T0258131

Finite Element Analysis: Biomedical Aspects

Finite Element Analysis: Biomedical Aspects

Edited by **Connie Mcguire**

New York

Published by NY Research Press,
23 West, 55th Street, Suite 816,
New York, NY 10019, USA
www.nyresearchpress.com

Finite Element Analysis: Biomedical Aspects
Edited by Connie Mcguire

International Standard Book Number: 978-1-63238-197-2 (Hardback)

Printed in the United States of America.

Contents

Preface

This book has been an outcome of determined endeavour from a group of educationists in the field. The primary objective was to involve a broad spectrum of professionals from diverse cultural background involved in the field for developing new researches. The book not only targets students but also scholars pursuing higher research for further enhancement of the theoretical and practical applications of the subject.

Finding approximate solutions to partial differential equations and integral equations, allowing numerical assessment of complicated structures based on their material properties is best represented by the mathematical method of Finite Element Analysis. This book presents varied topics on the utilization of Finite Elements in biomedical engineering under two sections on "Dentistry, Dental Implantology and Teeth Restoration" and "Cardiovascular and Skeletal Systems". The structure and language of the book has been so written that it is useful for graduate students learning applications of finite element and also encompasses topics and reference material useful for research and professionals who want to gain a deeper knowledge of finite element analysis.

It was an honour to edit such a profound book and also a challenging task to compile and examine all the relevant data for accuracy and originality. I wish to acknowledge the efforts of the contributors for submitting such brilliant and diverse chapters in the field and for endlessly working for the completion of the book. Last, but not the least; I thank my family for being a constant source of support in all my research endeavours.

Editor

Part 1

Dentistry, Dental Implantology and Teeth Restoration

FEA in Dentistry: A Useful Tool to Investigate the Biomechanical Behavior of Implant Supported Prosthesis

Wirley Gonçalves Assunção, Valentim Adelino Ricardo Barão,
Érica Alves Gomes, Juliana Aparecida Delben and Ricardo Faria Ribeiro
Univ Estadual Paulista (UNESP), Aracatuba Dental School,
Univ of Sao Paulo (USP), Dental School of Ribeirao Preto,
Brazil

1. Introduction

The use of dental implants is widespread and has been successfully applied to replace missing teeth (Amoroso et al., 2006). Although high success rate has been reported by several clinical studies, early or late dental implants failures are still inevitable. During mastication, overstress around dental implants may cause bone resorption, which leads to infection on the peri-implant region and failure of oral rehabilitation (Kopp, 1990). The way in which bone is loaded may influence its response (Koca et al., 2005). The results of cyclic loading into the bone differ from those of static loading (Papavasiliou et al., 1996). In case of repetitive cyclic load application, stress microfractures in bone may occur (Koca et al., 2005) and may induce osteoclastic activity to remove the damaged bone (Papavasiliou et al., 1996). So far, it is imperative to understand where the maximum stresses occur during mastication around the implants in order to avoid these complications (Nagasao et al., 2003).

Considering that stress/strain distribution at bone level is hard to be clinically assessed, the finite element analysis (FEA) has been extensively used in Dentistry to understand the biomechanical behavior of implant-supported prosthesis. To date, FEA was first used in the Implant Dentistry field by Weinstein et al. (1976) to evaluate the stress distribution of porous rooted dental implants. Nowadays, owing to the geometric complexity of implant-bone-prosthesis system, FEA has been viewed as a suitable tool for analyzing stress distribution into this system and to predict its performance clinically. Such analysis has the advantage of allowing several conditions to be changed easily and allows measurement of stress distribution around implants at optional points that are difficult to be clinically examined.

Therefore, this chapter provides the current status of using FEA to investigate the biomechanical behavior of implant-supported prosthesis. The modeling of complex structures that represents the oral cavity is described, and comparisons between two-dimensional (2D) and three-dimensional (3D) modeling techniques are discussed. Additionally, the application of microcomputer tomography to develop complex and more realistic FE models are assessed. Some sensitive cases are also illustrated.

2. Biomechanical behavior of implant-supported prosthesis

In order to enhance treatment longevity, it is important to understand the biomechanics of implant-supported prosthesis during masticatory loading. And the way that the stress/strain is transmitted and distributed to the bone tissue dictates whether the implant treatments will failure or succeed (Geng et al., 2001). Several variables affect the stress/strain distribution on the implant/bone complex such as prosthesis type, implant type, veneering and framework materials, bone quality, and presence of misfit.

2.1 Prosthesis and implant types

The implant-supported prosthesis can be classified as single- or multi- unit prosthesis. From a biomechanical point of view, the multi-unit prosthesis is subdivided into implant-supported overdentures and implant-supported fixed prosthesis (cantilevered design or not). The nature of FEA studies for these prosthesis designs is much more complex than for single-unit design (Geng et al., 2001).

Implant-retained overdentures are considered a simple, cost-effective, viable, less invasive and successful treatment option for edentulous patients (Assuncao et al., 2008; Barao et al., 2009). However, controversies toward the design of attachment systems for overdentures still exist (Bilhan et al., 2011). Our previous study (Barao et al., 2009) used a 2D FEA to investigate the effect of different designs of attachment systems on the stress distribution of implant-retained mandibular overdentures. The bar-clip attachment system showed the greatest stress values followed by bar-clip associated with two distally placed o'ring attachment systems, and o'ring attachment system (Fig. 1). Other 2D (Meijer et al., 1992) and 3D FEA studies (Menicucci et al., 1998) also showed stress optimization in overdenture with unsplinted implants (e.g. o'ring attachment system). The flexibility and resiliency provided by the o'ring rubber and the spacer in the o'ring system assembly may be the driven force toward the lower stress values with o'ring attachment system. Additionally, the stress breaking effect of the o'ring rubber can also decrease the stress in implants, prosthetics components and supporting tissues (Tokuhisa et al., 2003).

Tanino et al. (2007) evaluated the effect of stress-breaking attachments at the connections between maxillary palateless overdentures and implants using 3D models with two and four implants. Stress-breaking materials (with elastic modulus ranging from 1 to 3,000 MPa) connecting the implants and denture were included around each abutment. As the elastic modulus of the stress-breaking materials increased, the stress increased at the implant-bone interface and decreased at the cortical bone surface. Additionally, the 3-mm-thick stress-breaking material decreased the stress values at the implant-bone interface when compared to the 1-mm-thick material. Knowing that overdentures are retained by implants but are still supported by the mucosa, and facing the difference in displacement between implants (20-30 μm) and soft tissue (about 500 μm), our previous study (Barao et al., 2008) investigated the influence of different mucosa thickness and resiliency on stress distribution of implant-retained overdentures using a 2D FEA. Two models were designed: two-splinted-implants connected with bar-clip system and two-splinted-implants connected with bar-clip system associated with two-distally placed o'ring system. For each design, mucosa assumed three characteristics of thickness (1, 3 and 5 mm) varying its resiliencies (based on its Young's modulus) in hard (680 MPa), resilient (340 MPa) and soft (1 MPa), respectively. In general, the stress decreased at the supporting tissues as mucosa thickness and resiliency increased (Fig. 2).

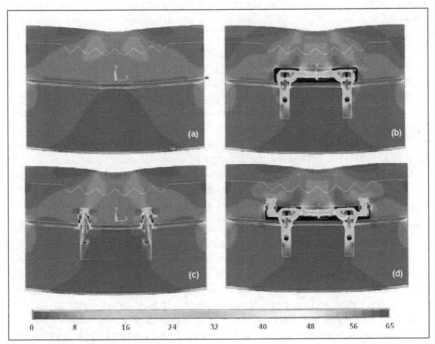

Fig. 1. First principal stress distribution (in MPa). (a) conventional complete denture. (b) overdenture – bar-clip system. (c) overdenture – o'ring system). (d) overdenture – bar-clip associated with distally placed o'ring system. Colors indicate level of stress from dark blue (lowest) to red (highest).

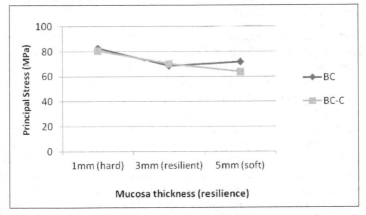

Fig. 2. Distribution of first principal stress (MPa) in supporting tissues for groups BC (bar-clip) and BC-C (bar-clip associated with two-distally placed o'rings) considering different mucosa thickness (1, 3 and 5mm) and resilience (hard, resilient and soft).

In relation to the implant-supported fixed prosthesis, the variety of factors that affect the stress distribution into the bone-implant complex comprise implant inclination, implant number and position, framework/veneering material properties, and cross-sectional design of the framework (Geng et al., 2001). The use of tilted implants mostly affected the stress concentration in the peri-implant bone tissue when compared to vertical implants (Canay et al., 1996). However, tilted implants have been used in case of atrophic jaw, to avoid maxillary sinus, and to reduce the cantilever extension (Silva et al., 2010). Caglar et al. (2006) investigated the effects of mesiodistal inclination of implants on the stress distribution of posterior maxillary implant-supported fixed prosthesis using a 3D FEA. Inclination of the implant in the molar region resulted in increased stress. Similar results were found by a Iplikcioglu & Akca (2002) who investigated the effect of buccolingual inclination in implant-supported fixed prosthesis applied to the posterior mandibular region using a 3D FEA. Bevilacqua et al. (2011) investigated the influence of cantilever length (13, 9, 5 and 0 mm) and implant inclination (0, 15, 30 and 45 degrees) on stress distribution in maxillary fixed dentures. This 3D FEA study showed that tilted implants, with consequent reduction of the posterior cantilevers, reduced the stress values in the peri-implant cortical bone.

Zarone et al. (2003) evaluated the relative deformations and stress distributions in six different designs of full-arch implant-supported fixed mandibular denture (six or four implants, cantilevered designed or not, cross-arch or midline-divided bar into two free-standing bridges) by means of 3D FEA. When the implants were rigidly connected by one-piece framework, the free bending of the mandible was hindered. The flexibility of the mandible was increased as the more distal implant supports were more mesially located. The use of two free-standing bars also reduced the overall stress on the bone/implant interface, fixtures and superstructure. Contradicting these findings, Yokoyama et al. (2005) observed that the use of single-unit superstructure was more effective in relining stress concentration in the edentulous mandibular bone than 3-unit superstructure. Other study (Silva et al., 2010), using a 3D FEA, assessed the biomechanical behavior of the "All-on-four" system with that of six-implant-supported maxillary prosthesis with tilted implants. The stress values were greater to the "All-on-four" concept, and the presence of cantilever increased the stress values about 100% in both models.

It is believed that loading distribution pattern in implant-retained overdentures differs from those in implant-supported fixed restorations (Tokuhisa et al., 2003). Our ongoing project has compared the effect of different designs of implant-retained overdentures and fixed full-arch implant-supported prosthesis on stress distribution in edentulous mandible by using a 3D-FEA based on a computerized tomography (CT). Four 3D FE models of an edentulous human mandible with mucosa and four implants placed in the interforamina area were constructed and restored with different designs of dentures. In the OR group, the mandible was restored with an overdenture retained by four unsplinted implants with O'ring attachment; in the BC-C and BC groups, the mandibles were restored with overdentures retained by four splinted implants with bar-clip anchor associated or not with two distally placed cantilevers, respectively; in the FD group, the mandible was restored with a fixed full-arch four-implant-supported prosthesis. The masticatory muscles and temporomandibular joints supported the models. A 100-N oblique load (30 degrees) was applied on the left first molar of each denture in a buccolingual direction. Qualitative and quantitative analysis based on the von Mises stress (σ_{vM}), the maximum (σ_{max}) (tensile) and minimum (σ_{min}) (compressive) principal stresses (in MPa) were obtained. BC-C group exhibited the highest stress values (σ_{vM} = 398.8, σ_{max} = 580.5 and σ_{min} = -455.2) while FD group showed the lowest one (σ_{vM} = 128.9, σ_{max} = 185.9 and σ_{min} = -172.1) in the implant/prosthetic components. Within overdenture groups, the use of unsplinted implants (OR group) reduced the stress level in the implant/prosthetic components (59.4% for σ_{vM}, 66.2% for σ_{max} and 57.7% for σ_{min} versus BC-C group) and supporting tissues (maximum stress reduction of 72% and 79.5% for σ_{max}, and 15.7% and 85.7% for σ_{min} on the cortical bone and the trabecular bone, respectively). The cortical bone exhibited greater stress concentration than the trabecular bone for all groups. We concluded that the use of fixed implant dentures and removable dentures retained by unsplinted implants to rehabilitate completely edentulous mandible reduced the stresses in the peri-implant cortical bone tissue (Fig. 3), mucosa and implant/prosthetic components.

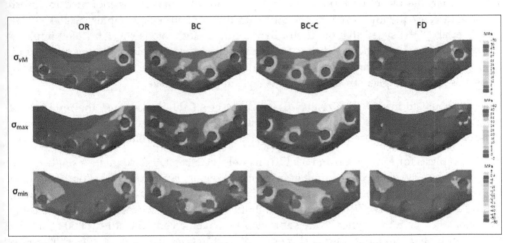

Fig. 3. von Mises stress (σ_{vM}), maximum (σ_{max}) and minimum (σ_{min}) principal stress distributions (in MPa) within cortical bone for o'ring (OR), bar-clip (BC), bar-clip with distally placed cantilever (BC-C) and fixed denture (FD) groups.

Concerning the implant design, Ding et al. (2009) analyzed the stress distribution around immediately loaded implants of different diameters (3.3; 4., and 4.8 mm) using an accurate complete mandible model. The authors observed that with the increase of implant diameter, stress/strain on the implant-bone interface decreased, mainly when the diameter increased from 3.3 to 4.1 mm for both axial and oblique loading conditions. Other studies also showed more favorable stress distribution with the use of wide-diameter implants (Himmlova et al., 2004; Matsushita et al., 1990). Huang et al. (2008) analyzed the peri-implant bone stress and the implant-bone sliding as affected by different implant designs and implant sizes of immediately loaded implant with maxillary sinus augmentation. Twenty-four 3D FE models with four implant designs (cylindrical, threaded, stepped and step-thread implants) and three dimensions (standard, long and wide threaded implants) with a bonded and three levels of frictional contact of implant-bone interfaces were analyzed. The use of threaded implants decreased the bone stress and sliding distance about 30% as compared with non-threaded (cylindrical and stepped) implants. With the increase of implant's length or diameter, the bone stress reduced around 13-26%. The immediately loaded implant with smooth machine surface increased the bone stress by 28-63% versus osseointegrated implants. The increase of implant's surface roughness did not reduce the bone stress but decrease the implant-bone interfacial sliding.

2.2 Veneering and framework material

The literature is scarce about the best material to fabricate superstructures of implant-supported prosthesis (Gomes et al., 2011). Originally, the protocol consisted of gold alloy framework and acrylic resin for denture base and acrylic resin or composite resin for artificial denture teeth (Zarb & Jansson, 1985). Rigid occlusal material such as porcelain on metal may increase the load transfer to the implant and surrounding bone tissue (Skalak, 1983). So far, the use of occlusal veneering based on resin material is indicated to absorb shock and consequently to reduce the stress on the implant-bone complex (Skalak, 1983). Gracis et al. (1991) stated that the use of harder and stiffer materials to fabricated implant-supported restorations increased the stress transmitted to the implant. On the other hand, some studies (Ciftci & Canay, 2001; Sertgoz, 1997) showed that the use of softer restorative materials lead to a higher stress on implants and supporting tissues.

Our previous studies (Delben et al., 2011; Gomes et al., 2011) evaluated the influence of different superstructures on preload maintenance of retention screw of single implant-supported crowns submitted to mechanical cycling and stress distribution through 3D FEA.

Twelve replicas for each group and 3D FEA models were created to simulate a single crown supported by external hexagon implant in premolar region. Five groups were obtained: gold abutment veneered with ceramic (GC) and resin (GR), titanium abutment veneered with ceramic (TC) and resin (TR), and zirconia abutment veneered with ceramic (ZC). During mechanical cycling, the replicas were submitted to dynamic vertical loading of 50 N at 2 Hz for detorque measurement after each period of 1×10^5 cycles up to 1×10^6 cycles. The FEA software generated the stress maps after vertical loading of 100 N on the contact points of the crowns. Significant difference (P<.05) between group TC (21.4 ± 1.78) and groups GC (23.9 ± 0,91), GR (24.1 ± 1.34) and TR (23.2 ± 1.33); and between group ZC (21,9 ± 2,68) and groups GC and GR for initial detorque mean (in N.cm) was noted. After mechanical cycling, there was significant difference (P<.05) between groups GR (23.8 ± 1.56) and TC (22.1 ± 1.86), and between group ZC (21.7 ± 2.02) and groups GR and TR (23.6 ± 1.30) (Fig. 4). The stress values and distribution in bone tissue were similar for groups GC, GR, TC and ZC (1574.3 MPa, 1574.3 MPa, 1574.3 MPa and 1574.2 MPa, respectively), except for group TR (1838.3 MPa) (Fig. 5). Group ZC transferred lower stress to the retention screw (785 MPa) than the other groups (939 MPa for GC, 961 MPa for GR, 1010 MPa for TC, and 1037 MPa for TR) We concluded that detorque reduction occurred for all superstructure materials but torque maintenance was enough to maintain joint stability in this study. The different materials did not affect stress distribution in bone. However, group ZC presented the best stress distribution for the retention screw. Previous study conducted by our research group also found similar stress distribution to single implant-supported prosthesis regardless of the type of veneering/framework material through a 2D FEA (Assuncao et al., 2010).

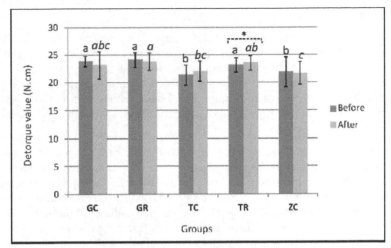

Fig. 4. Detorque mean value (N.cm) before and after mechanical cycling for all groups. Within each period, mean followed by different letters represent statistically significant difference (P<.05, Fisher's exact test).*denotes statistically significant difference within the same group (P<.05, Student's t-test).

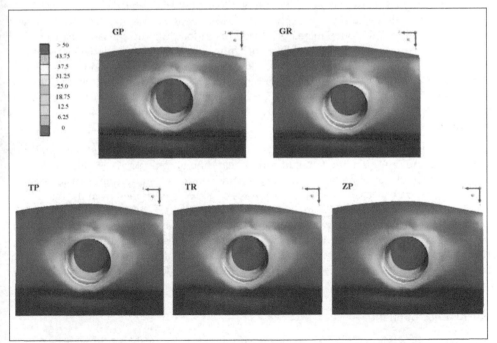

Fig. 5. von Mises stress distribution (MPa) within supporting bone for GP, GR, TP, TR and ZP groups. Colors indicate level of stress from dark blue (lowest) to red (highest).

2.3 Bone quality

The bone quality is strongly correlated with the implant success as pointed out by several longitudinal clinical studies (Friberg et al., 1991; Jemt & Lekholm, 1995; van Steenberghe et al., 1990). The biomechanical behavior among the different types of bone (I, II, III or IV) differs substantially, which affect the ability of bone to support physiological loads (de Almeida et al., 2010). The poor quality bone type 4 has promoted greater failures of dental implants (Jaffin et al., 2004) owing to its reduced capability to bond the implant to the bone (Drage et al., 2007; Shapurian et al., 2006).

de Almeida et al. (2010) investigated the influence of different types of bone (types I to IV) on the stress distribution on the supporting tissue of a fixed full-arch implant-supported mandibular prosthesis based on a prefabricated bar by using a 3D FEA. Three unilateral posterior loads of 150 N were applied on the prosthesis: L1 – axial loading; L2 – oblique loading (buccolingual direction, 30 degrees); L3 – oblique loading (linguobuccal direction, 30 degrees). Type III and IV bones displayed the greatest stress values in the axial and buccolingual loading conditions, while stiffer bones (type I and II) exhibited the lowest. For the linguobuccal loading condition, the poorest quality cortical bone (type IV) had the highest stress concentration followed by types III, II and I.

Tada et al. (2003) evaluated whether bone quality affect the stress/strain distribution of single-unit implant-supported mandibular prosthesis with different implant type and length. Screw and cylinder implants with 9.2, 10.8, 12.4 and 14.0 mm length were used and virtually placed in 4 types of bone. Two different loads (axial and buccolingual forces) were applied to the occlusal surface at the center of the abutment. As the bone density decreased, the stress/strain into the bone increased. Under axial loading, the stress in the cancellous bone was lower with the screw-type implant when compared with cylinder-type implant. Additionally, longer implants displayed lower stress values. The bone quality also influences the stress distribution under buccolingual load. According to the authors, low-density bone presents reduced stiffness, which increases implant displacement. Under greater displacement, the bone is deformed and consequently higher stresses in the cortical and cancellous bone are expected.

Another 3D FEA study (Sevimay et al., 2005) examined the effect of the bone quality on stress distribution for an implant-supported mandibular crown. A 3D FE model of a mandibular section of bone with a missing second premolar and an implant to receive a crown restoration was used. A total vertical force of 300 N was applied from the buccal cusp (150 N) and distal fossa (150 N) in centric occlusion. Low-density bone (types III and IV) displayed the greatest stress values (163 and 180 MPa, respectively) mainly at the peri-implant cortical bone. On the other hand, type I and II bones exhibited the lowest levels of stresses (150 and 152 MPa, respectively). Other similar study (Holmes & Loftus, 1997) found that the placement of implants in type I bone resulted in less micromotion and reduced stress concentration.

2.4 Presence of misfit

An increase of clinical failures has been correlated with misfit of implant-supported prostheses (Klineberg & Murray, 1985; Skalak, 1983). In order to prevent mechanical (i.e. retention screw and abutment screw loosening and fracture, superstructure mobility, and implant fracture) (Carlson & Carlsson, 1994; Dellinges & Tebrock, 1993) and biological (i.e. sensorial disturbances, soft tissue injuries, peri-implantitis, and bone loss) (Berglundh et al., 2002) complications of the implant treatment, a passive fit between the crown and implant should be achieved (Sahin & Cehreli, 2001). Previous studies (Assuncao et al., 2011; Kunavisarut et al., 2002; Natali et al., 2006) have showed an increase of stress on peri-implant bone tissue under the presence of misfit in implant-supported prostheses.

Our previous study (Assuncao et al., 2011) used a 3D FEA to investigate the effect of vertical and angular misfit in three-piece implant-supported screwed crown on the biomechanical behavior of peri-implant bone, implants, and prosthetic components. A total of four 3D models were fabricated to represent a posterior mandibular section with one implant in the region of the second premolar and another in the region of the second molar. The implants were splinted by a three-piece implant-supported metal-ceramic prosthesis and differed according to the type of misfit, as represented by four different models: control - prosthesis with complete fit to the implants; unilateral angular misfit - prosthesis presenting unilateral angular misfit of 100 µm in the mesial region of the second molar; unilateral vertical misfit - prosthesis presenting unilateral vertical misfit of 100 µm in the mesial region of the second molar; and total vertical misfit - prosthesis presenting total vertical misfit of 100 µm in the platform of the framework in the second molar (Fig. 6). A vertical load of 400 N was distributed and applied on 12 centric points (a vertical load of 150 N was applied to each molar in the prosthesis and a vertical load of 100 N was applied at the second premolar). We observed that stress on the peri-implant cortical bone was slightly affected by the presence of misfit. Each type of misfit overloaded a specific region of the implant-supported system. The unilateral angular misfit was most harmful for the implant body and retention screw, the unilateral vertical misfit placed the most stress on the framework, and the total vertical misfit added stress to the implant hexagon.

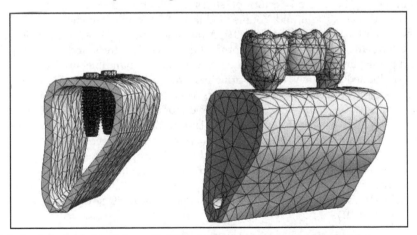

Fig. 6. Mesh of the main model: cortical bone, trabecular bone, implant and crown (framework, veneering material and retention screw).

Another study conducted by our group (Gomes et al., 2009) assessed the influence misfit on the displacement and stress distribution in the bone-implant-prosthesis complex using a 2D FEA. A single-unit mandibular implant-supported prosthesis was fabricated. Different unilateral angular misfit (0 μm – control, 50 μm, 100 μm, and 200 μm) was represented on the contact region between the implant and the crown. An oblique (30 degrees) load of 133 N was applied at the opposite direction of misfit on the models. The greater the angular misfit, the higher the stress and displacement values in the bone-implant-prosthesis assembly. On the other hand, Spazzin et al. (2011) investigated the effect of different levels of vertical misfit (5, 25, 50, 100, 200 and 300 μm) between implant and bar framework in overdenture and showed that the presence of misfit did not influence the stress level at the peri-implant bone tissue, but stress on the prosthetic components (bar framework, retention screw) and implants increased with greater misfit levels.

Spazzin et al. (2011) also assessed the influence of horizontal misfit (10, 50, 100 and 200 μm) and bar framework material (gold alloy, silver-palladium alloy, commercially pure titanium and cobalt-chromium alloy) on the stress distribution in implant-retained mandibular overdenture associated with bar attachment system using a 3D FEA. The increase in horizontal misfit promoted an enhancement of stress levels in the inferior region of the bar, retention screw neck, cervical and medium third of the implant, and peri-implant cortical bone. The stiffer the bar material was, the greater the stress on the framework.

3. Modeling complex structures

As the oral cavity is very complex in nature, it is very hard to represent this structure with high accuracy by means FE models. For this reason, several simplifications are necessary to our reality. Most of the FEA studies in Dental field consider the materials as isotropic, homogeneous and linearly elastic. The modeling of biological tissues (e.g. bone) is a very difficult task because of their inherent heterogeneous and anisotropic character (Cowin & editor, 2001). The use of isotropic properties instead of anisotropic properties for bone tissue may affect the overall results of stress distribution (O'Mahony et al., 2001). In addition, the ultimate strain and Young modulus of bone under compression is different than those under tension (Geng et al., 2001). A previous study (Liao et al., 2008) investigated in what extend the anisotropic elastic properties affect the stress and strain distribution around implants under physiologic load in a complete mandible model based on CT. Models were loaded obliquely, and the principal stress and strain values in the peri-implant bone tissue were recorded. The authors observed an increase of up to 70% of stress/strain values for the anisotropic model versus isotropic model. O'Mahony et al. (2001) compared implant-bone interface stresses and peri-implant principal strains in anisotropic versus isotropic 3D models of the posterior mandible. Anisotropy increased by 20 to 30% the stress and strain in the cortical bone when compared with isotropic case. In the trabecular bone, the anisotropy enhanced by 3- to 4-fold the stress level versus isotropic condition. So far, anisotropy has significant effects on peri-implant stress and strain; therefore, careful consideration should be given to its use in biomechanical FE studies (Liao et al., 2008; O'Mahony et al., 2001).

The bone-implant contact in most of FEA studies is considered at a 100%; however, in the clinical scenario the implant-bone contact ranges from 30% to 70% (Geng et al., 2001). Additionally, the bone is not homogeneous and porosities are presented. Nowadays, it is possible to insert contact algorithms to simulate contacts (friction coefficients). In Dentistry, this factor is very important because it allows the representation of different degrees of osseointegration and the scenario of immediate loading. Huang et al. (2008) compared two conditions of implant-bone contact (bonded interface and non-bonded interface – friction coefficient of 0.3) on peri-implant bone stress distribution and implant-bone interfacial sliding of different implant designs and implant sizes. The use of friction coefficient to represent the immediate loading condition of the implants increased the bone stress by 28-63% when compared with the osseointegrated condition (bonded contact) for all implant designs and sizes. The interfacial sliding between bone and implant decreased with the presence of friction coefficient.

Thread configuration is an important objective in biomechanical optimization of dental implants (Valen & Locante, 2000). However, several 2D and 3D FEA studies have not considered the threads modeling of implants and retention screws or have modeled them using only concentric rings owing to the difficulty in modeling the thread helix (Sertgoz, 1997). Additionally, some justify the use of simplified model due to the difficulty in constructing a 3D complex model and the enormous increase in element numbers (Assuncao et al., 2009). As the oversimplication of implant complex geometry may affect the results of several FEA studies (Al-Sukhun et al., 2007), some authors (Lang et al., 2003; Sakaguchi & Borgersen, 1995) believe that the modeling of the perfect geometry of the implant, including the thread helix of the screw and the screw bore, is essential to finite element analysis simulation. Therefore, our previous study (Assuncao et al., 2009) investigated whether or not the representation of implant's threads would affect the outcome of a 2D FEA. Two models reproducing a frontal section of edentulous mandibular posterior bone were constructed. In the first models, the implants threads were accurately simulated (precise model) and, on the other implants with a smooth surface (press-fit implant) were used (simplified model). A load of 133 N was obliquely (30 degrees) applied with on the models. Precise model (1,45 MPa) showed higher maximum stress values than simplified model (1,2 MPa). Stress distribution and stress values in the cortical bone (292.95 MPa for precise model and 401.14 MPa for simplified model) and trabecular bone (19.35 MPa for precise model and 20.35 MPa for simplified) were similar, and the stresses were mostly located around implant neck and implant apex. We concluded that considering implant and screw analysis, remarkable differences in stress values were found between the models. Although models have showed differences on absolute stress values, the stress distribution was similar.

4. Two-dimensional (2D) versus three-dimensional (3D) analysis

The biomechanical performance of mandibular and maxillary bones associated to other natural (teeth, periodontal ligament, gingiva, etc.) and artificial (prostheses, dental implants, etc.) structures has been considered clinically and biologically relevant for modern Dentistry. In this sense, several tools have been developed for biomechanical evaluations, such as the finite element method.

The first studies reported in the literature using the Finite Element Method in Dentistry presented simple 2D models (Takahashi et al., 1978; Weinstein et al., 1979; Yettram et al., 1976). However, the complexity of the oral environment required the development of 3D models. Khera et al. (1988) were pioneers on 3D modeling for human mandible.

Although the technological advancement allowed the use of accurate and fast software and hardware to obtain the images, the decision about 2D or 3D modeling for FEA remains uncertain. It is important to understand that the biomechanical performance of complex oral structures depends on several factors. Therefore, the accuracy of the results may be influenced by: 1- complex geometry of the model; 2 – materials properties, such as isotropy or anisotropy; 3 – type, size and quantity of elements in the mesh; 4 – boundary and loading conditions similar to clinical scenario; and 5 – analysis mode (static or dynamic). In addition, the choice between 2D or 3D models should be guided by the expectances and applicability of the results. Thus, the researcher should understand the advantages and limitations of both modeling types (Romeed et al., 2006).

The 2D modeling has been continually applied in Dentistry since it is a simple, fast and low-cost approach. On the other hand, the 3D model is more accurate and may represent the details of a real condition (Fig. 7). However, a complex model is not worthy if it is misinterpreted considering that the higher the complexity of the model, the higher the density of the elements mesh. Thus, the following question should be answered before starting a FEA study "How can I represent the model accurately to obtain results within the ideal parameters?"

Fig. 7. Different types of FE implant models: Two-dimensional model and Three-dimensional model from left to right, respectively.

Several studies (Poiate et al., 2011; Romeed et al., 2006) were conducted to compare the different models applied from Engineering to Dentistry to verify the effect of 2D or 3D modeling on the analysis of the biomechanical performance of complex structures. Romeed et al. (2006) evaluated the mechanical behavior of a second maxillary premolar restored with full-coverage crown under different occlusal schemes and observed similar stress distribution and minor differences for the stress values. On the other hand, Poiate et al. (2011) compared the biomechanical performance of a maxillary central incisor between 2D and 3D modeling and concluded that 2D models can be safely applied only for qualitative analysis since these models showed overestimated values for the quantitative analysis of the stress. The differences between 2D and 3D models have been attributed to the geometric representation of the model. Although 2D models are simplified and easier to be obtained than 3D models, the biaxial state may influence the reliability of the results since some important biomechanical aspects may be not reproduced (Gao et al., 2006).

4.1 Case sensitive

Two- and 3- D models were constructed to evaluate the behavior of different veneering materials of single implant-supported prostheses (Assuncao et al., 2010; Gomes et al., 2011). For both models, five types of finite element (FE) models were simulated according to the different framework (gold alloy, titanium, and zirconia) and veneering (porcelain and modified composite resin) materials. However, several differences between the models representation were introduced according to the limitations of each modeling process. First of all, it was observed differences about the geometry, mainly for the superstructure (veneering material and framework) and bone tissue (Fig. 8). In the 2D model, the superstructure was modeled with 8-mm in height and 8-mm in diameter and the surrounding bone model assumed the characteristic of a block. In the 3D model, the superstructure was modeled based on the characteristics of a left first premolar tooth and the bone tissue reproduced a segment of the maxilla with a missing left first premolar tooth. In relation to the loading, the load was applied at a 30-degree inclination and 2-mm off-axis in 2D simulation while a 100-N vertical force was applied to the contact points of the crowns in the 3D model. Additionally, the 3D model represented a contact element between the abutment and the implant to simulate the clinical situation. At the end, it was observed difference between the 2D and 3D models for both qualitative and quantitative analyses.

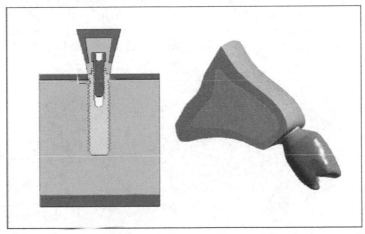

Fig. 8. Differences about the geometry representation of the superstructure and bone tissue for single implant-supported prostheses: 2D and 3D model from left to right, respectively.

Thus, the selection between 2D or 3D modeling should be guided by the researcher knowledge about both methods. The 3D models are similar to real structures but are time-consuming for modeling and data processing even when powerful computers are used. The higher the model complexity, the higher the number of elements and the complexity of the analysis. Thus, a simple geometry may generate an accurate mesh with satisfactory results in a faster way.

Two-dimensional models offer excellent access for pre- and post-processing, and because of the reduced dimensions, computational capacity can be preserved for improvements in element and simulation quality. On the other hand, 3D models, although more realistic with respect to the dimensional properties, are generally more coarse, with elements that are far from their ideal shapes. Moreover, examination of the model is more difficult. Depending on the investigated structure and boundary conditions, 2D modeling may be justified as a reasonable or even sensible simplification (Korioth & Versluis, 1997). Additionally, combinations of 2D or 3D FEA may offer the best understanding of the biomechanical behavior of complex dental structures in certain situations (Romeed et al., 2006).

5. Microcomputer tomography in FE models

In the beginning, the FE models were obtained from sectional images of bone tissue, tooth, surrounding structures and other elements related with the study that would be executed. Khera et al. (1988) constructed a 3D human mandible based on a 2D model. Initially, a 2D model was obtained and, using the projection of several pictures in a magnifying monitor, a 3D model was generated and an axial z-axis was defined.

Another widely used technique to obtain 2D and 3D models is the embedment of structures in acrylic resin. Gomes et al. (2009) embedded a prosthesis/retention screw/implant system in resin and sectioned it longitudinally to investigate the effect of different levels of unilateral angular misfit prostheses in the assembly and surrounding bone using a 2D FEA. The embedded model was scanned to produce digitalized images that were imported into CAD image analysis software and placed within the supporting tissue based on literature data. The outline of the images was manually quoted and each point was converted into x and y coordinates. At the end, the coordinates were finally imported into the FE software as key points of the final images (Fig. 9).

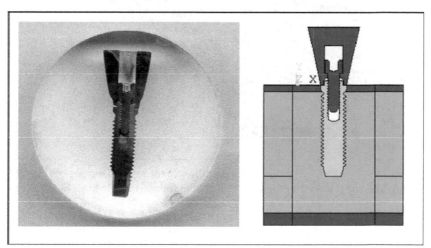

Fig. 9. Scanned embedded model and finite element model.

Recently, microcomputer tomography (CT) images have been obtained as a useful tool to model the bone complex in FE models and have gained general consensus among researches. The 3D model simulated from CT images provides high fidelity to the anatomical dimensions and configuration of all oral structures because it is possible to define the geometry and the local tissue properties of the bone segment to be modeled.

However, some caution should be taken when these data are used. The X-ray images in grey scale must be recorded by a CT in DICOM (Digital Imaging Communications in Medicine) format. Thus, the procedure should be carefully indicated for the patients due to the exposition to radiation and the research project should be approved by the ethics committee ensuring that physical and geometrical parameters within safety limits will be adopted. Furthermore, the patient should sign an informed consent form to authorize the reproduction of images and results.

The processing techniques used to extract this information from the CT data may be also frequently affected by no negligible errors that propagate in an unknown way through the various steps of the model generation, affecting the accuracy of the model (Taddei et al., 2006). The first source of geometric error and distortion is the resolution of the dataset that depends on the scan parameters setting (Taddei et al., 2006). The ones that yielded the best results for image quality were obtained in the regime of 120 kV, 150 mA, 512 × 512 matrix, 14 cm × 14 cm field of view, and slice thickness of 0.5 mm (Poiate et al., 2011). The second error may result from the segmentation process of the region of interest. Several segmentation algorithms have been proposed, with various level of automation, starting from complete manual contours extraction to complex fully automatic algorithms (Taddei et al., 2006). Considering that several softwares are currently available in the market, the professional should be trained to accurately use the tools to convert CT images into FE models.

5.1 Converting CT images into FEA models

Some steps to convert CT images into solid models will be presented in this section considering the availability of softwares, such as Mimics (Materialise, Leuven, Belgium), Simpleware (Simpleware Ltd, United Kingdom), InVesalius (Brazilian Public Software, Brazil), and ITK Snap (General Public License, USA).

The use of Simpleware software to convert a maxillary CT images into a maxillary FEA model will be discussed. After submission by the Ethic Committee in Research of Dental School of Ribeirão Preto, University of São Paulo (Process CAAE – 0038.0.138.000-11 and SISNEP – FR – 430209), the CT images obtained from patient imaging with 219 cross sections were imported into the Simpleware 4.1 software (Simpleware Ltd, United Kingdom) (Fig. 10) at the *ScanIP* segment.

In the *ScanIP*, the segmentation tool was used to identify the pixels value of the tomographic image. Then, it was possible to separate an object from other adjacent anatomical structures in different masks, such as cortical bone and trabecular bone (Fig. 11). According to its radio-density, expressed in Hounsfield unities, the program picked the values up and automatically created different masks. After that, fine adjustments were executed to further improve the quality of the model masks (Fig. 12). The program also provides tools to eliminate any interference of the tomographic image. By examining the image, a certain level of noise in the data could be corrected by filters. After obtaining the masks for the cortical and trabecular bone, the soft tissue was constructed with 2.0 mm in thickness for the whole model using a morphological filter tool in structuring element. Then, the step to convert the CT image into a solid model was completed (Figs. 13 and 14).

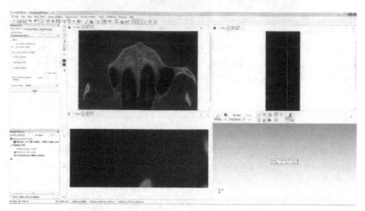

Fig. 10. Tomographic image of an edentulous maxilla imported into *ScanIP*.

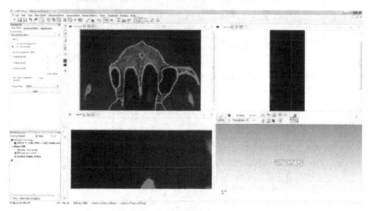

Fig. 11. Tomographic image of an edentulous maxilla processed in *ScanIP*. Determination of the cortical bone tissue.

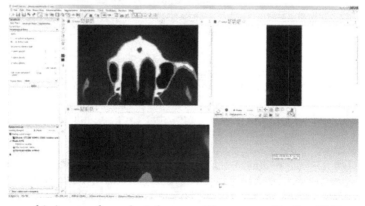

Fig. 12. Tomographic image of an edentulous maxilla processed in *ScanIP*. Determination of the trabecular bone tissue.

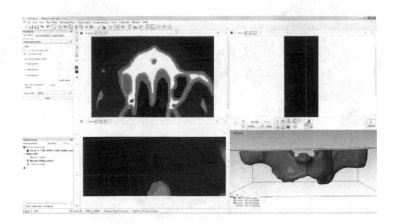

Fig. 13. Tomographic image of an edentulous maxilla processed in *ScanIP*. Determination of the soft tissue with 2.0mm in thickness in the whole extension.

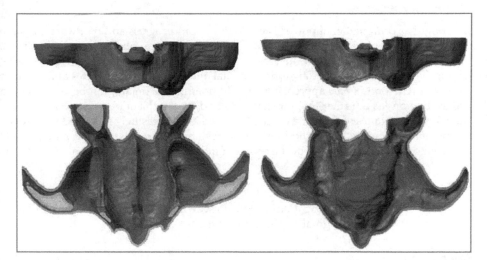

Fig. 14. Edentulous maxilla generated in the *ScanIP* representing the cortical and medullary bone and the soft tissue.

After obtaining the solid model, the finite elements mesh was generated. The mesh can be created either in the software for image conversion or in the FE software. In this study, the Simpleware software generated the mesh. A mix of tetrahedral and hexahedral elements was obtained using gallery elements + FE Free (Fig. 15). Afterwards, the meshed model is ready to be exported to a FEA software in order to conduct stress and displacement analysis.

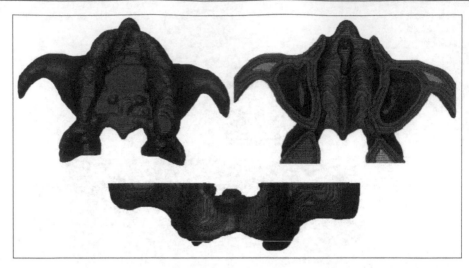

Fig. 15. Generated mesh with parabolic tetrahedral interpolation solid elements by the Simpleware software. The meshed model is ready to be imported by the finite element analysis software to investigate de stress distribution into the bone tissue.

6. Future perspectives

Considering that computational power is exhibiting rapid progress and hardware costs are decreasing, the numerical techniques probably will increase its application over time. Thus, the finite element method will be increasingly applied in Dentistry to generate reliable results for the biomechanical investigation of dental and supporting structures at lower cost than other *in vitro* and *in vivo* approaches. In addition, this technique can be associated to clinical evaluations as a further tool for diagnosis and/or treatment planning. For instance, the numerical techniques of the finite element method are increasingly indicated to simulate dental movements induced by orthodontic force systems. Thus, this method may provide information to the orthodontist about the choice of individual therapy (Clement et al., 2004).

7. Acknowledgment

Authors express gratitude to Sao Paulo State Research Foundation (FAPESP – Process number 2010/09857-3) for the grant support provided.

8. References

Al-Sukhun, J.; Lindqvist, C. & Helenius, M. (2007). Development of a Three-Dimensional Finite Element Model of a Human Mandible Containing Endosseous Dental Implants. Ii. Variables Affecting the Predictive Behavior of a Finite Element Model of a Human Mandible. *Journal of biomedical materials research Part A*, Vol.80, No.1, (Jan), pp. 247-256, ISSN 1549-3296

Amoroso, P. F.; Adams, R. J.; Waters, M. G. & Williams, D. W. (2006). Titanium Surface Modification and Its Effect on the Adherence of Porphyromonas Gingivalis: An in

Vitro Study. *Clinical oral implants research,* Vol.17, No.6, (Dec), pp. 633-637, ISSN 0905-7161

Assuncao, W. G.; Tabata, L. F.; Barao, V. A. & Rocha, E. P. (2008). Comparison of Stress Distribution between Complete Denture and Implant-Retained Overdenture-2d Fea. *Journal of oral rehabilitation,* Vol.35, No.10, (Oct), pp. 766-774, ISSN 1365-2842

Assuncao, W. G.; Gomes, E. A.; Barao, V. A. & de Sousa, E. A. (2009). Stress Analysis in Simulation Models with or without Implant Threads Representation. *The International journal of oral & maxillofacial implants,* Vol.24, No.6, (Nov-Dec), pp. 1040-1044, ISSN 0882-2786

Assuncao, W. G.; Gomes, E. A.; Barao, V. A.; Delben, J. A.; Tabata, L. F. & de Sousa, E. A. (2010). Effect of Superstructure Materials and Misfit on Stress Distribution in a Single Implant-Supported Prosthesis: A Finite Element Analysis. *The Journal of craniofacial surgery,* Vol.21, No.3, (May), pp. 689-695, ISSN 1536-3732

Assuncao, W. G.; Barao, V. A.; Delben, J. A.; Gomes, E. A. & Garcia, I. R., Jr. (2011). Effect of Unilateral Misfit on Preload of Retention Screws of Implant-Supported Prostheses Submitted to Mechanical Cycling. *Journal of prosthodontic research,* Vol.55, No.1, (Jan), pp. 12-18, ISSN 1883-1958

Barao, V. A.; Assuncao, W. G.; Tabata, L. F.; de Sousa, E. A. & Rocha, E. P. (2008). Effect of Different Mucosa Thickness and Resiliency on Stress Distribution of Implant-Retained Overdentures-2d Fea. *Computer methods and programs in biomedicine,* Vol.92, No.2, (Nov), pp. 213-223, ISSN 0169-2607

Barao, V. A.; Assuncao, W. G.; Tabata, L. F.; Delben, J. A.; Gomes, E. A.; de Sousa, E. A. & Rocha, E. P. (2009). Finite Element Analysis to Compare Complete Denture and Implant-Retained Overdentures with Different Attachment Systems. *The Journal of craniofacial surgery,* Vol.20, No.4, (Jul), pp. 1066-1071, ISSN 1536-3732

Berglundh, T.; Persson, L. & Klinge, B. (2002). A Systematic Review of the Incidence of Biological and Technical Complications in Implant Dentistry Reported in Prospective Longitudinal Studies of at Least 5 Years. *Journal of clinical periodontology,* Vol.29 Suppl 3, pp. 197-212; discussion 232-193, ISSN 0303-6979

Bevilacqua, M.; Tealdo, T.; Menini, M.; Pera, F.; Mossolov, A.; Drago, C. & Pera, P. (2011). The Influence of Cantilever Length and Implant Inclination on Stress Distribution in Maxillary Implant-Supported Fixed Dentures. *The Journal of prosthetic dentistry,* Vol.105, No.1, (Jan), pp. 5-13, ISSN 1097-6841

Billian, H.; Mumcu, E. & Arat, S. (2011). The Comparison of Marginal Bone Loss around Mandibular Overdenture-Supporting Implants with Two Different Attachment Types in a Loading Period of 36 Months. *Gerodontology,* Vol.28, No.1, (Mar), pp. 49-57, ISSN 1741-2358

Caglar, A.; Aydin, C.; Ozen, J.; Yilmaz, C. & Korkmaz, T. (2006). Effects of Mesiodistal Inclination of Implants on Stress Distribution in Implant-Supported Fixed Prostheses. *The International journal of oral & maxillofacial implants,* Vol.21, No.1, (Jan-Feb), pp. 36-44, ISSN 0882-2786

Canay, S.; Hersek, N.; Akpinar, I. & Asik, Z. (1996). Comparison of Stress Distribution around Vertical and Angled Implants with Finite-Element Analysis. *Quintessence Int,* Vol.27, No.9, (Sep), pp. 591-598, ISSN 0033-6572

Carlson, B. & Carlsson, G. E. (1994). Prosthodontic Complications in Osseointegrated Dental Implant Treatment. *The International journal of oral & maxillofacial implants,* Vol.9, No.1, (Jan-Feb), pp. 90-94, ISSN 0882-2786

Ciftci, Y. & Canay, S. (2001). Stress Distribution on the Metal Framework of the Implant-Supported Fixed Prosthesis Using Different Veneering Materials. *The International journal of prosthodontics,* Vol.14, No.5, (Sep-Oct), pp. 406-411, ISSN 0893-2174

Clement, R.; Schneider, J.; Brambs, H. J.; Wunderlich, A.; Geiger, M. & Sander, F. G. (2004). Quasi-Automatic 3d Finite Element Model Generation for Individual Single-Rooted Teeth and Periodontal Ligament. *Computer methods and programs in biomedicine,* Vol.73, No.2, (Feb), pp. 135-144, ISSN 0169-2607

Cowin, S. C. & editor (2001). *Bone Mechanics Handbook,* CRC Press, ISBN 0849391172, Boca Raton, USA

de Almeida, E. O.; Rocha, E. P.; Freitas, A. C., Jr. & Freitas, M. M., Jr. (2010). Finite Element Stress Analysis of Edentulous Mandibles with Different Bone Types Supporting Multiple-Implant Superstructures. *The International journal of oral & maxillofacial implants,* Vol.25, No.6, (Nov-Dec), pp. 1108-1114, ISSN 0882-2786

Delben, J. A.; Gomes, E. A.; Barao, V. A.; Kuboki, Y. & Assuncao, W. G. (2011). Evaluation of the Effect of Retightening and Mechanical Cycling on Preload Maintenance of Retention Screws. *The International journal of oral & maxillofacial implants,* Vol.26, No.2, (Mar-Apr), pp. 251-256, ISSN 1942-4434

Dellinges, M. A. & Tebrock, O. C. (1993). A Measurement of Torque Values Obtained with Hand-Held Drivers in a Simulated Clinical Setting. *Journal of prosthodontics : official journal of the American College of Prosthodontists,* Vol.2, No.4, (Dec), pp. 212-214, ISSN 1059-941X

Ding, X.; Zhu, X. H.; Liao, S. H.; Zhang, X. H. & Chen, H. (2009). Implant-Bone Interface Stress Distribution in Immediately Loaded Implants of Different Diameters: A Three-Dimensional Finite Element Analysis. *Journal of prosthodontics : official journal of the American College of Prosthodontists,* Vol.18, No.5, (Jul), pp. 393-402, ISSN 1532-849X

Drage, N. A.; Palmer, R. M.; Blake, G.; Wilson, R.; Crane, F. & Fogelman, I. (2007). A Comparison of Bone Mineral Density in the Spine, Hip and Jaws of Edentulous Subjects. *Clinical oral implants research,* Vol.18, No.4, (Aug), pp. 496-500, ISSN 0905-7161

Friberg, B.; Jemt, T. & Lekholm, U. (1991). Early Failures in 4,641 Consecutively Placed Branemark Dental Implants: A Study from Stage 1 Surgery to the Connection of Completed Prostheses. *The International journal of oral & maxillofacial implants,* Vol.6, No.2, (Summer), pp. 142-146, ISSN 0882-2786

Gao, J.; Xu, W. & Ding, Z. (2006). 3d Finite Element Mesh Generation of Complicated Tooth Model Based on Ct Slices. *Computer methods and programs in biomedicine,* Vol.82, No.2, (May), pp. 97-105, ISSN 0169-2607

Geng, J. P.; Tan, K. B. & Liu, G. R. (2001). Application of Finite Element Analysis in Implant Dentistry: A Review of the Literature. *The Journal of prosthetic dentistry,* Vol.85, No.6, (Jun), pp. 585-598, ISSN 0022-3913

Gomes, E. A.; Assuncao, W. G.; Tabata, L. F.; Barao, V. A.; Delben, J. A. & de Sousa, E. A. (2009). Effect of Passive Fit Absence in the Prosthesis/Implant/Retaining Screw

System: A Two-Dimensional Finite Element Analysis. *The Journal of craniofacial surgery*, Vol.20, No.6, (Nov), pp. 2000-2005, ISSN 1536-3732

Gomes, E. A.; Assuncao, W. G.; Barao, V. A.; Rocha, E. P. & de Almeida, E. O. (2011). Effect of Metal-Ceramic or All-Ceramic Superstructure Materials on Stress Distribution in a Single Implant-Supported Prosthesis: Three-Dimensional Finite Element Analysis. *The International journal of oral & maxillofacial implants*, Vol.26, No.6, (Nov-Dec), pp. in press, ISSN 1942-4434

Gracis, S. E.; Nicholls, J. I.; Chalupnik, J. D. & Yuodelis, R. A. (1991). Shock-Absorbing Behavior of Five Restorative Materials Used on Implants. *The International journal of prosthodontics*, Vol.4, No.3, (May-Jun), pp. 282-291, ISSN 0893-2174

Himmlova, L.; Dostalova, T.; Kacovsky, A. & Konvickova, S. (2004). Influence of Implant Length and Diameter on Stress Distribution: A Finite Element Analysis. *The Journal of prosthetic dentistry*, Vol.91, No.1, (Jan), pp. 20-25, ISSN 0022-3913

Holmes, D. C. & Loftus, J. T. (1997). Influence of Bone Quality on Stress Distribution for Endosseous Implants. *The Journal of oral implantology*, Vol.23, No.3, pp. 104-111, ISSN 0160-6972

Huang, H. L.; Hsu, J. T.; Fuh, L. J.; Tu, M. G.; Ko, C. C. & Shen, Y. W. (2008). Bone Stress and Interfacial Sliding Analysis of Implant Designs on an Immediately Loaded Maxillary Implant: A Non-Linear Finite Element Study. *Journal of dentistry*, Vol.36, No.6, (Jun), pp. 409-417, ISSN 0300-5712

Iplikcioglu, H. & Akca, K. (2002). Comparative Evaluation of the Effect of Diameter, Length and Number of Implants Supporting Three-Unit Fixed Partial Prostheses on Stress Distribution in the Bone. *Journal of dentistry*, Vol.30, No.1, (Jan), pp. 41-46, ISSN 0300-5712

Jaffin, R. A.; Kumar, A. & Berman, C. L. (2004). Immediate Loading of Dental Implants in the Completely Edentulous Maxilla: A Clinical Report. *The International journal of oral & maxillofacial implants*, Vol.19, No.5, (Sep-Oct), pp. 721-730, ISSN 0882-2786

Jemt, T. & Lekholm, U. (1995). Implant Treatment in Edentulous Maxillae: A 5-Year Follow-up Report on Patients with Different Degrees of Jaw Resorption. *The International journal of oral & maxillofacial implants*, Vol.10, No.3, (May-Jun), pp. 303-311, ISSN 0882-2786

Khera, S. C.; Goel, V. K.; Chen, R. C. & Gurusami, S. A. (1988). A Three-Dimensional Finite Element Model. *Operative dentistry*, Vol.13, No.3, (Summer), pp. 128-137, ISSN 0361-7734

Klineberg, I. J. & Murray, G. M. (1985). Design of Superstructures for Osseointegrated Fixtures. *Swedish dental journal Supplement*, Vol.28, pp. 63-69, ISSN 0348-6672

Koca, O. L.; Eskitascioglu, G. & Usumez, A. (2005). Three-Dimensional Finite-Element Analysis of Functional Stresses in Different Bone Locations Produced by Implants Placed in the Maxillary Posterior Region of the Sinus Floor. *The Journal of prosthetic dentistry*, Vol.93, No.1, (Jan), pp. 38-44, ISSN 0022-3913

Kopp, C. D. (1990). Overdentures and Osseointegration. Case Studies in Treatment Planning. *Dental clinics of North America*, Vol.34, No.4, (Oct), pp. 729-739, ISSN 0011-8532

Korioth, T. W. & Versluis, A. (1997). Modeling the Mechanical Behavior of the Jaws and Their Related Structures by Finite Element (Fe) Analysis. *Critical reviews in oral biology and medicine : an official publication of the American Association of Oral Biologists,* Vol.8, No.1, pp. 90-104, ISSN 1045-4411

Kunavisarut, C.; Lang, L. A.; Stoner, B. R. & Felton, D. A. (2002). Finite Element Analysis on Dental Implant-Supported Prostheses without Passive Fit. *Journal of prosthodontics : official journal of the American College of Prosthodontists,* Vol.11, No.1, (Mar), pp. 30-40, ISSN 1059-941X

Lang, L. A.; Kang, B.; Wang, R. F. & Lang, B. R. (2003). Finite Element Analysis to Determine Implant Preload. *The Journal of prosthetic dentistry,* Vol.90, No.6, (Dec), pp. 539-546, ISSN 0022-3913

Liao, S. H.; Tong, R. F. & Dong, J. X. (2008). Influence of Anisotropy on Peri-Implant Stress and Strain in Complete Mandible Model from Ct. *Computerized medical imaging and graphics : the official journal of the Computerized Medical Imaging Society,* Vol.32, No.1, (Jan), pp. 53-60, ISSN 0895-6111

Matsushita, Y.; Kitoh, M.; Mizuta, K.; Ikeda, H. & Suetsugu, T. (1990). Two-Dimensional Fem Analysis of Hydroxyapatite Implants: Diameter Effects on Stress Distribution. *The Journal of oral implantology,* Vol.16, No.1, pp. 6-11, ISSN 0160-6972

Meijer, H. J.; Kuiper, J. H.; Starmans, F. J. & Bosman, F. (1992). Stress Distribution around Dental Implants: Influence of Superstructure, Length of Implants, and Height of Mandible. *The Journal of prosthetic dentistry,* Vol.68, No.1, (Jul), pp. 96-102, ISSN 0022-3913

Menicucci, G.; Lorenzetti, M.; Pera, P. & Preti, G. (1998). Mandibular Implant-Retained Overdenture: Finite Element Analysis of Two Anchorage Systems. *The International journal of oral & maxillofacial implants,* Vol.13, No.3, (May-Jun), pp. 369-376, ISSN 0882-2786

Nagasao, T.; Kobayashi, M.; Tsuchiya, Y.; Kaneko, T. & Nakajima, T. (2003). Finite Element Analysis of the Stresses around Fixtures in Various Reconstructed Mandibular Models--Part Ii (Effect of Horizontal Load). *Journal of cranio-maxillo-facial surgery : official publication of the European Association for Cranio-Maxillo-Facial Surgery,* Vol.31, No.3, (Jun), pp. 168-175, ISSN 1010-5182

Natali, A. N.; Pavan, P. G. & Ruggero, A. L. (2006). Evaluation of Stress Induced in Peri-Implant Bone Tissue by Misfit in Multi-Implant Prosthesis. *Dental materials : official publication of the Academy of Dental Materials,* Vol.22, No.4, (Apr), pp. 388-395, ISSN 0109-5641

O'Mahony, A. M.; Williams, J. L. & Spencer, P. (2001). Anisotropic Elasticity of Cortical and Cancellous Bone in the Posterior Mandible Increases Peri-Implant Stress and Strain under Oblique Loading. *Clinical oral implants research,* Vol.12, No.6, (Dec), pp. 648-657, ISSN 0905-7161

Papavasiliou, G.; Kamposiora, P.; Bayne, S. C. & Felton, D. A. (1996). Three-Dimensional Finite Element Analysis of Stress-Distribution around Single Tooth Implants as a Function of Bony Support, Prosthesis Type, and Loading During Function. *The Journal of prosthetic dentistry,* Vol.76, No.6, (Dec), pp. 633-640, ISSN 0022-3913

Poiate, I. A.; Vasconcellos, A. B.; Mori, M. & Poiate, E., Jr. (2011). 2d and 3d Finite Element Analysis of Central Incisor Generated by Computerized Tomography. *Computer methods and programs in biomedicine,* Vol.104, No.2, (Nov), pp. 292-299, ISSN 1872-7565

Romeed, S. A.; Fok, S. L. & Wilson, N. H. (2006). A Comparison of 2d and 3d Finite Element Analysis of a Restored Tooth. *Journal of oral rehabilitation,* Vol.33, No.3, (Mar), pp. 209-215, ISSN 0305-182X

Sahin, S. & Cehreli, M. C. (2001). The Significance of Passive Framework Fit in Implant Prosthodontics: Current Status. *Implant dentistry,* Vol.10, No.2, pp. 85-92, ISSN 1056-6163

Sakaguchi, R. L. & Borgersen, S. E. (1995). Nonlinear Contact Analysis of Preload in Dental Implant Screws. *The International journal of oral & maxillofacial implants,* Vol.10, No.3, (May-Jun), pp. 295-302, ISSN 0882-2786

Sertgoz, A. (1997). Finite Element Analysis Study of the Effect of Superstructure Material on Stress Distribution in an Implant-Supported Fixed Prosthesis. *The International journal of prosthodontics,* Vol.10, No.1, (Jan-Feb), pp. 19-27, ISSN 0893-2174

Sevimay, M.; Turhan, F.; Kilicarslan, M. A. & Eskitascioglu, G. (2005). Three-Dimensional Finite Element Analysis of the Effect of Different Bone Quality on Stress Distribution in an Implant-Supported Crown. *The Journal of prosthetic dentistry,* Vol.93, No.3, (Mar), pp. 227-234, ISSN 0022-3913

Shapurian, T.; Damoulis, P. D.; Reiser, G. M.; Griffin, T. J. & Rand, W. M. (2006). Quantitative Evaluation of Bone Density Using the Hounsfield Index. *The International journal of oral & maxillofacial implants,* Vol.21, No.2, (Mar-Apr), pp. 290-297, ISSN 0882-2786

Silva, G. C.; Mendonca, J. A.; Lopes, L. R. & Landre, J., Jr. (2010). Stress Patterns on Implants in Prostheses Supported by Four or Six Implants: A Three-Dimensional Finite Element Analysis. *The International journal of oral & maxillofacial implants,* Vol.25, No.2, (Mar-Apr), pp. 239-246, ISSN 0882-2786

Skalak, R. (1983). Biomechanical Considerations in Osseointegrated Prostheses. *The Journal of prosthetic dentistry,* Vol.49, No.6, (Jun), pp. 843-848, ISSN 0022-3913

Spazzin, A. O.; Abreu, R. T.; Noritomi, P. Y.; Consani, R. L. & Mesquita, M. F. (2011). Evaluation of Stress Distribution in Overdenture-Retaining Bar with Different Levels of Vertical Misfit. *Journal of prosthodontics : official journal of the American College of Prosthodontists,* Vol.20, No.4, (Jun), pp. 280-285, ISSN 1532-849X

Tada, S.; Stegaroiu, R.; Kitamura, E.; Miyakawa, O. & Kusakari, H. (2003). Influence of Implant Design and Bone Quality on Stress/Strain Distribution in Bone around Implants: A 3-Dimensional Finite Element Analysis. *The International journal of oral & maxillofacial implants,* Vol.18, No.3, (May-Jun), pp. 357-368, ISSN 0882-2786

Taddei, F.; Martelli, S.; Reggiani, B.; Cristofolini, L. & Viceconti, M. (2006). Finite-Element Modeling of Bones from Ct Data: Sensitivity to Geometry and Material Uncertainties. *IEEE transactions on bio-medical engineering,* Vol.53, No.11, (Nov), pp. 2194-2200, ISSN 0018-9294

Takahashi, N.; Kitagami, T. & Komori, T. (1978). Analysis of Stress on a Fixed Partial Denture with a Blade-Vent Implant Abutment. *The Journal of prosthetic dentistry,* Vol.40, No.2, (Aug), pp. 186-191, ISSN 0022-3913

Tanino, F.; Hayakawa, I.; Hirano, S. & Minakuchi, S. (2007). Finite Element Analysis of Stress-Breaking Attachments on Maxillary Implant-Retained Overdentures. *The International journal of prosthodontics,* Vol.20, No.2, (Mar-Apr), pp. 193-198, ISSN 0893-2174

Tokuhisa, M.; Matsushita, Y. & Koyano, K. (2003). In Vitro Study of a Mandibular Implant
        Overdenture Retained with Ball, Magnet, or Bar Attachments: Comparison of Load
        Transfer and Denture Stability. *The International journal of prosthodontics*, Vol.16,
        No.2, (Mar-Apr), pp. 128-134, ISSN 0893-2174

Valen, M. & Locante, W. M. (2000). Laminoss Immediate-Load Implants: I. Introducing
        Osteocompression in Dentistry. *The Journal of oral implantology*, Vol.26, No.3, pp.
        177-184, ISSN 0160-6972

van Steenberghe, D.; Lekholm, U.; Bolender, C.; Folmer, T.; Henry, P.; Herrmann, I.;
        Higuchi, K.; Laney, W.; Linden, U. & Astrand, P. (1990). Applicability of
        Osseointegrated Oral Implants in the Rehabilitation of Partial Edentulism: A
        Prospective Multicenter Study on 558 Fixtures. *The International journal of oral &
        maxillofacial implants*, Vol.5, No.3, (Fall), pp. 272-281, ISSN 0882-2786

Weinstein, A. M.; Klawitter, J. J.; Anand, S. C. & Schuessler, R. (1976). Stress Analysis of
        Porous Rooted Dental Implants. *Journal of Dental Research*, Vol.55, No.5, (Sep-Oct),
        pp. 772-777, ISSN 0022-0345

Weinstein, A. M.; Klawitter, J. J. & Cook, S. D. (1979). Finite Element Analysis as an Aid to
        Implant Design. *Biomaterials, medical devices, and artificial organs*, Vol.7, No.1, pp.
        169-175, ISSN 0090-5488

Yettram, A. L.; Wright, K. W. & Pickard, H. M. (1976). Finite Element Stress Analysis of the
        Crowns of Normal and Restored Teeth. *Journal of Dental Research*, Vol.55, No.6,
        (Nov-Dec), pp. 1004-1011, ISSN 0022-0345

Yokoyama, S.; Wakabayashi, N.; Shiota, M. & Ohyama, T. (2005). Stress Analysis in
        Edentulous Mandibular Bone Supporting Implant-Retained 1-Piece or Multiple
        Superstructures. *The International journal of oral & maxillofacial implants*, Vol.20, No.4,
        (Jul-Aug), pp. 578-583, ISSN 0882-2786

Zarb, G. A. & Jansson, T. (1985). Prosthodontic Procedures, In: *Tissue-Integrated Prostheses:
        Osseointegration in Clinical Dentistry*, Branemark, P.-I., Zarb, G. A. & Albrektsson, T.
        editors, 241-282, Quintessence, ISBN 0867151293 / 0-86715-129-3, Chicago, USA

Zarone, F.; Apicella, A.; Nicolais, L.; Aversa, R. & Sorrentino, R. (2003). Mandibular Flexure
        and Stress Build-up in Mandibular Full-Arch Fixed Prostheses Supported by
        Osseointegrated Implants. *Clinical oral implants research*, Vol.14, No.1, (Feb), pp. 103-
        114, ISSN 0905-7161

Finite Element Analysis in Dentistry – Improving the Quality of Oral Health Care

Carlos José Soares[1], Antheunis Versluis[2],
Andréa Dolores Correia Miranda Valdivia[1],
Aline Arêdes Bicalho[1], Crisnicaw Veríssimo[1],
Bruno de Castro Ferreira Barreto[1] and Marina Guimarães Roscoe[1]

[1]*Federal University of Uberlândia,*
[2]*University of Tennessee,*
[1]*Brazil*
[2]*USA*

1. Introduction

The primary function of the human dentition is preparation and processing of food through a biomechanical process of biting and chewing. This process is based on the transfer of masticatory forces, mediated through the teeth (Versluis & Tantbirojn, 2011). The intraoral environment is a complex biomechanical system. Because of this complexity and limited access, most biomechanical research of the oral environment such as restorative, prosthetic, root canal, orthodontic and implant procedures has been performed in vitro (Assunção et al., 2009). In the in vitro biomechanical analysis of tooth structures and restorative materials, destructive mechanical tests for determination of fracture resistance and mechanical properties are important means of analyzing tooth behavior. These tests, however, are limited with regard to obtaining information about the internal behavior of the structures studied. Furthermore, biomechanics are not only of interest at the limits of fracture or failure, but biomechanics are also important during normal function, for understanding property-structure relationships, and for tissue response to stress and strain. For a more precise interrogation of oral biomechanical systems, analysis by means of computational techniques is desirable.

When loads are applied to a structure, structural strains (deformation) and stresses are generated. This is normal, and is how a structure performs its structural function. But if such stresses become excessive and exceed the elastic limit, structural failure may result. In such situations, a combination of methodologies will provide the means for sequentially analyzing continuous and cyclic failure processes (Soares et al., 2008). Stresses represent how masticatory forces are transferred through a tooth or implant structure (Versluis & Tantbirojn, 2011). These stresses cannot be measured directly, and for failure in complex structures it is not easy to understand why and when a failure process is initiated, and how we can optimize the strength and longevity of the components of the stomatognathic system. The relationship between stress and strain is expressed in constitutive equations according to universal physical laws. When dealing with physically and geometrically complex systems, an engineering concept that uses a numerical analysis to solve such equations becomes inevitable. Finite Element Analysis (FEA) is a widely used numerical analysis that has been applied successfully in many engineering and bioengineering areas since the 1950s. This computational numerical analysis can be considered the most comprehensive method currently available to calculate the complex conditions of stress distributions as are encountered in dental systems (Versluis & Tantbirojn, 2009).

The concept of FEA is obtaining a solution to a complex physical problem by dividing the problem domain into a collection of much smaller and simpler domains in which the field variables can be interpolated with the use of shape functions. The structure is discretized into so called "elements" connected through nodes. In FEA choosing the appropriate mathematical model, element type and degree of discretization are important to obtain accurate as well as time and cost effective solutions. Given the right model definition, FEA is capable of computationally simulating the stress distribution and predicting the sites of stress concentrations, which are the most likely points of failure initiation within a structure or material. Other advantages of this method compared with other research methodologies are the low operating costs, reduced time to carry out the investigation and it provides information that cannot be obtained by experimental studies (Soares et al. 2008).

However, FEA studies cannot replace the traditional laboratory studies. FEA needs laboratory validation to prove its results. The properties and boundary conditions dentistry is dealing with are complex and often little understood, therefore requiring assumptions and simplifications in the modeling of the stress-strain responses. Furthermore, large anatomical variability precludes conclusions based on unique solutions. The most powerful application of FEA is thus when it is conducted together with laboratory studies. For example, the finite element method can be performed before a laboratory study as a way to design and conduct the experimental research, to predict possible errors, and serve as a pilot study for the standardization protocols. The use of this methodology can also occur after laboratory experimental tests in order to explain ultra-structural phenomena that cannot be detected or isolated. The identification of stress fields and their internal and external distribution in the specimens may therefore help answer a research hypothesis (Ausiello et al., 2001).

The complexity of a FEA can differ depending on the modeled structure, research question, and available knowledge or operator experience. For example, FEA can be performed using two-dimensional (2D) or three-dimensional (3D) models. The choice between these two models depends on many inter-related factors, such as the complexity of the geometry, material properties, mode of analysis, required accuracy and the applicability of general findings, and finally the time and costs involved (Romeed et al., 2004; Poiate et al., 2011). 2D FEA is often performed in dental research (Soares et al., 2008; Silva et al., 2009; Soares et al., 2010). The advantage of a 2D-analysis is that it provides significant results and immediate insight with relatively low operating cost and reduced analysis time. However, the results of 2D models also have limitations regarding the complexity of some structural problems. In contrast, 3D FEA has the advantage of more realistic 3D stress distributions in complex 3D geometries (Fig. 1). However, creating a 3D model can be considered more costly, because it is more labor-intensive and time-consuming and may require additional technology for acquiring 3D geometrical data and generation of models (Santos-Filho et al., 2008).

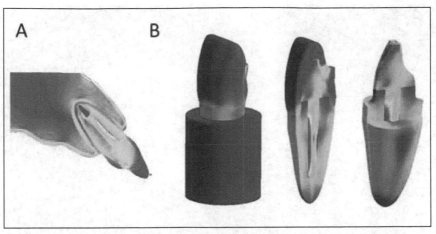

Fig. 1. Upper central incisor restored with cast post-and-cores. A) 2D FEA model, B) 3D FEA model with different cutting planes, showing internal stress distributions (ANSYS 12 Workbench - Ansys Inc., Houston, USA). (Santos-Filho, 2008).

In dental research, FEA has been used effectively in many research studies. For example, FEA has been used to analyze stress generation during the polymerization process of composite materials and stress analyses associated with different restorative protocols like tooth implant, root post canal, orthodontic approaches (Versluis et al., 1996; Versluis et al., 1998; Ausiello et al., 2001; Lin et al., 2001; Ausiello et al., 2002; Versluis et al., 2004; Misra et al., 2005; Meira et al., 2007; Witzel et al., 2007; Meira et al., 2010). This chapter will discuss the application and potential of finite element analysis in biomechanical studies, and how this method has been instrumental in improving the quality of oral health care.

2. Application of finite element analysis in dentistry - Modeling steps: Geometry, properties, and boundary conditions

The FEA procedure consists of three steps: pre-processing, processing and post-processing.

2.1 Pre-processing: Building a model

Pre-processing involves constructing the "model". A model consists of: (1) the geometrical representation, (2) the definition of the material properties, and (3) the determination of what loads and restraints are applied and where. Model construction is often difficult, because biological structures have irregular shapes, consist of different materials and/or compositions, and the exact loading conditions can have a large effect on the outcome. Therefore, the correct construction of a model to obtain accurate results from a FEA is very important. The development of FEA models can follow different protocols, depending on the aim of the study. Models used to analyze laboratory test parameters, like microtensile bond tests, flexural tests, or push-out tests usually have the simplest geometries and can be generated directly into the FEA software (Fig. 2.). Modeling of 2D and 3D biological structures are often more intricate, and may have to be performed with Computer Aided Design (CAD) or Bio-CAD software. This chapter mainly discusses 3D Bio-CAD modeling.

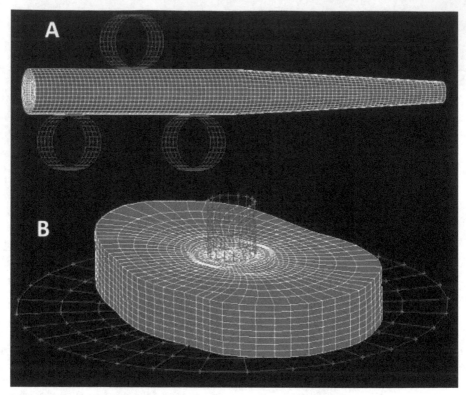

Fig. 2. Finite element models of test specimens made directly in FEA software.

2.2 Bio-CAD protocol for 3D modeling of organic structures

The modeling technique often used in bioengineering studies is called Bio-CAD, and consists of obtaining a virtual geometric model of a structure from anatomical references (Protocol developed in the Center for Information Technology Renato Archer, Travassos, 2010). The obtained geometrical model consists of closed volumes or solid shapes, in which a mesh distribution of discrete elements can be generated. The shape of the object of study can be reconstructed as close to reality as possible, for example, by reducing the size of the elements in regions that require more details. However, higher detail and thus reducing the element sizes will increase the total number of elements and consequently, the computational requirements. Modeling Bio-CAD involves the stages of obtaining the base-geometry, creation of reference curves, construction of surface areas, union of surfaces for generation of solids and exportation of the model to FEA software.

2.2.1 Obtaining the base-geometry

References for model creation, whether 2D or 3D, are images of the structure that is modeled. Modeling of biological structures for the finite element method usually requires CAD techniques. For 2D models, the modeling is made from the images or planar sections of a structure (photograph, tomography or radiograph) (Fig. 3).

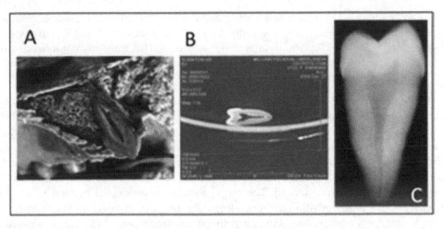

Fig. 3. Images used to build a two-dimensional model. A) Photographs, B) Section plane of computed tomography, C) Radiograph.

These images can be imported into different software programs that can digitize reference points of the structure, such as Image J (available at http://imagej.nih.gov). These points can be exported as a list of coordinates, which can subsequently be imported into a finite element program, for example MENTAT-MARC package (MSC. Software Corporation, Santa Ana, CA, USA), or CAD software, such as Mechanical Desktop (Autodesk, San Rafael, CA, USA) that can generate IGES-files that can be read by most FEA software. The imported reference points can be used to outline the shape of the modeled structure or materials, and hence the finite element mesh.

NURBS Modeling (Non Uniform Rational Bezier Spline) is one of several methods applied for building 3D models. This methodology involves a model creation from a base geometry in STL (stereolithography) format. Obtaining an STL-file, consisting of a mesh of triangular surfaces created from a distribution of surface points, is a critical step for 3D modeling. Several methods have been described in the literature (Magne, 2007; Soares et al., 2008). The

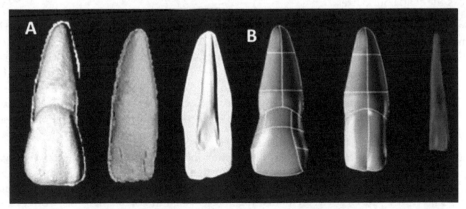

Fig. 4. NURBS modeling. A) STL model (stereolithography), B) NURBS-based geometry created from the STL (Santos-Filho, 2008).

STL file can be obtained by computed tomography, Micro-CT, magnetic resonance imaging (MRI) or optical, contact or laser scanning. Using CAD software, NURBS curves can be defined that follow the anatomical details of the structure. This transformation from surface elements to a NURBS-based representation allows for greater control of the shape and quality of the resulting finite element mesh (Fig. 4).

Our research group has used this strategy to create models of the tooth. First the outer shape of an intact tooth is scanned using a laser scanner (LPX 600, Roland DG, Osaka, Japan). Next, the enamel is removed by covering the root surface with a thin layer of nail polish, and immersing the tooth in 10% citric acid for 10 minutes in an ultra-sound machine. Using a stereomicroscope (40X) the complete removal of enamel can be confirmed. Then the tooth is scanned again and the two shapes (sound tooth and dentin) are fit using PixForm Pro II software (Roland DG, Osaka, Japan). The pulp geometry is generated by two X-ray images obtained from the tooth positioned bucco-lingually and mesio-distally. These images are exported to Image J software where the pulp shapes are traced and digitized, and eventually merged with the scanned tooth and dentin surfaces.

2.2.2 NURBS Modeling: Creation of the curves, surfaces and solids

NURBS Modeling or irregular surface modeling begins with planning the number and position of curves that will represent the main anatomic landmarks of the models, justifying the level of detail in each case. From these curves surfaces and volumes (solids) will be created. The NURBS curves will determine the quality of the model, and consequently, the quality of the finite element mesh. The modeling strategy begins with knowledge of the anatomy of the structure to be studied. The curves should be as regular as possible, and should not form a very small or narrow area with sharp angles, as this would hinder the formation of meshes. The boundary conditions, defined by external restrictions, contact structures and loading definition, must already be defined at the time of construction of lines and surfaces. The curves should provide continuity to ensure that the model will result in closed volumes. If models are made up of multiple solids, NURBS curves from adjacent solids should have the same point of origin to facilitate the formation of a regular mesh across the solid boundaries.

After curves have been defined, surfaces can be created using three or four curves each. The formation of surfaces should follow a chess pattern to prevent wrinkling of the end surfaces caused by the assigned tangency between the surfaces (Fig. 5). This makes it possible to choose the form of tangency between the surfaces and avoid creases in the models that would become areas of mesh complications and consequently locations of erroneous stress concentrations in the final finite element model. It is recommended that there is continuity of curvature between the surfaces. Finally the surfaces should be joined to form a closed NURBS volume.

Most cases involve more than one solid, with different materials and contact areas defined, among other features. In these cases, a classification is assigned to *multi-bodies*. Another important requirement is that there can be no intersection between bodies. There should also be no empty spaces between solids in contact, which in contact analysis would cause single contacts with associated stress peaks, or would cause gaps for intended bonded interfaces. In order to avoid these problems, it is recommended that the contact surfaces of both bodies are identical and coincide using commands such as *Boolean Operations*, or copying common surfaces.

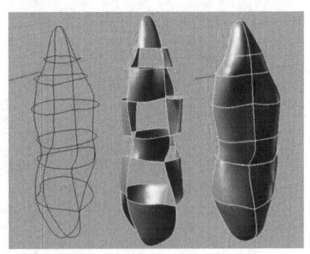

Fig. 5. Creation of surfaces and solids from reference curves in Rhinoceros 3D (Robert McNeel & Associates, USA).

2.2.3 Exporting the solids

The export model is usually saved in STEP (STP) format or IGES (IGS) format. The choice of format depends on the compatibility with the pre-processing software of the FEA program. Also be aware of the units (chosen at the beginning of modeling, usually in millimeters) before importing the model into the pre-processing software. Before exporting, it is recommended to carefully re-check the model to avoid rework: ensure that solids are closed, check for acute angles in surfaces or discontinuities, check for very short edges, check for surfaces that are too narrow or small, and inspect intersections between solids (Travassos, 2010).

2.3 Material properties

Material properties can be determined by means of mechanical tests and applied for any material with the same characteristics. Specimens and procedures can be carried out following agreed testing standards (ASTM - American Society for Testing and Materials). The minimum properties required for most linear elastic isotropic finite element analyses are the elastic modulus and Poisson's ratio.

2.3.1 Methods for obtaining material properties used in FEA

The elastic modulus (E) represents the inherent stiffness of a material within the elastic range, and describes the relationship between stress and strain. The elastic modulus can thus be determined from the slope of a stress/strain curve. Such relationship can be acquired by means of a uniaxial tensile test in the elastic regime (Chabrier et al., 1999). The modulus of elasticity is defined as:

$$E = \sigma/\varepsilon \tag{1}$$

where (σ) is the stress and (ε) is the strain (ratio between amount of deformation and original dimension).

Various methods have been used to measure the elastic modulus (Chung et al., 2005; Vieira et al., 2006; Boaro et al., 2010; Suwannaroop et al., 2011). For dental materials and tissues, the classical uniaxial tensile test is often problematic due to small specimen dimensions dictated by size, cost, and/or manufacturing limitations. Therefore, other methods such as 3-point bending, indentation, nanoindentation and ultrasonic waves have been used to determine the elastic modulus. Using a Knoop hardness setup, the elastic modulus of composites can be estimated with an empirical relationship, yielding a simple and low cost method (Marshall et al., 1982). Using the dimensions of the short and long diagonals of the indentation, the elastic modulus (GPa) can be determined using the following equation:

$$E = 0.45 \; KHN/((0.140647-d/D) \; 100) \tag{2}$$

where KHN is the Knoop Hardness (kg/mm^2), d is the short diagonal of the indentation, D is the long diagonal of the indentation, and 0.140647 is the ratio of the short and long diagonals of the Knoop indenter (1/7.11). Nanoindentation systems have also been used for this purpose. The elastic modulus from nanoindentation is obtained from the data generated in the load-displacement curve by means of the equation (Suwannaroop et al., 2011):

$$1/E^* = (1-v2)/E + (1-v2)/E' \tag{3}$$

where E* is the reduced modulus from the nanoindenter, E is the modulus of the Berkovich diamond indenter (1,050 GPa), E' is the modulus of the specimen, υ is the Poisson's ratio for the indenter (0.07)28, and υ' is the Poisson's ratio for the specimens.

The ISO 4049 (Dentistry - Resin based dental fillings) provides a standard for the use of three-point bending tests for determining the flexural modulus (elastic modulus) for composites. Generally, the preparation of specimens for microindentation tests is easier, specimen size is smaller and it has been suggested that their results are more consistent than with an ISO 4049 three-point bending test (Chung et al., 2005).

The analysis of anisotropic materials (i.e., materials with different stress-strain responses in different directions) requires the application of elastic moduli and Poisson's ratios in 3 directions (2 in case of orthotropy), as well as shear moduli in those directions. It is well accepted that enamel is not isotropic, but the anisotropy of dentin is less well established. Analyzing the effect of anisotropy in dentin, the presence and direction of dentinal tubules were not found to affect the mechanical response, indicating that dentin behaved homogeneous and isotropic (Peyton et al., 1952). More recently, some heterogeneity and anisotropy was demonstrated for dentin. However, the stiffness response seems to be only mildly anisotropic (Wang & Weiner, 1998; Kinney et al., 2004; Huo, 2005). Therefore, dentin properties in FEA are usually assumed to be isotropic. Potential simplifications such as the assumption of linear-elastic isotropic material behavior may be necessary in FEA simulations due to the difficulty of obtaining the correct directional properties, or the need to reduce the complexity of an analysis. As in other research approaches, some simplifications and assumptions are also common in FEA, and are permissible provided that their impact on the conclusions is carefully taken into account. It has been shown, for example, that the assumption of isotropic properties for enamel did not change the conclusions of a shrinkage stress analysis (Versluis & Tantbirojn, 2011).

The other mandatory property for a FEA, the Poisson's ratio, is the ratio of lateral contraction and longitudinal elongation of a material subjected to a uniaxial load (Chabrier et al., 1999). Among the static methods are tensile and compression tests, in which a uniaxial stress is applied to the material and the Poisson's ratio is calculated from the resulting axial and transverse strains. Another method uses ultrasound (resonance), where the Poisson's ratio is obtained from the speed or natural frequency of the generated longitudinal and transverse waves.

2.3.2 Type of structural analysis: Linear and nonlinear analysis

The type of structural analysis depends on the subject that is being modeled. Depending on the model, the FEA can be linear or nonlinear. Linear or nonlinear analysis refers to the proportionality of the solutions. A solution is linear if the outcome is independent of its loading history. For example, an analysis is linear if the outcome will be the same irrespective of if the load is applied in one or multiple increments. Some conditions are inherently nonlinear, such as nonlinear material responses (e.g., rate-dependent properties or viscoelasticity, plastic deformation), time-dependent boundary conditions (e.g., contact analysis where independent bodies interact), or geometric instabilities (e.g., buckling). Sometimes linear conditions become nonlinear when general assumptions become invalid. For example, the stress-strain responses are generally based on the assumption of small displacements. When large deformations occur, the numerical solution procedures must be adjusted. Most high-end FEA software programs have the capability to resolve nonlinear equation systems. For the end-user, the difference between submitting a linear or nonlinear analysis is minimal, and usually only involves the prescription of multiple increments or invoking an alternative solver for the ensuing nonlinear solution. Since nonlinear systems potentially have multiple solutions, nonlinear analyses should also be checked more thoroughly for the convergence to the correct solution. Nonlinear solutions require more computational iterations to converge to a final solution, therefore nonlinear analyses are more costly in terms of computation and time.

Nonlinear FEA is a powerful tool to predict stress and strain within structures in situations that cannot be simulated in conventional linear static models. However the determination of elastic, plastic, and viscoelastic material behavior of the materials involved requires accurate mechanical testing prior to FEA. The experimental determination of mechanical properties continues to be a major challenge and impediment for more accurate FEA. For example, periodontal ligament (PDL) is a dental tissue structure with significant viscoelastic behavior, and simulation using nonlinear analysis would be more realistic. However, due to its complex structure; the exact mechanical properties of PDL must still be considered poorly understood. It such case it can be argued that using incorrect or questionable nonlinear mechanical properties in a FEA may be more obscuring than a well defined and understood simplification.

An example of the need for a nonlinear analysis is the simulation of the mechanical behavior across an interface. Interfacial areas are among the most important areas for the performance of materials or structures. Interfaces between different materials can often be modeled as a perfect bond, where nodes are shared across the interface (Wakabayashi et al., 2008). The simulations of such interface can normally be conducted in a linear analysis. However, depending the actual conditions at a simulated interface, such perfect fusion can occasionally lead to unrealistic results. Fig. 6 shows a 2D FEA model of a root filled tooth restored with a cast post and core and a fiberglass post and composite core. When perfect bonding was assumed in a linear analysis, the stress distribution indicated higher stress concentrations in the cast post and core compared to the fiberglass post, while the stress distribution in the root dentin was nearly identical between these two models. Experimental failure data showed, however, that the failure modes of the cast post and core group were more catastrophic and involved longitudinal root fractures while all fractures of the root with fiberglass post were coronal fractures. Simulating the interface more realistically with friction between resin cement and cast post and core (requiring a nonlinear analysis) rather than a perfect fusion, the stress distribution changed substantially between the two post types (Fig. 6), and yielded more realistic results when compared with the experimental observations.

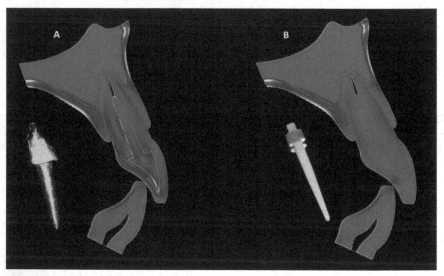

Fig. 6. Nonlinear FEA of endodontic treated tooth restored with A. Cast post and core and B. Fiberglass post.

2.4 Mesh generation

In FEA the whole domain is divided into smaller elements. The collection and distribution of these elements is called a mesh. Elements are interconnected by nodes, which are thus the only points though which elements interact with each other. The process of creating an element mesh is referred to as "discretization" of the problem domain (Geng et al., 2001).

There are many different types of elements. One of the differences can be their basic shape, such as triangular, tetrahedral, hexahedral, etc. Triangular or tetrahedral elements are popular because automatic meshing software routines are easier to develop and thus more advanced for those shapes. Automatic generation of element distributions is especially useful in bioengineering, which often deals with irregular geometries. Since few modeled geometries have perfectly square dimensions or even straight edges, element shapes must be adapted to fit. Note that the accuracy of elements deteriorates the further they are distorted from their ideal basic shape. Besides their basic geometrical shapes, elements can differ in the way they are solved, such as linear or quadratic interpolation. This refers to how stress and strain is interpolated within an element.

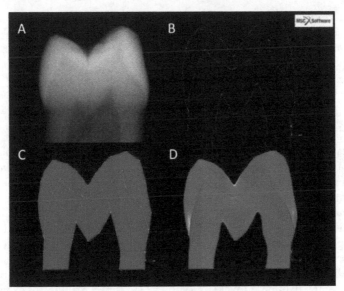

Fig. 7. Radiography of maxillary premolar (A); Lines plotted in the MARC/MENTAT software (B); Manual creation of mesh (C); Final subdivision for improving the mesh quality (D).

Most FEA software provides *automesh* or automatic mesh generation options. The program may suggests the size and number of elements or allow manual control for generating the element meshes. Manual mesh generation can give good results for 2D models (Fig. 7). However, most 3D models rely on automated mesh generators because manual creation of 3D models is very time-consuming. Still, various aspects of the 3D automeshing need to be controlled manually, such as the number of elements required in a given pre-selected area, the distribution of elements, the range of element sizes within a model, uniting or dividing elements, etc. The manual controls also allows selective distribution of elements, for example, more refined meshes in special regions of interest (contact interfaces, geometric discontinuities) while creating coarser mesh distributions in regions of less interest.

In finite element modeling, a finer (denser) mesh should allow a more accurate solution. However, as a mesh is made finer and the element count increases, the computation time also increases. How can you get a mesh that balances accuracy and computing resources? One way is to perform a convergence study. This process involves the creation and analysis of multiple mesh distributions with increasing number of elements or refinements. When results with the various models are plotted, a convergence to a particular solution can be found. Based on this convergence data, an estimation of the error can be made for the various mesh distributions. A mesh convergence study can thus be used to find a balance between an efficient mesh distribution and an acceptably accurate solution within the limitations of the computing resources. Moreover, a convergence test can verify if an obtained solution is true or if it was an artifact of a particular element distribution.

2.5 Boundary conditions

The boundary conditions define the external influences on a modeled structure, usually loading and constraints. Boundary conditions are associated with six degrees of freedom (DOF). The combination of all boundary conditions of a FEA model must represent the procedural conditions to which the actual structure that is simulated is subjected. The choice and application of boundary conditions is extremely important, because they determine the outcome of the FEA.

2.5.1 Prescribed displacement – Fixation and symmetry

In a simple way, restrictions can be summarized as the imposition of displacements and rotations on a finite element model, which can be either null or have fixed values or rates. These restrictions concern three rotations (around X, Y, Z-axes) and three translations (in X, Y, Z-directions). Static analysis requires sufficient fixation of a model to remain in place. Insufficient fixation will lead to instability and failure to reach a numerical solution in the FEA. Since nodes are the points through which elements communicate, boundary conditions are usually applied to nodes, where in a 3D model each free node has 6 degrees of freedom (3 translations and 3 rotations). Although some FEA software may allow application of boundary conditions to element edges or surfaces, they are extrapolated to the associated nodes. To achieve the fixation mimicking a support system in real life, for example complete immobilization of a modeled specimen in a test fixture, displacement constraints can be applied to nodes located in a region equivalent to those of the real support system. Symmetry can be viewed as a form of fixation. Since all displacements are mirrored, the displacement across the symmetry-axis is zero.

2.5.2 Load application

The application of loads in a FEA model must also represent the external loading situations to which the modeled structure is subjected. These loads can be tensile, compressive, shear, torque, etc. To simulate the masticatory forces, loads have been applied using different methods, for example point loads, distributed loads across a specific area, and by means of a simulated opposing cusp of the antagonist tooth (Fig. 8). A point load application may result in high stress concentrations around the loaded nodes, creating unrealistic stress concentrations. In reality, a masticatory contact force is likely to be distributed across certain contact areas on both the buccal and lingual cuspal inclines. However, the most realistic load application is not always the best choice for all research questions. Contact areas move depending on stiffness and thus deformation of both opposing teeth. If contact areas change, contact loads change also, which can have significant effect on the stress distribution. When a research question requires well-defined load conditions, point loads or prescribed distributed loading may be better choices than the seemingly more realistic simulated tooth contact.

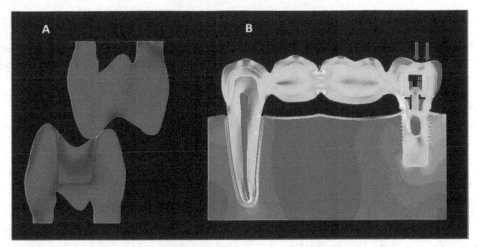

Fig. 8. Load application A. Contact analysis using antagonist tooth (Marc/Mentat software); B. Point load application (Ansys Inc., Houston, USA).

3. Evaluation of finite element analysis

FEA results can be intimidating, the amount of generated data (displacements, strains, stresses, temperatures, etc) is almost unlimited. However, one of the most powerful features of FEA is that results can be easily visualized and made accessible. Visualization of the results can be done by showing data distributions using a colour scale, where each colour corresponds to a range of values. Furthermore, deformations and displacements can be shown by comparing the original unstressed model outlines with the outline of the model under stress. Based on this immediate visualized output, the operator can investigate the displacement of the structure, the type of movement that was performed, which region has a higher dislodgement, or how to redistribute the stresses in the analyzed structure in either three dimensions or in two-dimensions. Such a structural analysis allows the determination of stress and strain resulting from external force, pressure, thermal change, and other factors. This section discusses how results from a finite element analysis should be evaluated, starting with a check of the model, followed by checking the outcome. Then the relationship between finite element and experimental will be discussed with respect to the limitations of either method.

3.1 Analysis of coherency

All finite element analyses should first be checked for coherence (or sanity). The first step of a coherence analysis is to visualize the displacements and deformations to verify that the simulated model moves in the expected direction. The second step of a coherence analysis is to analyze if the distribution stresses is as expected.

3.2 Validation of the outcome

After the model definition is confirmed to be sound, the validity of the outcome still has to be validated. A finite element analysis is modelled based on geometric, property, and boundary conditions, each of which may have required significant assumptions or simplifications. The purpose of validation is thus to confirm that the general response of the model is realistic. Unlike stress, which cannot be measured directly, deformation or displacement can be directly measured. Therefore, displacement is often a good choice for comparing the simulated behaviour with the behaviour observed in reality, even if the simulation was a "stress" analysis. Validation can be achieved by comparing the outcome with published results from validated analyses or with laboratory measurements. Examples are strain gauge measurement, cuspal flexure, bending displacements, etc. Effects of stresses can also be indirectly validated through the observation of their expected effects, such as crack initiation and fractures. It is not realistic to expect an exact fit between experimental and numerical results, because even between experimental results there will not be an exact fit due to natural and experimental variations. Therefore it is important to remember that it is not an exact fit that validates a finite element analysis but rather the similarity in general tendencies (Versluis et al., 1997).

3.3 Interpretation of the results

After a finite element analysis has been checked and validated, it can be used to interpret the research question. Most finite element analysis results should be interpreted qualitatively. Quantitative interpretation can sometimes be justified, provided all input is verifiable and quantitatively validated. The current state of the art of the use of finite element modelling in dentistry indicates that predictive results are still best viewed in a qualitative manner. It is the search for optimal balance between the objectives of a study, computational efforts (accuracy and efficiency), and practical limitations that ultimately determines the value of a finite element model. Since most finite element models are linear, errors in magnitudes of the loads will not have a direct effect on qualitative predictions. However, small changes in types of boundary conditions such as the location of the loading can substantially alter even the qualitative performance predictions.

Biomechanical performance involves efficient function as well as failure. One of the failure mechanisms is loss of structural integrity, which can eventually result in loss of function. FEA can be used in the determination of fracture mechanics parameters, and examination of experimental failure test methods. In the dental FEA literature, failure is usually extrapolated from maximum stress values, where stress concentrations are identified as possible locations for failure initiation and relative concentration values are interpreted as related to the failure risk. When using this process for interpreting failure behaviour, it is important to carefully assess the stress concentration locations because they may depend heavily on the chosen modelling and boundary condition options (Korioth & Versluis, 1997). Furthermore, stress distributions change when a crack propagates. Therefore, researchers should be extremely cautious about extrapolating crack behaviour based on the distribution of stress concentrations from a static analysis.

Interfacial stress was previously noted as an important area that needs careful interpretation in FEA. For example, stress analyses of the tooth–restoration complex have been performed to predict failure risks at the interfaces as well as stresses transferred across such interfaces. Usually such interfaces are modelled as perfectly bonded, where tooth and restoration elements share the same node. Depending on the accuracy of this assumption, this may lead to erroneous interpretation of the results of a finite element analysis (Srirekha & Bashetty,

2010). Sometimes interfacial interactions are more complex, such as areas where two materials may contact but do not bond. In such cases contact analysis needs to be simulated which will require FEA software that can perform nonlinear analyses.

Most research protocols, including finite element analyses, will have limitations. Some limitations in finite element modelling are a deliberately choice. For example, although teeth are 3D structures, often they are modelled in 2D. Two-dimensional models offer excellent visual access for pre- and post-processing, improving their didactic potential. Furthermore, because of the reduced dimensions, more computational capacity can be preserved for improvements in element and simulation quality or functional processes such as masticatory movements. On the other hand, 3D models, although geometrically more realistic, may give a false impression of accuracy, because they are generally more coarse, contain elements with compromised shapes, and examination or improvement of the model is far more difficult (Korioth & Versluis, 1997).

3.4 Relationship between finite element and experimental analysis

The finite element method is sometimes viewed as a less time-consuming process than experimental research, and therefore could minimize laboratory testing requirements. For some applications, finite element analysis may provide faster solutions, for example for the testing of parameters, which can be changed more easily in FEA than in laboratory experiments. However, due to the complexity of shape, properties, and boundary conditions of dental structures, comprehensive modelling can also quickly becomes very complex and time-consuming. Finite element analysis should be viewed in combination with experimental methods, not as a substitute. Finite element analysis can provide information that would be difficult or impossible to obtain with experimental observations, but at the same time, finite element analysis cannot be performed without experimental input and validation.

Compared with experiments, FEA has clear limitations. These limitations are mainly due to the many factors that contribute to the mechanical response but are still poorly understood. Such lack of understanding usually does not affect experiments, if their outcomes are simply considered as phenomena. However, FEA is the compilation of our understanding of physical laws and material properties, expressed in a theoretical model that describes the interactions between the various factors. Therefore, phenomena are no acceptable input for a theoretical model. Limitations in FEA therefore most often refer back to our own lack of understanding the reality. In other words, our own limitations in understanding are the cause of limitations in FEA. Our inability to accurately describe and simulate biomechanical dynamics and properties of a tooth and its supporting structures limits the accuracy of our FEA models. Fortunately, even imperfect experimental or FEA testing methods can improve our insight and continuously expand our understanding of reality. Therefore, although certain differences may remain between reality and the analyses we conduct using the finite element method, the numerical approach can approximate, for example, otherwise inaccessible stress distributions within a tooth-restoration complex. Furthermore, the ability to visualize many of the results from finite element analyses has also undoubtedly helped researchers to more clearly convey their data, and helped expand the discussion and dissemination of research findings that have contributed to improve oral health.

4. Impact of finite element analysis on dentistry - How FE analyses have contributed to improved oral health

Oral health is important to an individual's well-being and overall health. In dentistry, most oral diseases are neither self-limiting nor self-repairing (Vargas & Arevalo, 2009). Therefore, prompt professional care is fundamental, given that oral diseases follow a downward spiral: incipient diseases requiring minimum dental care, if untreated, progress into diseases that require increasingly more complex and expensive treatments; increases in complexity and cost usually make the treatment even more out of reach for a large proportion of the population (Vargas & Ronzio, 2002). In this context, finite element analysis has been applied in various areas in dentistry (1) to improve the understanding of these complex processes and (2) to help to design better procedures.

4.1 Non-Carious Cervical Lesions (NCCL)

FEA has been used in the investigation of NCCL (Michael et al., 2009). Although the etiology of NCCL remains a controversial subject, there is a general consensus that the process is multi-factorial, and that stress can be one of the factors. Goel et al. (1991) investigated stresses arising at the dentino-enamel junction during function and noted that the shape of the dentino-enamel junction was different under working cusps than non-working cusps. Tensile stresses were elevated toward cervical enamel where the mechanical inter-locking between enamel and dentin is weaker than in other areas of the tooth, making it susceptible to cracking, which could contribute to cervical caries (Goel et al., 1991). Finite element analyses have usually assumed the NCCL across the CEJ (Fig. 9). A recent study, however, did not find clinical evidence of enamel loss above the occlusal margin of NCCL,

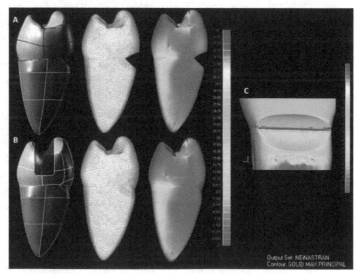

Fig. 9. A. 3D Model of FEA analysis of a non-carious cervical lesions not restored; B. 3D Model of FEA analysis of a non-carious cervical lesions restored with composite resin; C. Maximum principal stress distribution at the unrestored non-carious cervical lesion (Pereira FA, 2011).

except for fracture of enamel that was undermined by the NCCL (Hur et al., 2011). Since the location of the lesion will affect the stress conditions, combining clinical observations and finite element modeling will be essential to determine the stress factor in the initiation and development of NCCL.

4.2 Endodontic treatment

In the case of dental caries, the decay process can continue until the destruction of the tooth and the compromise of adjacent tissues. As the caries process progresses without some type of intervention, the pulp ultimately becomes involved and the root canal therapy is required (Vargas & Arevalo, 2009). One of the steps in root canal treatment is to completely fill the root canal system. During root canal preparation, many variables are outside the control of the clinician (natural root morphology, canal shape and size, dentine thickness) other factors can be addressed during treatment to reduce fracture susceptibility. Using finite element analysis, Versluis et al. (2006) demonstrated that the potential for fracture susceptibility may be reduced by ensuring round canal profiles and smooth canal taper (Fig. 10). Even when fins were not contacted by the instrument, stresses within the root were lower and more evenly distributed than before preparation. Rundquist & Versluis (2006) also used FEA to demonstrate that with increasing taper, root stresses decreased during root filling but tended to increase slightly during a masticatory load. Based on the simulation of vertical warm gutta-percha compaction and a subsequent occlusal load, they suggested that root fracture originating at the apical third was likely initiated during filling, whilst fracture originating in the cervical portion was likely caused by occlusal loads. Gutta-percha is the

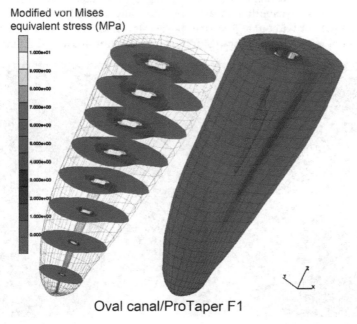

Fig. 10. Stress distribution during obturation pressure in a root with oval canal, cleaned with ProTaper F1 (Versluis et al., 2006).

most common core material used (Er et al., 2007). Although the softening of gutta-percha by heat is a widely used technique, the use of high levels of heat can lead to complications. When heat compaction techniques are used, the procedure should not harm the periodontal ligament (Budd et al., 1991). The use of the technique may result in an unintentional transmission of excessive heat to the surrounding periodontal tissues (Er et al., 2007). Excessive heat during obturation techniques may cause irreversible injury to tissues (Atrizadeh et al., 1971; Albrektsson et al., 1986). By using a three-dimensional thermal finite element analysis the distribution and temperatures were evaluated in a virtual model of a maxillary canine and surrounding tissues during a simulated continuous heat obturation procedure (Er et al., 2007).

4.3 Restoration of root filled teeth

Endodontically treated teeth are compromised by coronal destruction from dental caries (Ross, 1980), fractures (Soares et al., 2007), previous restorations (Schatz et al., 2001), and endodontic access (Soares et al., 2007). How these compromised teeth should be reconstructed to regain their original fracture resistance has been the subject of many studies investigating restoration types and benefits of posts (Fokkinga et al., 2005; Salameh et al., 2006; Salameh et al., 2007). It is not sufficient to only measure an endpoint such as fracture resistance to fully understand the effect of restoration type and post application. A more comprehensive analysis is thus needed to determine the optimal procedures for reconstructing endodontically treated teeth (Soares et al., 2008). The biomechanical conditions that lead to fracture are characterized by the stress state in a tooth, which can be assessed by finite element analysis (Fig. 11). Soares et al. (2008) therefore used FEA to investigate the stress distribution in an endodontically treated premolar restored with composite resin with or without a glass fiber post system and concluded that the use of glass fiber posts did not reinforce the tooth-restoration complex. Intraradicular retention should

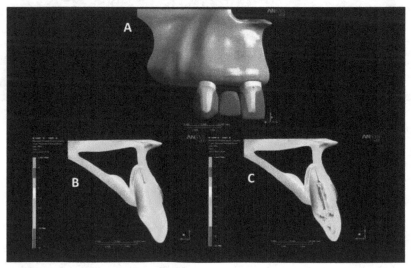

Fig. 11. A. 3D Model of FEA analysis of a 3 elements fixed prosthesis regarding the effect of post type (B. fiber glass post; C. Cast post and core) (Silva GR, 2011).

thus be indicated for endodontically treated teeth that have suffered excessive coronary structure loss (Yu et al., 2006). Research studies using FEA concluded that the use of post systems that have an elastic modulus similar to that of dentine result in a mechanically homogenous units with better biomechanical performance (Barjau-Escribano et al., 2006; Silva et al., 2009). Some studies have concluded that the attributes of carbon and glass fiber dowels make them suitable for dowel restoration (Glazer, 2000; Lanza et al., 2005). Dowel length, size, and design have also been shown to influence the biomechanics and stress distribution of restored teeth (Barjau-Escribano et al., 2006). Using finite element analysis it is possible to evaluate the influence of the type of material (carbon and glass fiber) and the external configuration of the dowel (smooth and serrated) on the stress distribution of teeth restored with varying dowel systems (Soares et al., 2009). Moreover, the difference in elastic modulus between dentin, intraradicular retainers, and cements could result in stress concentrations at the restoration interface when the tooth is in function (Soares et al., 2010, Silva et al., 2011).

4.4 Restorative procedures

In the field of operative dentistry, FEA seems to be an appropriate method for obtaining answers about the interferences caused by the restorative process in a complete structure, for optimizing the design of dental restorations and for evaluating stress distributions in relation to different designs. Many materials are available for dental restorations. The selection and indications for direct and indirect restorative materials involve esthetic, financial, and anatomic considerations, as well as important factors such as analysis of the biomechanical characteristics of the restorative materials, and the amount and state of remaining tooth structure (Soares et al., 2008). In recent years, the demand for nonmetal dental restorations has grown considerably. Metal-free reinforced restorative systems have become popular because of the less favorable esthetic appearance of metal ceramic crowns (Gardner et al., 1997). The primary advantages of nonmetal alternatives (composite resins and ceramics) are improved esthetics, the avoidance of mercury, and cost effectiveness (Stein et al., 2005). Composite resin and ceramic restorations retained with an adhesive resin are the most popular restorations currently used. Composite resins have mechanical properties similar to dentin (Willams et al., 1992) while ceramic has an elastic modulus similar to that of enamel (Albakry et al., 2003).

The conservation of dental structure is crucial to offering fracture resistance, since the removal of dentin reduces the structural integrity of a tooth and causes alteration in stress distributions (Soares et al., 2008b). In this context, the use of adhesive restorations is recommended for reinforcing remaining dental structure (Soares et al., 2008b, Versluis & Tantbirojn, 2011). By using the finite element analysis, stress distributions could be accessed within endodontically treated maxillary premolars that lost tooth structure and the effect of the type of restorative material used for restorations could be studied (Soares et al., 2008). The use of directly placed adhesive restorative materials, such as composite resin, and indirectly placed restorations, such as ceramic inlays, cemented with adhesive materials, generally reduced stress concentrations in comparison with amalgam restorations (Soares et al., 2008). Although indirect restorations may be recommended, the dentist still faces to the choice of geometric configuration of the cavity preparation (Soares et al., 2003).

Inlays and onlays are the 2 technical choices for indirect restorations (Fig. 12). Some studies have shown that after endodontic treatment, teeth restored with intracoronal restorations show more severe fracture patterns (Hannig et al., 2005; Soares et al., 2008c). However, it is unclear whether bonded intracoronal restorations should be used for large defects and which material is the most indicated. In this context, Soares et al. (2003) evaluated the cavity preparation influence on the stress distribution of molar teeth restored with esthetic indirect restorations. The stress distribution pattern of the sound tooth was compared to several different extensions of preparation for inlay, onlay and overlay restored with ceramic or ceromer materials. The cavity preparation extension was significant only for onlays covering one cusp and for overlays. Ceramic restorations had higher stress concentrations, while ceromer restorations caused higher stresses in the tooth structure (Soares et al., 2003).

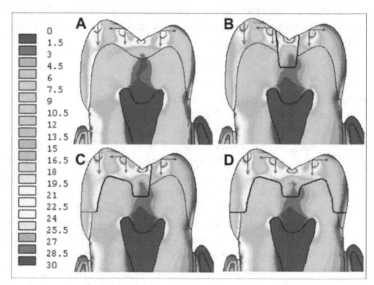

Fig. 12. FEA of different cavity restoration designs for ceramic indirect restorations. A. Intact tooth, B. Inlay restoration; C. Onlay convering buccal cusps; D. Overlay ceramic. (Von mises Stress distribution)

The routine use of metal-free crowns has resulted in an increasing number of fractured restorations (Bello & Jarvis, 1997). Increased fracture resistance of ceramic systems when metal reinforcement was eliminated, has been obtained by the addition of chemical components such as aluminum oxide, leucite, and lithium disilicate (Mak et al., 1997; Drummond et al., 2000). Considering that any restoration has a risk of fracture, the finite element analysis provides a method to evaluate stress distributions in different ceramic systems under occlusal forces. Various studies investigating the performance of ceramic restorations have been performed. Using the finite element analysis method, some investigators (Hubsch et al., 2000; Magne et al., 2002; Magne, 2007; Dejak & Mlotkowski, 2008) demonstrated that ceramic inlays reduced tension at the dentin-adhesive interface and may offer better protection against debonding at the dentin restoration interface, compared with the composite resin inlay. In this context, Reis et al. (2010) investigated, through a 3D finite element analysis, the biomechanical behavior of indirect restored maxillary premolars based on type of preparation (inlay or onlay), and restorative material (composite resin, resin laboratory, reinforced ceramic with lithium disilicate or reinforced ceramic with leucite). Materials with higher modulus of elasticity transfer less stress into the tooth structure. However, materials with modulus of elasticity much larger than the dental structure caused more severe stress concentrations. The models that used reinforced ceramics with leucite showed a behavior that was biomechanically closest to healthy teeth (Reis et al., 2010).

4.5 Composite and resin cement shrinkage

Resin-composite materials have been widely and increasingly used today in adhesive dental restorative procedures (Fagundes et al., 2009). An important advantage over metallic filling materials is the well-known possibility of bonding the restoration to dental tissues (Marques de Melo et al., 2008) and a significant disadvantage of many of these materials are still the polymerization shrinkage (Pereira et al., 2008). The clinical concern about polymerization shrinkage is evident from the large number of publications and large number of controversial opinions about this topic (Versluis et al., 2004). Shrinkage stress has been associated with various clinical symptoms, including fracture propagation, microleakage and post-operative sensitivity, none of which are direct measures of shrinkage stress. Since stress cannot be measured directly, the presence of shrinkage stresses can only be quantified through indirect manifestations, in particular tooth deformation (Tantbirojn et al., 2004).

Various methods have been used to estimate residual shrinkage stresses, ranging from extrapolated shrinkage or load measurements in vitro to stress analyses in tooth shaped anatomies using photoelastic or finite element methods (Kinomoto et al., 1999; Ausiello et al., 2001). Determination of shrinkage stress is difficult, because it is a transient and nonlinear process. The amount of stress after polymerization therefore depends on the correct description of all changes in mechanical properties and their sequence. Moreover, stress is not a material property or even a structural value, because stress is a three-dimensional local tensor (system of related vectors) that is determined by the combination of multiple material properties and local conditions. Since finite element analysis performs its calculation based on such input (mechanical properties, geometry, boundary conditions), it is eminently suitable for studying residual shrinkage stress in dental systems. On the other hand, as the input for especially the mechanical properties remains to be determined more comprehensively, any polymerization shrinkage predicted by finite element analysis should be validated experimentally using indirect factors that can be measured, such as displacement.

Using such validated finite element analyses, shrinkage stresses in restored teeth (enamel and dentin) were found to increase with increasing restoration size, while stresses in the restoration and along the tooth-restoration interface decreased (Versluis et al., 2004). This outcome was explained by the change in tooth stiffness: removal of dental hard tissue decreases the stiffness of the tooth, causing the tooth to be deformed more by the shrinkage stresses (higher stress in the tooth) and causing less resistance to the composite shrinkage (lower stress in the composite). As this example shows, shrinkage stresses are generated in the adhesive interface as well as in the composite and in the residual tooth structure (Versluis et al., 2004; Ausiello et al., 2011).

Restoration placement, techniques are widely recognized as a major factor in the modification of shrinkage stresses. Various techniques, ranging from incremental composite placement to light-exposure regimes, have been advocated to reduce shrinkage stress effects on a restored tooth. Using finite element analysis, it was shown that even during restoration, cavities deform, and thus that incremental application of composite may end up with a higher tooth deformation than a bulk filling (Versluis et al., 1996). Recently the interaction between incremental filling technique, elastic modulus, and post-gel shrinkage of different dental composites was investigated in a restored premolar. Sixteen composites, indicated for restoring posterior teeth, were analyzed. Two incremental techniques, horizontal or oblique, were applied in a finite element model using experimentally determined properties. The calculated shrinkage stress showed a strong correlation with post-gel shrinkage and a weaker correlation was found with elastic modulus. The oblique incremental filling technique resulted in slightly lower residual shrinkage stress along the enamel/composite interface compared to the horizontal technique. However horizontal incremental filling resulted in slightly lower stresses along the dentin/composite interface compared to the oblique technique (Soares et al., 2011). FEA has been used also to analyze the residual shrinkage stress of resin cement used to cement a ceramic inlay, recently we proved that resin cement polymerized immediately after cementation produced significantly more residual stress than when was delayed for 5 minutes after setting ceramic inlay and polymerization (Fig. 13).

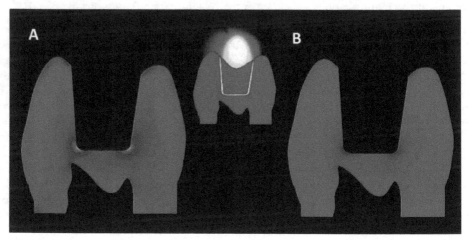

Fig. 13. FEA of residual shrinkage stress of resin cement used to cement a ceramic inlay. A. resin cement polymerized immediately after cementation; B. Reduction of shrinkage stress with delay for 5 minutes after setting ceramic inlay and polymerization.

An often used experimental test for measuring shrinkage forces uses a cylindrical composite specimen bonded between two flat surfaces of steel, glass, composite, or acrylic rods. Even for such seemingly simple experimental tests, understanding the outcome can be difficult. Although one may expect that for a specific experimental set-up, differences in the measured force could be attributed to the composite properties, particularly shrinkage and elastic modulus, it was found that the relative ranking of a series of materials was affected by differences in system compliance. As a result, different studies may show different rankings and may draw contradictory conclusions about polymerization stress, shrinkage or modulus (Meira et al., 2011). Finite element analysis can help to better understand the test mechanics that cause such divergences among studies. Using an FEA approach, a commonly used test apparatus was simulated with different compliance levels defined by the bonding substrate (steel, glass, composite, or acrylic). The authors showed that when shrinkage and modulus increased simultaneously, stress increased regardless of the substrate. However, if shrinkage and modulus were inversely related, their magnitudes and interaction with rod material determined the stress response (Meira et al., 2011).

4.6 Periodontology and implantology

Another oral problem with high prevalence, mainly in adults, is periodontal disease. "Periodontal disease" is a generic term describing diseases affecting the gums and tissues that support the teeth (Thomson et al., 2004). A periodontal compromised tooth can be diagnosed from probing depth, mobility, supporting bone volume, crown-to-root ratio, and root form (Grossmann & Sadan, 2005). It is generally accepted that a reduction of periodontal support worsens the prognosis of a tooth. However, the morphology of the periodontum with reduced structural support has not been well understood in relation to clinical functions, such as load-bearing capability (Ona & Wakabayashi, 2006). To determine the interaction of reduced periodontal support with mechanical function, one must determine the stress and strain created in the periodontum in accordance with the morphologic alteration of the structures (Ona & Wakabayashi, 2006). Finite element analysis can be used for such assessment, and of the influence of progressive reduction of alveolar support on stress distributions in periodontal structures (Ona & Wakabayashi, 2006). The stress in the periodontum could also predict the potential pain and damage that may occur under functional bite force (Kawarizadeh et al., 2004).

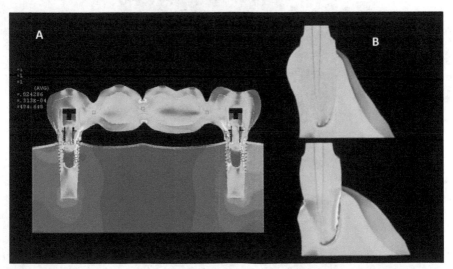

Fig. 14. FEA analysis of implant prosthesis demonstrating the stress concentration on the mesial region of the interface between implant and prosthesis. B. FEA analysis of canine restored with fiber glass post and its effect on bone loss (Roscoe MG, 2010).

Historically, periodontal disease is one of the main causes of tooth loss (Deng et al., 2010). Traditionally, patients with severe periodontitis have ultimately had all teeth removed due to severe alveolar bone resorption and high risks for systemic infections (Deng et al., 2010). In this context implant therapy has been applied successfully for three decades, and proven to be a successful means for oral rehabilitation (Albrektsson et al., 1986). The knowledge of physiologic values of alveolar stresses provides a guideline reference for the design of dental implants and it is also important for the understanding of stress-related bone remodeling and osseointegration (Srirekha & Bashetty, 2010). Stiffness of the tissue-implant interface and implant-supporting tissues is considered the main determinant factor in osseointegration (Ramp & Jeffcoat, 2001; Turkyilmaz et al., 2009). Finite element analysis has been used extensively in the field of implant research over the past 2 decades (Geng et al., 2001). It has been used to investigate the impact of implant geometry (Himmlova et al., 2004), material properties of implants (Yang & Xiang, 2007), quality of implant-supporting tissues (Petrie & Williams, 2007), fixture-prosthesis connections (Akca et al., 2003), and of implant loading conditions (Natali et al., 2006).

4.7 Trauma and orthodontics

Beyond caries and periodontal disease, orofacial trauma is also considered a public health problem (Ferrari & Ferreria de Mederios, 2002). Finite element analysis has also been widely used for dental trauma analysis (Huang et al., 2005). In the real world traumatic injuries to teeth typically result from a dynamic force (Huang et al., 2005). Therefore, for traumatic analysis of a tooth, it has been recommended to simulate time-dependent behavior and analyze different rates of loading (Natali et al., 2004). Finite element analysis can provide insight into the process of impact stresses and fracture propagation in teeth subjected to dynamic impact loads in various directions.

Fig. 15. FEA analysis of orthodontic intrusion movement of a maxillary canine.

Finite element analysis has also been used to study the biomechanics of tooth movement, which allows accurate assessment of appliance systems and materials without the need to go to animal or other less representative models (Srirekha & Bashetty, 2010). Orthodontic tooth movement is a biomechanical process, because the remodeling processes of the alveolar support structures that result in the tooth movement are triggered by orthodontic forces and moments and their consequences for the stress / strain distribution in the periodontium. The redistribution of stresses and strains causes site-specific resorption and formation of the alveolar bone and with it the translation and rotation of the associated tooth (Cattaneo et al., 2009). Finite element analysis can provide insight into the stress and strain distributions around teeth with orthodontic loading to help orthodontists define a loading regime that results in a maximal rate of tooth movement with a minimum of adverse side-effects. The main challenges for the application of finite element analysis in orthodontics has been the definition of the mechanical properties of the periodontal ligament (Toms et al., 2002) and to move beyond the currently most common practice of static finite element models.

4.8 Summary: How FE analysis contributes to improve oral health

It is often commented that finite element analysis is a powerful tool for the interpretation of complex biomechanical systems. Yet, all clinicians and dental researchers are acutely aware of the complexity of oral tissues and their interactions, and hence of the limitations of any theoretical model that depends on input from our incomplete knowledge. The reason why FEA is nonetheless considered such a powerful tool is that it does not need perfect input to be already extremely useful. FEA helps researchers and clinicians formulate the right research questions, design appropriate experiments, and through the underlying universal physics that form the basis of FEA it provides an almost instant insight into complex biomechanical relationships (cause and effect) that cannot be easily obtained or communicated with any other method. The expanded insight and understanding of mechanical responses have undeniably been of direct significance for justifying experimental questions and improving clinical treatments.

As the preceding examples show, finite element analysis not only offers solutions for the engineering problems, but it has been instrumental in the progress in many areas of dentistry. Finite element analysis has improved the understanding of complex processes and has assisted researchers and clinicians in designing better procedures to maintain oral health. Finite element simulation provides unique advantages for dental research, such as its precision and its ability to solve complex biomechanical problems for which other research methods are too cumbersome or even impossible (Ersoz, 2000). Finite element simulation allows more comprehensive prediction and analysis of medical processes or treatments because in a process where many variables need to be considered, it allows for manipulation of single parameters, making it possible to isolate and study the influence of each parameter with more precision (Sun et al., 2008). Thanks to the highly graphic pre- and post-processing features, finite element analysis has also brought researchers and clinicians closer together. It can be argued that without such visualization, stress and strain development would remain mostly academic. The visual interface has improved the communication and collaboration between clinical and research expertise, and is likely to have had a significant impact on the current state of the art in dentistry.

Finite element analysis is not perfect. But we should not expect our theoretical models to be perfect because our understanding of dental properties and processes is still developing. Finite element analysis, however, will continue to improve along with our own understanding about reality. Such continuous improvement will happen as long as we keep comparing reality with theory, and use the insight we gain from these comparisons for improving the theory. The past decades have shown how finite element simulation, which is an expression of our theoretical understanding of biomechanics, has moved from mainly static and linear conditions to more dynamic or transient and nonlinear conditions (Wakabayashi et al., 2008; Srirekha & Bashetty, 2010), thus reflecting the gains that were made in dental science with support from finite element analysis.

5. Acknowledgment

The authors are indebted to financial support granted by FAPEMIG and CNPq.

6. References

Akca K, Cehreli MC, Iplikcioglu H. Evaluation of the mechanical characteristics of the implant-abutment complex of a reduced-diameter morse-taper implant. A nonlinear finite element stress analysis. Clin Oral Implants Res. 2003;14(4):444-54.

Albakry M, Guazzato M, Swain MV. Biaxial flexural strength, elastic moduli, and x-ray diffraction characterization of three pressable all-ceramic materials. J Prosthet Dent. 2003;89(4):374-80.

Albrektsson T, Zarb G, Worthington P, Eriksson AR. The long-term efficacy of currently used dental implants: a review and proposed criteria of success. Int J Oral Maxillofac Implants. 1986;1(1):11-25.

Assunção WG, Barão VA, Tabata LF, Gomes EA, Delben JA, dos Santos PH. Biomechanics studies in dentistry: bioengineering applied in oral implantology. J Craniofac Surg. 2009 Jul;20(4):1173-7.

Atrizadeh F, Kennedy J, Zander H. Ankylosis of teeth following thermal injury. J Periodontal Res. 1971;6(3):159-67.

Ausiello P, Apicella A, Davidson CL, Rengo S. 3D-finite element analyses of cusp movements in a human upper premolar, restored with adhesive resin-based composites. J Biomech. 2001;34(10):1269-77.

Ausiello P, Apicella A, Davidson CL. Effect of adhesive layer properties on stress distribution in composite restorations--a 3D finite element analysis. Dent Mater. 2002;18(4):295-303.

Ausiello P, Franciosa P, Martorelli M, Watts DC. Numerical fatigue 3D-FE modeling of indirect composite-restored posterior teeth. Dent Mater. 2011;27(5):423-30.

Barjau-Escribano A, Sancho-Bru JL, Forner-Navarro L, Rodriguez-Cervantes PJ, Perez-Gonzalez A, Sanchez-Marin FT. Influence of prefabricated post material on restored teeth: fracture strength and stress distribution. Oper Dent. 2006;31(1):47-54.

Bello A, Jarvis RH. A review of esthetic alternatives for the restoration of anterior teeth. J Prosthet Dent. 1997;78(5):437-40.

Boaro LC, Goncalves F, Guimaraes TC, Ferracane JL, Versluis A, Braga RR. Polymerization stress, shrinkage and elastic modulus of current low-shrinkage restorative composites. Dent Mater. 2010;26(12):1144-50.

Budd CS, Weller RN, Kulild JC. A comparison of thermoplasticized injectable gutta-percha obturation techniques. J Endod. 1991;17(6):260-4.

Cattaneo PM, Dalstra M, Melsen B. Strains in periodontal ligament and alveolar bone associated with orthodontic tooth movement analyzed by finite element. Orthod Craniofac Res. 2009;12(2):120-8.

Chabrier F, Lloyd CH, Scrimgeour SN. Measurement at low strain rates of the elastic properties of dental polymeric materials. Dent Mater. 1999;15(1):33-8.

Chung SM, Yap AU, Tsai KT, Yap FL. Elastic modulus of resin-based dental restorative materials: a microindentation approach. J Biomed Mater Res B Appl Biomater. 2005;72(2):246-53

Dejak B, Mlotkowski A. Three dimensional finite element analysis of strength and adhesion of composite resin versus ceramic inlays in molars. J Prosthet Dent. 2008;99(2):131-40.

Deng F, Zhang H, Shao H, He Q, Zhang P. A comparison of clinical outcomes for implants placed in fresh extraction sockets versus healed sites in periodontally compromised patients: a 1-year follow-up report. Int J Oral Maxillofac Implants. 2010;25(5):1036-40.

Drummond JL, King TJ, Bapna MS, Koperski RD. Mechanical property evaluation of pressable restorative ceramics. Dent Mater. 2000;16(3):226-33.

Er O, Yaman SD, Hasan M. Finite element analysis of the effects of thermal obturation in maxillary canine teeth. Oral Surg Oral Med Oral Pathol Oral Radiol Endod. 2007;104(2):277-86.

Ersoz E. Evaluation of stresses caused by dentin pin with finite elements stress analysis method. J Oral Rehabil. 2000;27(9):769-73.

Fagundes TC, Barata TJ, Carvalho CA, Franco EB, van Dijken JW, Navarro MF. Clinical evaluation of two packable posterior composites: a five-year follow-up. J Am Dent Assoc. 2009;140(4):447–54.

Ferrari CH, Ferreria de Mederios JM. Dental trauma and level of information: mouthguard use in different contact sports. Dent Traumatol. 2002;18(3):144-7.

Fokkinga WA, Le Bell AM, Kreulen CM, Lassila LV, Vallittu PK, Creugers NH. Ex vivo fracture resistance of direct resin composite complete crowns with and without posts on maxillary premolars. Int Endod J. 2005;38(4):230 –7.

Gardner FM, Tillman-McCombs KW, Gaston ML, Runyan DA. In vitro failure load of metal-collar margins compared with porcelain facial margins of metal-ceramic crowns. J Prosthet Dent. 1997;78(1):1-4.

Geng JP, Tan KB, Liu GR. Application of finite element analysis in implant dentistry: a review of the literature. J Prosthet Dent. 2001;85(6):585-98.

Glazer B. Restoration of endodontically treated teeth with carbon fiber posts: prospective study. J Can Dent Assoc. 2000;66(11):613-8.

Goel VK, Khera SC, Ralston JL, Chang KH. Stresses at the dentinoenamel junction of human teeth--a finite element investigation. J Prosthet Dent. 1991;66(4):451-9.

Grossmann Y, Sadan A. The prosthodontic concept of crown-to-root ratio: a review of the literature. J Prosthet Dent. 2005;93(6):559-62.

Hannig C, Westphal C, Becker K, Attin T. Fracture resistance of endodontically treated maxillary premolars restored with CAD/CAM ceramic inlays. J Prosthet Dent. 2005;94(4):342-9.

Himmlova L, Dostalova T, Kacovsky A, Konvickova S. Influence of implant length and diameter on stress distribution: a finite element analysis. J Prosthet Dent. 2004;91(1):20-5.

Huang HM, Ou KL, Wang WN, Chiu WT, Lin CT, Lee SY. Dynamic finite element analysis of the human maxillary incisor under impact loading in various directions. J Endod. 2005;31(10):723-7.

Hubsch PF, Middleton J, Knox J. A finite element analysis of the stress at the restoration-tooth interface, comparing inlays and bulk fillings. Biomaterials. 2000;21(10):1015-9.

Huo B. An inhomogeneous and anisotropic constitutive model of human dentin. J Biomech. 2005;38(3):587-94.

Hur B, Kim HC, Park JK, Versluis A. Characteristics of non-carious cervical lesions--an ex vivo study using micro computed tomography. J Oral Rehabil. 2011. Jun;38(6):469-74. doi: 10.1111/j.1365-2842.2010.02172.x. Epub 2010 Oct 19.

Kawarizadeh A, Bourauel C, Zhang D, Gotz W, Jager A. Correlation of stress and strain profiles and the distribution of osteoclastic cells induced by orthodontic loading in rat. Eur J Oral Sci. 2004;112(2):140-7.

Kinomoto Y, Torii M, Takeshige F, Ebisu S. Comparison of polymerization contraction stresses between self- and light- curing composites. J Dent. 1999;27(5):383–9.

Kinney JH, Balooch M, Marshall GW, Marshall SJ. A micromechanics model of the elastic properties of human dentine. Archives of Oral Biology 2004; 44: 813–822.

Korioth TW, Versluis A. (1997). Modeling the mechanical behavior of the jaws and their related structures by finite element (FE) analysis. Critical Reviews in Oral Biology & Medicine, Vol.8, No.1, (January 1997), pp. 90-104, ISSN 1544-1113

Lanza A, Aversa R, Rengo S, Apicella D, Apicella A. 3D FEA of cemented steel, glass and carbon posts in a maxillary incisor. Dent Mater. 2005;21(8):709-15.

Lin CL, Chang CH, Ko CC. Multifactorial analysis of an MOD restored human premolar using auto-mesh finite element approach. J Oral Rehabil. 2001;28(6):576-85.

Magne P, Perakis N, Belser UC, Krejci I. Stress distribution of inlay-anchored adhesive fixed partial dentures: a finite element analysis of influence of restorative materials and abutment preparation design. J Prosthet Dent. 2002;87(5):516-27.

Magne P. Efficient 3D finite element analysis of dental restorative procedures using micro-CT data. Dent Mater. 2007;23(5):539-48.

Mak M, Qualtrough AJ, Burke FJ. The effect of different ceramic materials on the fracture resistance of dentin-bonded crowns. Quintessence Int. 1997;28(3):197-203.

Marques de Melo R, Galhano G, Barbosa SH, Valandro LF, Pavanelli CA, Bottino MA. Effect of adhesive system type and tooth region on the bond strength to dentin. J Adhes Dent. 2008;10(2):127–33.

Marshall DB, Noma T, Evans AG. A simple method for determining elastic-modulus-to-hardness ratios using Knoop indentation measurements. J Am Ceram Soc 1982;65: C175-6.

Meira JB, Braga RR, de Carvalho AC, Rodrigues FP, Xavier TA, Ballester RY. Influence of local factors on composite shrinkage stress development--a finite element analysis. J Adhes Dent. 2007;9(6):499-503.

Meira JB, Braga RR, Ballester RY, Tanaka CB, Versluis A. Understanding contradictory data in contraction stress tests. J Dent Res. 2011;90(3):365-70.

Michael JA, Townsend GC, Greenwood LF, Kaidonis JA. Abfraction: separating fact from fiction. Aust Dent J. 2009;54(1):2-8.

Misra A, Spencer P, Marangos O, Wang Y, Katz JL. Parametric study of the effect of phase anisotropy on the micromechanical behaviour of dentin-adhesive interfaces. J R Soc Interface. 2005;2(3):145-57.

Natali AN, Pavan PG, Ruggero AL. Analysis of bone-implant interaction phenomena by using a numerical approach. Clin Oral Implants Res. 2006;17(1):67-74.

Natali AN, Pavan PG, Scarpa C. Numerical analysis of tooth mobility: formulation of a non-linear constitutive law for the periodontal ligament. Dent Mater. 2004;20(7):623-9.

Ona M, Wakabayashi N. Influence of alveolar support on stress in periodontal structures. J Dent Res. 2006;85(12):1087-91.

Pereira RA, Araujo PA, Castaneda-Espinosa JC, Mondelli RF. Comparative analysis of the shrinkage stress of composite resins. J Appl Oral Sci. 2008;16(1):30-4.

Pereira FA, Influência do material restaurador, lesão cervical e tipo de contato no comportamento biomecânico de pré-molares superiores tratados endodonticamente. [Faculdade de Odontologia – UFU] Dissertação de Mestrado 2011

Petrie CS, Williams JL. Probabilistic analysis of peri-implant strain predictions as influenced by uncertainties in bone properties and occlusal forces. Clin Oral Implants Res. 2007;18(5):611-9.

Peyton FA, Mahler DB, Hershenov B. Physical properties of dentin. J Dent Res. 1952;31(3):366-70.

Poiate IA, Vasconcellos AB, Mori M, Poiate E, Jr. 2D and 3D finite element analysis of central incisor generated by computerized tomography. Comput Methods Programs Biomed. 2011;

Ramp LC, Jeffcoat RL. Dynamic behavior of implants as a measure of osseointegration. Int J Oral Maxillofac Implants. 2001;16(5):637-45.

Reis BR. Influência da configuração cavitária e tipo de material restaurador no comportamento biomecânico de pré-molar superior. Análise por elementos finitos. [Dissertação]. Uberlândia: Universidade Federal de Uberlândia; 2010. Mestrado em Odontologia.

Romeed SA, Fok SL, Wilson NH. Finite element analysis of fixed partial denture replacement. J Oral Rehabil. 2004;31(12):1208-17.

Roscoe MG, Influência da perda óssea, tipo de retentor e presença de férula no comportamento biomecânico de caninos superiores tratados endodonticamente. [Dissertação de Mestrado], Universidade Federal de Uberlândia, Uberlândia, MG, 2010.

Ross IF. Fracture susceptibility of endodontically treated teeth. J Endod. 1980;6(5):560 –5.

Rundquist BD, Versluis A. How does canal taper affect root stresses? Int Endod J. 2006 Mar;39(3):226-37. PubMed PMID: 16507077.

Salameh Z, Sorrentino R, Ounsi HF, Goracci C, Tashkandi E, Tay FR, Ferrari M. Effect of different all-ceramic crown system on fracture resistance and failure pattern of endodontically treated maxillary premolars restored with and without glass fiber posts. J Endod. 2007;33(7):848-51.

Salameh Z, Sorrentino R, Papacchini F, Ounsi HF, Tashkandi E, Goracci C, Ferrari M. Fracture resistance and failure patterns of endodontically treated mandibular molars restored using resin composite with or without translucent glass fiber posts. J Endod. 2006;32(8):752-5.

Santos-Filho PC. Biomecânica restauradora de dentes tratados endodonticamente – Análise por elementos Finitos. Piracicaba: Faculdade de Odontologia da Universidade Estadual de Campinas; 2008.

Schatz D, Alfter G, Goz G. Fracture resistance of human incisors and premolars: morphological and patho-anatomical factors. Dent Traumatol. 2001;17(4):167-73.

Silva NR, Castro CG, Santos-Filho PC, Silva GR, Campos RE, Soares PV, et al. Influence of different post design and composition on stress distribution in maxillary central incisor: Finite element analysis. Indian J Dent Res. 2009;20(2):153-8.

Silva GR, Efeito do tipo de retentor e da presença de férula na distribuição de tensões em prótese fixas em cerâmica pura – Análise por elementos finitos [Tese de Doutorado]. Piracicaba: Faculdade de Odontologia da Universidade Estadual de Campinas; 2011.

Soares CJ. Influência da configuração do preparo cavitário na distribuição de tensões e resistência à fratura de restaurações indiretas estéticas. Piracicaba: Universidade Estadual de Campinas; 2003. Doutorado em Clínica Odontológica, área de concentração em dentística.

Soares CJ, Martins LR, Fonseca RB, Correr-Sobrinho L, Fernandes-Neto AJ. Influence of cavity preparation design on fracture resistance of posterior Leucite-reinforced ceramic restorations. J Prosthet Dent. 2006;95(6):421-9.

Soares CJ, Santana FR, Silva NR, Preira JC, Pereira CA. Influence of the endodontic treatment on mechanical properties of root dentin. J Endod. 2007;33(5): 603- 6.

Soares CJ, Soares PV, Santos-Filho PC, Armstrong SR. Microtensile specimen attachment and shape--finite element analysis. J Dent Res. 2008 Jan;87(1):89-93.

Soares CJ, Castro CG, Santos Filho PC, Soares PV, Magalhaes D, Martins LR. Two-dimensional FEA of dowels of different compositions and external surface configurations. J Prosthodont. 2009;18(1):36-42.

Soares CJ, Raposo LH, Soares PV, Santos-Filho PC, Menezes MS, Soares PB, et al. Effect of different cements on the biomechanical behavior of teeth restored with cast dowel-and-cores-in vitro and FEA analysis. J Prosthodont. 2010;19(2):130-7.

Soares CJ, Versluis A, Tantbirojn D. Polymerization shrinkage stresses in a premolar restored with different composites and different incremental techniques. Dent Mater., In press.

Soares PV, Santos-Filho PC, Gomide HA, Araujo CA, Martins LR, Soares CJ. Influence of restorative technique on the biomechanical behavior of endodontically treated maxillary premolars. Part II: strain measurement and stress distribution. J Prosthet Dent. 2008;99(2):114-22.

Soares PV, Santos-Filho PC, Martins LR, Soares CJ. Influence of restorative tech- nique on the biomechanical behavior of endodontically treated maxillary premolars. Part I: fracture resistance and fracture mode. J Prosthet Dent. 2008;99(1):30-7.

Soares PV, Santos-Filho PC, Queiroz EC, Araújo TC, Campos RE, Araújo CA, Soares CJ. Fracture resistance and stress distribution in endodontically treated maxillary premolars restored with composite resin. J Prosthodont. 2008;17(2):114-9.

Soares PV. Análise do complexo tensão-deformação e mecanismo de falha de pré-molares superiores com diferentes morfologias radiculares e redução seqüencial de estrutura dental. [Tese de Doutorado]. Piracicaba: Faculdade de Odontologia da Universidade Estadual de Campinas; 2008.

Srirekha A, Bashetty K. Infinite to finite: an overview of finite element analysis. Indian J Dent Res. 2010;21(3):425-32.

Stein PS, Sullivan J, Haubenreich JE, Osborne PB. Composite resin in medicine and dentistry. J Long Term Eff Med Implants. 2005;15(6):641-54.

Sun X, Witzel EA, Bian H, Kang S. 3-D finite element simulation for ultrasonic propagation in tooth. J Dent. 2008;36(7):546-53.

Suwannaroop P, Chaijareenont P, Koottathape N, Takahashi H, Arksornnukit M. In vitro wear resistance, hardness and elastic modulus of artificial denture teeth. Dent Mater J. 2011. 30(4):461-8.

Tantbirojn D, Versluis A, Pintado MR, DeLong R, Douglas WH. Tooth deformation patterns in molars after composite restoration. Dent Mater. 2004; 20: 535-542.

Thomson WM, Slade GD, Beck JD, Elter JR, Spencer AJ, Chalmers JM. Incidence of periodontal attachment loss over 5 years among older South Australians. J Clin Periodontol. 2004;31(2):119-25.

Toms SR, Lemons JE, Bartolucci AA, Eberhardt AW. Nonlinear stress-strain behavior of periodontal ligament under orthodontic loading. Am J Orthod Dentofacial Orthop. 2002;122(2):174-9.

Travassos AB. Protocolo BioCAD para modelagem de estruturas orgânicas. Campinas: Centro de Tecnologia da Informação Renato Archer; 2010.

Turkyilmaz I, Sennerby L, McGlumphy EA, Tozum TF. Biomechanical aspects of primary implant stability: a human cadaver study. Clin Implant Dent Relat Res. 2009;11(2):113-9.

Vargas CM, Arevalo O. How dental care can preserve and improve oral health. Dent Clin North Am. 2009;53(3):399-420.

Vargas CM, Ronzio CR. Relationship between children's dental needs and dental care utilization: United States, 1988-1994. Am J Public Health. 2002;92(11):1816-21.

Versluis A, Douglas WH, Cross M, Sakaguchi RL. Does an incremental filling technique reduce polymerization shrinkage stresses? J Dent Res. 1996;75(3):871-8.

Versluis A, Tantbirojn D, Douglas WH. Distribution of transient properties during polymerization of a light-initiated restorative composite. Dent Mater. 2004;20(6):543-53.

Versluis A, Tantbirojn D, Douglas WH (1997). Why do shear bond tests pull out dentin? Journal of Dental Research 76: 1298-1307.

Versluis A, Tantbirojn D, Douglas WH. Do dental composites always shrink toward the light? J Dent Res. 1998;77(6):1435-45.

Versluis A, Tantbirojn D, Pintado MR, DeLong R, Douglas WH. Residual shrinkage stress distributions in molars after composite restoration. Dent Mater. 2004;20(6):554-64.

Versluis A, Tantbirojn D. Filling cavities or restoring teeth? J Tenn Dent Assoc. 2011;91(2):36-42; quiz 42-3.

Versluis A, Tantbirojn D. Relationship between shrinkage and stress. A, D, editor. Hershey, PA: IGI Global; 2009.

Versluis A, Messer HH, Pintado MR. Changes in compaction stress distributions in roots resulting from canal preparation. Int Endod J. 2006 Dec;39(12):931-9.

Vieira AP, Hancock R, Dumitriu M, Limeback H, Grynpas MD. Fluoride's effect on human dentin ultrasound velocity (elastic modulus) and tubule size. Eur J Oral Sci. 2006;114(1):83-8.

Wakabayashi N, Ona M, Suzuki T, Igarashi Y. Nonlinear finite element analyses: advances and challenges in dental applications. J Dent. 2008;36(7):463-71.

Wang R, Weiner S. Human root dentin: Structural anisotropy and Vickers microhardness isotropy. Connective Tissue Research 1998; 39, 269–279.

Willams G, Lambrechts P, Braem M, Celis JP, Vanherle G. A classification of dental composites according to their morphological and mechanical characteristics. Dent Mater. 1992;8(5):310-9.

Witzel MF, Ballester RY, Meira JB, Lima RG, Braga RR. Composite shrinkage stress as a function of specimen dimensions and compliance of the testing system. Dent Mater. 2007;23(2):204-10.

Yang J, Xiang HJ. A three-dimensional finite element study on the biomechanical behavior of an FGBM dental implant in surrounding bone. J Biomech. 2007;40(11):2377-85.

Yu WJ, Kwon TY, Kyung HM, Kim KH. An evaluation of localized debonding between fibre post and root canal wall by finite element simulation. Int Endod J. 2006;39(12):959-67.

3

Critical Aspects for Mechanical Simulation in Dental Implantology

Erika O. Almeida[1,2] et al.*
1Department of Biomaterials and Biomimetics,
New York University College of Dentistry, New York, NY,
2Department of Dental Materials and Prosthodontics,
Sao Paulo State University College of Dentistry, Araçatuba, SP,
1USA
2Brazil

1. Introduction

Dental implants have been widely used for the rehabilitation of completely and partially edentulous patients. (Branemark et al. 1969; Branemark et al. 1977; Adell et al. 1990; van Steenberghe et al. 1990) Despite the high success rates reported by a vast number of clinical studies, early or late implant failures are still unavoidable. (Esposito et al. 1998) Mechanical complications and failures have frequently been reported during prosthetic treatment. (Roberts 1970; Randow et al. 1986; Walton et al. 1986; Levine et al. 1999)

From a biomechanical point of view, forces occurring either from functional or parafunctional occlusal contact may result in a physiologic adaptation of the supporting tissues since implants are rigidly anchored to the bone. However, if the stress generated is beyond the adaptive capacity of the host, the response of the supporting tissues and prosthetic components may result in failures. Therefore, the load magnitude and duration employed to implant-restoration systems play a significant role in biomechanical stress dissipation on the implant-prosthesis system and surrounding tissues. (Menicucci et al. 2002) Other factors that are known to affect the stress/strain distribution on bone surrounding implants are the implant position and angulations, implant-abutment connection and the magnitude and direction of the occlusal load. (Stanford and Brand 1999; Kozlovsky et al. 2007; Lin et al. 2009)

*Amilcar C. Freitas Júnior[3], Eduardo P. Rocha[2], Roberto S. Pessoa[4], Nikhil Gupta[5], Nick Tovar[1] and Paulo G. Coelho[1]
1Department of Biomaterials and Biomimetics, New York University College of Dentistry, New York, USA
2Department of Dental Materials and Prosthodontics, São Paulo State University College of Dentistry, Araçatuba, SP, Brazil
3Postgraduate Program in Dentistry, Potiguar University, College of Dentistry – UnP, Natal, RN, Brazil
4Department of Mechanical Engineering - Federal University at Uberlândia,
FEMEC - Uberlândia, Uberlândia, MG, Brazil
5Department of Mechanical and Aerospace Engineering, Polytechnic Institute of New York University, New York, USA

While randomized controlled clinical studies are strongly suggested (Bozini et al. 2011; Pieri et al. 2011; Turkyilmaz 2011) as the optimal approach to evaluate the performance of biomaterials and biomechanical aspects of dental implants and prosthetic components, often times these studies are not economically viable (especially in cases where multiple variables are to be included). Thus, well designed *in vitro* studies utilizing virtual models via finite element analysis (FEA) should be considered, as this method allows researchers to address a range of questions that are otherwise intractable due to the number of variations within clinical trials or the difficulty in solving analytically (Ross 2005).

Although there are several implant dentistry studies using the FEA method, (Rayfield 2004; Yokoyama et al. 2005; Galantucci et al. 2006; Huang et al. 2007; Assuncao et al. 2008; Lim et al. 2008; de Almeida et al. 2011) each step of the method deserves discussion in order to facilitate its understanding when applied to implant/restoration/bone system. Thus, the aim of this chapter is to conduct a critical review of simulated biomechanical scenarios describing essential basic aspects and approaches currently used in implantology.

2. Model creation

The first step in creating a finite element (FE) model is to determine the appropriate number of dimension to be utilized for evaluation (1, 2, or 3D). While three dimensions are more realistic for the complex anatomy/implant/restoration biomechanical interaction, it is proportionally more challenging when CAD modeling, solving, and software output interpretation is considered. (Richmond et al. 2005) For selected situations, 2D analyses are often adequate for the questions at hand. (Rayfield 2004; Assuncao et al. 2008; Freitas et al. 2010; Freitas Junior et al. 2010)

The choice between 2D and 3D FEA for investigating the biomechanical behavior of complex structures depends on several factors, including the complexity of the geometry, the type of analyses required, expectations in terms of accuracy as well as the general applicability of the results. (Romeed et al. 2006)

Two-dimensional FEA has previously been used in different areas of dental research. (Burak Ozcelik et al. ; Freitas, Rocha et al. 2010; Freitas Junior, Rocha et al. 2010) However, its main limitation is known to be the lower accuracy and reliability relative to 3D and its utilization has drastically fallen from favor when dental treatment biomechanical simulations are considered. (Yang et al. 1999; Yang et al. 2001) In contrast, 3D FEA has acceptable accuracy/reliability while properly capturing the geometry of complex structures. However, the higher the complexity of 3D FE models the higher the difficulty in generating appropriate mesh refinement for simulation, which is more easily achievable for 2D models. (Romeed, Fok et al. 2006)

It is general consensus that 3D reconstruction is essential to study the interaction between anatomy and mechanical behavior of restorative components and dental implants. (Galantucci, Percoco et al. 2006) For that purpose, models may be manually constructed or may be generated from imaging methods such as computed tomography (Figures 1 A and B). The choice for building a model using either a manual or automatic technique depends on the purpose of the study and the structure of interest. Considering the specific morphology of biological structures, the non-uniform rational basis spline (NURBS) has been commonly used for 3D modeling as it allows the achievement of reliable analytic or freeform parts based on an efficient management of curves and surfaces.

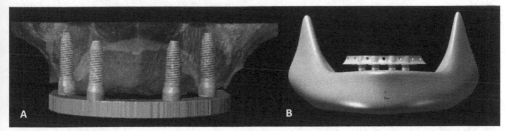

Fig. 1. Models of maxilla (A) and mandible (B) implant supported prosthesis based on (A) model reconstruction from imaging techniques and (B) manual construction on CAD software (Solid Works Corp).

The manual input technique generate structures in appropriate aided design software such as AutoCAD (Autodesk Inc, San Rafael, CA, USA), SolidWorks (SolidWorks Corp., Concord, MA, USA), Pro/Engineer (Wildfire, PTC, Needham, MA, USA), Rhino 3D (McNeel North America, Seatle, WA, USA) (Figure 1B).

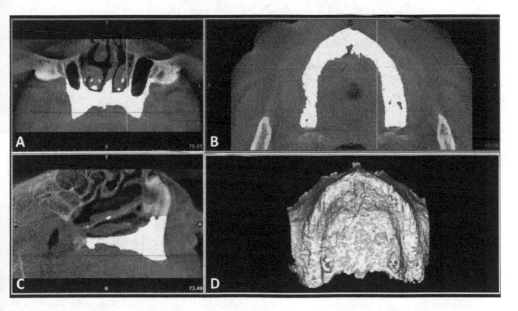

Fig. 2. CT scan data as seen in Mimics 13.0 (Materialise, Leuven, Belgium). (A-C) The maxilla is presented in three different cross-sectional views. Masks have been applied according to voxel density thresholding to determine the regions of interest. (D) 3D representation of maxilla as a result of segmentation in Mimics.

The imaging approach involves transforming available medical imaging files from computed tomography (CT) scans, magnetic resonance images (MRI), ultrasound, and laser digitizers into wireframe models that are then converted into FE models. (Romeed, Fok et al. 2006) While laser scans offer a high-resolution representation of the outer surface, these lack information about internal geometry. (Kappelman 1998) Creating models from CT or MRI is often time-consuming but can provide accurate models with fine structural details based on image density thresholding. (Cohen et al. 1999; Ryan and van Rietbergen 2005) The 3D object is automatically created in the form of masks by thresholding the region of interest on the entire stack of scans (Figure 2A-D). The degree of automation and high resolution make this model creation method attractive, but determining the appropriate thresholding algorithms to the bone-air boundary reliably throughout a structure with varying bone thicknesses and density can be challenging. (Fajardo et al. 2002)

When advanced imaging data is used to generate solid models, surface smoothing is advised in order to decrease the number of nodes and elements in the discretized FE model as such an approach generally decreases computation time. (Wang et al. 2005; Magne 2007) Fig. 3A shows an excessive number of elements in a dental implant that was subsequently reduced (Fig. 3B) by computer software (Materialise, Leuven, Belgium). However, it is also advisable that when surface smoothing is performed it does not over simplify the geometries, causing a decrease in solution accuracy.

Fig. 3. (A) 3D CAD of the Nobel Speed ™ RP Implant (Nobel Biocare, CA, USA) based on the micro-CT (μCT; CT40, Scanco Medical, Bassersdorf, Switzerland). (B) The Mimics Remesh (Materialise) function "quality preserving reduce triangle" and "reduce triangles" were used to reduce the number of elements at the implant CAD.

3. Material properties and software input limitation

Material properties such as heat conductivity, linear and nonlinear elastic properties, and temperature-dependent elastic properties may be utilized in FEA. (Richmond, Wright et al. 2005) In dentistry, the majority of previous FE scientific communications in implant prosthodontics has considered material properties to be isotropic, homogeneous, and linear elastic. (Huang et al. 2008; Canay and Akca 2009; Eser et al. 2009; Li et al. 2009; Chang et al. 2010; Chou et al. 2010; Okumura et al. 2010; Sagat et al. 2010; Wu et al. 2010; Burak Ozcelik et al. 2011) An isotropic material indicates that the mechanical response is similar regardless of the stress field direction, requiring Young's modulus (E) and Poisson's ratio (v) values for the FE calculation. (Richmond, Wright et al. 2005)

The elastic, or Young's modulus (E), is defined as stress/strain (σ/ϵ) and is measured in simple extension or compression. It is a measure of material deformation under a given axial load. In other words, a numerical description of its stiffness. Poisson's ratio (v) is the lateral strain divided by axial strain, thus representing how much the sides of a material as it is tensile tested. (Richmond, Wright et al. 2005)

Since bone is one of the structures to be simulated in FEA, (Peterson and Dechow 2003; Bozkaya et al. 2004; Danza et al. 2010; de Almeida et al. 2010) it is often treated as an anisotropic structure and three elastic modulus (E), three Poisson's ratio (v) and three shear modulus (G) are required (Table 1). (Chen et al. 2010; Sotto-Maior et al. 2010) Anisotropic materials are characterized by different stress responses under forces applied in varied directions within the structure. (O'Mahony et al. 2001; Natali et al. 2009; Eraslan and Inan 2010) The elastic behavior in cortical bone approximates to orthotropic, which is a type of anisotropy in which the internal structure of the material results in unique elastic behavior along each of the three orthogonal axes of the material. In this case, three elastic (E) and shear modulus (G) and six Poisson's ratios (v) are necessary for model input. (Richmond, Wright et al. 2005; Natali et al. 2010)

MATERIAL PROPERTIES	TRABECULLAR BONE	CORTICAL BONE
E_X (MPa)	1,148	12,600
E_Y (MPa)	210	12,600
E_Z (MPa)	1,148	19,400
G_{XY} (MPa)	68	4,850
G_{YZ} (MPa)	68	5,700
G_{XZ} (MPa)	434	5,700
v_{YX}	0,010	0,300
v_{ZY}	0,055	0,390
v_{ZX}	0,322	0,390
v_{XY}	0,055	0,300
v_{YZ}	0,010	0,253
v_{XZ}	0,322	0,253

Table 1. Material properties used in an anisotropic model. The material axes correspond to the global coordinate system. E = Young's modulus. G = shear modulus. v_{yx} = Poisson's ratio for strain in the y-direction when loaded in the x-direction.

Since in dental implantology bone quality has been related to the structural efficiency of the cortical and trabecular bone architecture and ratio (lower bone quality results in biomechanically challenged treatments), mechanical simulations of poor bone quality critical as clinical studies have shown that dental implants placed in regions of the jaw bones with lower density have a higher chance to fail than implants placed at regions with higher bone density. (Genna 2003)

To date, no consensus regarding the mechanical properties that are appropriate for simulating the different bone density scenarios clinically encountered in implant dentistry has been reached. For instance, the value of trabecular bone elastic modulus observed in the literature range from 0.3 to 9.5 GPa. (Zarone et al. 2003; Eskitascioglu et al. 2004; Sevimay et al. 2005; Yokoyama, Wakabayashi et al. 2005) A different approach has been employed by Tada and coworkers (Tada et al. 2003) who assigned different elastic moduli to bone depending on its density from most dense (Type 1) to least dense (Type IV). The moduli utilized were 9.5 GPa, 5.5 GPa, 1.6 GPa and 0.69 GPa for bone types I, II, III, and IV, respectively. Recent work has also been carried out using these bone property values for different simulations (Table 2).

MATERIAL	YOUNG'S MODULUS (GPa)	POISSON'S RATIO	REFERENCES
Bone type I	9.5	0.3	Tada et al. (2003)
Bone type II	5.5	0.3	Tada et al. (2003)
Bone type III	1.6	0.3	Tada et al. (2003)
Bone type III	0.69	0.3	Tada et al. (2003)
Cortical bone	13.70	0.3	Shunmugasamy et al. (2011)
Titanium	110	0.35	Huang et al. (2008)

Table 2. Mechanical properties of the most used structures in implant prosthodontics research through finite element analysis.

Considering the fact that the most common bone quality types in mandible are type I or II in the anterior region and type III in the posterior region, the elastic modulus that would best represent the mandible would be 9.5 GPa or 5.5 GPa for the anterior region and 1.6 GPa for the posterior region. For the maxillary bone, the most prevalent types of bone are type II or III in the anterior region and type IV in the posterior region. Therefore, the Young's modulus that would best represent these sites would be 5.5 GPa or 1.6 GPa for the anterior region and 0.69 GPa for the posterior region, respectively. For cortical bone, studies typically use the elastic modulus (E) of 13.7 GPa and the Poisson ratio (v) similar to 0.3 for the trabecular and cortical bone (Table 2). (Sevimay, Turhan et al. 2005; Huang, Hsu et al. 2008)

4. Bone-implant interface

In implant therapy, the fact that bone quantity and related biomechanical behavior differs for each patient implies a challenge to model the percentage of osseointegration. A critical issue when evaluating a study is to analyze the conditions specified for the interface. What is to be specified is the ability of the interface to resist three different types of stresses: compressive stresses at a right angle to the interface; tensile stresses, also at a right angle to the interface; and shear stresses in parallel with the interface (Figure 4). (Hansson and Halldin 2009)

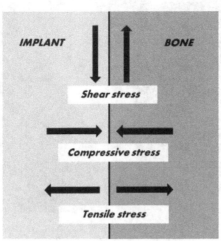

Fig. 4. The different types of stresses occurring at the implant–bone interface.

Linear static models have been employed extensively in previous FEA studies. (Kohal et al. 2002; Menicucci, Mossolov et al. 2002; Romanos 2004; Van Staden et al. 2006; Lim, Chew et al. 2008; Eser, Akca et al. 2009; Hsu et al. 2009; Li, Kong et al. 2009; Faegh and Muftu 2010; Freitas Junior et al. 2010; Hasan et al. 2011) These analyses usually assumed that all modeled volumes were bonded as one unit, which indicates that the trabecular and cortical bone are perfectly bonded to the implant interface. (Winter et al. 2004; Maeda et al. 2007; Fazel et al. 2009; Chang, Chen et al. 2010; Okumura, Stegaroiu et al. 2010). However, the validity of a linear static analysis may be questionable when the investigation aims to explore more realistic situations that are generally encountered in the dental implant field. Some actual implant clinical situations will give rise to nonlinearities, mainly related to the chang of interrelations between the simulated structures of a FE model. (Wakabayashi et al. 2008) Moreover, frictional contact mode potentially provides an improved accuracy with respect to the relative component's micromotion within the implant system, and, therefore, a more reasonable representation of the real implant clinical condition. (Merz et al. 2000) This configuration allows minor displacements between all components of the model without interpenetration. Under these conditions, the contact zones transfer pressure and tangential forces (i.e. friction), but not tension. Some FE analyses have shown remarkable differences in the values and even in the distribution of stresses between "fixed bond" and "non-linear contact" interface conditions. (Brunski 1992; Van Oosterwyck et al. 1998; Huang, Hsu et al. 2008) Not only the stress and strain levels but also the stress and strain highly affected by the interface state (Figure 5). Van Oosterwyck et al. (Van Oosterwyck, Duyck et al. 1998) argued that through the bonded interface the force was dissipated evenly in both the compressive and the tension site. However, on the contact interfaces, tensions are not transferred and force is only passed on through the compressive site, which results in excessive stresses. Additionally, the condition of the bone to implant interface also influences the strain distribution and level inside of the implant system. (Pessoa et al. 2010)

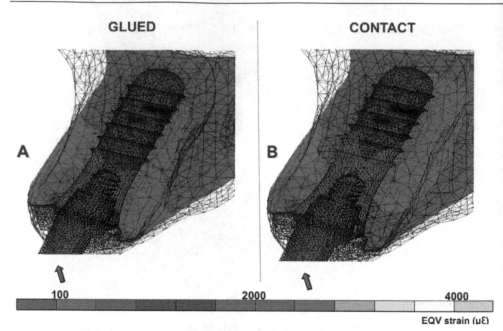

GLUED   **CONTACT**

A   B

100   2000   4000

EQV strain (µε)

Fig. 5. Equivalent strain (µe) distribution for 100N loaded implant in a superior central incisor region, in a median buccopalatal plane. The bone to implant interfaces were assumed as fixed bond ("glued") (A) and frictional contact (B). The arrows indicate the loading direction for clarity.

Conventionally, the osseointegrated bone to implant interface is treated as fully bonded. (Kohal, Papavasiliou et al. 2002; Menicucci, Mossolov et al. 2002; Romanos 2004; Van Staden, Guan et al. 2006; Lim, Chew et al. 2008; Eser, Akca et al. 2009; Hsu, Fuh et al. 2009; Li, Kong et al. 2009; Faegh and Muftu 2010; Freitas Junior, Rocha et al. 2010; Hasan, Rahimi et al. 2011)

This assumption is supported by experimental investigations in which removal of rough implants frequently resulted in fractures in bone distant from the implant surface, (Gotfredsen et al. 2000) suggesting the existence of an implant-bone "bond". On the other hand, frictional contact elements are used to simulate a nonintegrated bone to implant interface (i.e. in immediately loaded protocols), which allows minor displacements between the implant and the bone. (Pessoa, Muraru et al. 2010) The occurrence of relative motion between implant and bone introduces a source of non-linearity in FEA, since the contact conditions will change during load application.

The friction coefficient (µ) to be used in such simulations depends on many factors including mechanical properties and the roughness of the contact interface, exposure to interfacial contaminants (Williams 2000) and in some cases the normal load. A µ = 0.3 was measured for interfaces between a smooth metal surface and bone, while a µ = 0.45, for interfaces between a rough metal surface and bone. (Rancourt et al. 1990) Frequently, the friction coefficient between bone and implant is assumed as being µ = 0.3. Huang et al. (Huang, Hsu et al. 2008) investigated the effects of different frictional coefficients (µ = 0.3, 0.45 and 1) on the stress and displacement of an immediately loaded implant simulation. The authors demonstrated that the value of μ shows no significant influence for increasing or decreasing the tensile and compressive stresses of bone. Nevertheless, increasing μ from 0.3 to 1, the interfacial sliding between implant and bone was mainly reduced from 20% to 30–60%, depending on the implant design. (Natali, Carniel et al. 2009)

Moreover, when an implant is surgically placed into the jawbone, the implant is mechanically screwed into a drilled hole of a smaller diameter. Large stresses will occur due to the torque applied in the process of implant insertion. As the implant stability and stress state around an immediately loaded implant may be influenced by such conditions, this should be also considered in FE simulation of immediately loaded implants. However, this phenomenon has not been thoroughly investigated to date. The implementation of such implant insertion stresses in FEA is still unclear and should therefore be a matter of further investigations.

Another commonly observed assumption in dental implantology FE models, perfect bonding between implant, abutment and abutment screw also is not the most realistic scenario. Non-linear contact analysis was proven to be the most effective interface condition for realistically simulating the relative micromotions occurring between different components within the implant system. (Williams 2000; Pessoa, Muraru et al. 2010; Pessoa et al. 2010) Therefore, for correct simulation of an implant-abutment connection, frictional contact should be defined between the implant components. Accordingly, between implant, abutment and abutment screw regions in contact, a frictional coefficient of 0.5 was generally assumed in non-linear simulations of implant-abutment connection. (Merz, Hunenbart et al. 2000; Lin et al. 2007; Pessoa, Muraru et al. 2010) When using a contact interface between implant components, for lateral or oblique loading conditions, specific parts can separate, or new parts that were initially not in contact can come into contact. Consequently, higher stress levels may be expected to occur in an implant-abutment connection simulated with contact interfaces, compared to a glued connection. In this regard, the pattern and magnitude of deformation in both periimplant bone and implant components will be influenced by the implant connection design. (Pessoa, Vaz et al. 2010)

5. Mesh and convergence analysis

Since the components in a dental implant-bone system are complex from a geometric standpoint, FEA has been viewed as the most suitable tool for mechanically analyzing them. A mesh is needed in FEA to divide the whole domain into elements. The process of creating the mesh, elements, their respective nodes, and defining boundary conditions is called "discretization" of the problem domain. (Geng et al. 2001; Richmond, Wright et al. 2005)

The 2D structures are typically meshed with triangular or quadrilateral elements (Figure 6A). These elements may possess two or more nodes per side, with one node at each vertex. If nodes are only placed at the vertices, the element is called linear because a line function describes the element geometry and how the displacement field will vary along an element edge. (Richmond, Wright et al. 2005)

Typical 3D elements include eight-node bricks, six-node wedges, five-node pyramids, and four-node tetrahedral (Figure 6B). With increasing numbers of elements and nodes, the model becomes more complex and computationally more difficult and lengthy. (Assuncao et al. 2009)

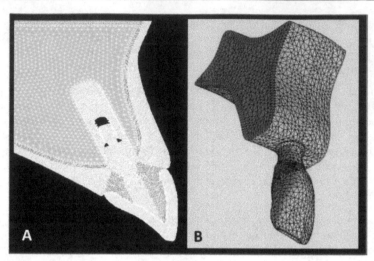

Fig. 6. (A) 2D Finite element model meshed with triangular elements. (B) 3D Finite element model meshed with tetrahedral elements.

To address this problem, researchers often devise strategies to minimize computational expense, such as taking advantage of symmetry when possible by modeling only half the structure, using a generally coarse mesh with finer elements only near regions of geometric complexity or high stress and strain, and/or using a 2D model when it suffices. (Richmond, Wright et al. 2005)

A convergence study of the model is always important to verify the mesh quality. A measurement of convergence is the degree of difference in the total strain energy between two successive mesh refinements. (Hasan, Rahimi et al. 2011) When the difference in energy is less than some tolerable limit specified by the user, the solution is considered converged

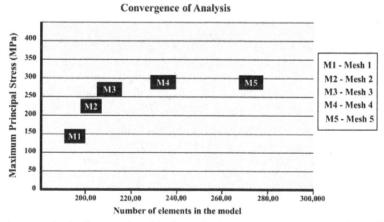

Fig. 7. Maximum principal stress (σ_{max}) in hypothetic convergence testing of mandible FE models meshed by 0.6 mm elements with no refinement (mesh 1) and with refinement levels 2, 3, 4 and 5 at a posterior implant region (meshes 2, 3, 4 and 5, respectively).

(Figure 7). Some authors considered that the convergence criterion between meshes refinement was a change of less than 5 (Li, Kong et al. 2009; Lin et al. 2010) or 6% (Huang, Hsu et al. 2008; Huang et al. 2009) in the maximum simulated stress of the bone-implant edge. Alternatively, a mesh can be considered converged when the rate of density change was less than 10^{-5} between subsequent iterations. (Hasan, Rahimi et al. 2011)

6. Loading and boundary conditions

A attempting to successfully replicate the clinical situation that an implant might encounter in the oral environment, it is also important to understand and correctly reproduce the natural forces that are exerted throughout the system. These forces are mainly the result of the masticatory muscles action, and are related to the amount, frequency and duration of the masticatory function.

Forces acting on dental implants possess both magnitude and direction, and are referred to as vector quantities. For accurate predictions on the implant-bone behavior, it is essential to determine realistic in vivo loading magnitudes and directions. However, at each specific bite point, bite forces can be generated in a wide range of directions. Also, although bite forces are generally act downward, toward the apex of the implant and thereby tending to compress the implant into the alveolar bone, tensile forces and bending moments may also be present depending on where the bite force is applied relative to the implant-supported prosthesis. This fact is even more important when the investigation aims to simulate multiple-implant treatment modalities, because of the geometric factors involving the restorations that are linking the implants, such as the existence of distal cantilevers. (Mericske-Stern et al. 1996; Fontijn-Tekamp et al. 1998) Nevertheless, although for implants used in single-tooth replacement simulations the in vivo forces ought to replicate the forces exerted on natural teeth, factors such as the width of the crown occlusal table, the height of the abutment above the bone level, and the angulation of the implant with respect to the occlusal plane will affect the value of the moment on the single-tooth implant.

A significant amount of investigations have assumed the direction of the load applied to the implant to be horizontal, vertical and oblique. However, the rationale for use of an oblique loading condition is based on finding the vertical (axial) forces directed to the implant system that are relatively low and well tolerated in comparison to oblique forces, which generate bending moments (Figure 8A). The combination of axial and transverse loading, termed as mixed loading, simulates practical conditions where the actual applied force may be inclined with respect to the implant axis and components can be solved in the longitudinal and transverse directions. (Shunmugasamy et al. 2011) These oblique forces have been considered more clinically realistic in FEA than vertical ones. (Chun et al. 2002; Pierrisnard et al. 2002)

Several investigators have tried to gain insight into implant loading magnitudes by performing tests using experimental, analytical, and computer-based simulations of various implant-supported prosthesis types. (Mericske-Stern, Piotti et al. 1996; Fontijn-Tekamp, Slagter et al. 1998; Duyck et al. 2000; Morneburg and Proschel 2002) Bite forces ranging from 50 to 400 N in the molar regions and 25 to 170 N in the incisor areas have been reported. These variations are influenced by patient's gender, muscle mass, exercise, diet, bite location, parafunction, number of teeth and implants, type of implant-supported prosthesis, physical status and age. (Duyck, Van Oosterwyck et al. 2000; Morneburg and Proschel 2002; Eser, Akca et al. 2009)

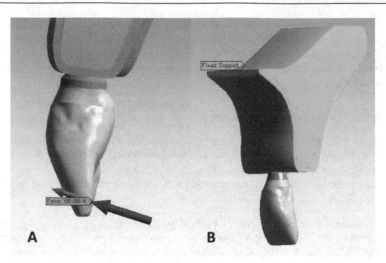

Fig. 8. (A) Transversal loading being applied on the lingual surface of the anterior crown. (B) Boundary conditions of the anterior maxilla at the upper and lateral sides. The green region represent the fixed support constrained of x, y and z directions (displacement = 0).

Ideally, the entire jawbone structure should be evaluated for its contribution to the force exerted onto the dental implant. However, since the simulation of the whole mandibular and maxillary bone is very elaborate, smaller models have been proposed. (Lin, Chang et al. 2007; Lin et al. 2008; Natali, Carniel et al. 2010; Pessoa et al. 2010). (Pierrisnard et al. 2003; Tada, Stegaroiu et al. 2003) It can be explained by the Saint Venant principles, the actual force system may be replaced by a equivalent load system and the distribution of stress and strains are only affected near the region of loading. (Ugural 2003) Thus, if the interest in the study is on the biomechanical peri-implant environment, the modeling of no more than the relevant segment of the bone is required (Figure 9). This procedure allows saving computing and modeling time. In addition, Teixeira et al. (Teixeira et al. 1998) demonstrated by a 3D FEA that modeling the mandible at a distance greater than 4.2 mm medially or distally from the implant did not result in any significant improvement in accuracy. Hence, besides the application of a proper implant loading, the determination of restrictions to the model displacement compatible with the anatomic segment to be simulated is advised. Obviously, if necessary, expanding the domain of the model could reduce the effect of inaccurate modeling of the boundary condition. (Zhou et al. 1999)

It has been common to apply fixed constraints to the uper region of the maxilla and to its lateral sides during the construction of the FE model to simulate the continuity of the bone (Figure 8B). (Van Staden, Guan et al. 2006) These fixed supports represent the constrained of x, y and z directions (displacement = 0). (Chaichanasiri et al. 2009; Hsu, Fuh et al. 2009; Faegh and Muftu 2010; Wu, Liao et al. 2010)

Ishigaki and coworkers (Ishigaki et al. 2003) determined the directions of displacement constrains, which were applied to the jawbone according to the angles of the closing pathways of chopping type and grinding type chewing patters, where the models were constrained at the base of the maxillary first molar to avoid sliding of the entire model.

Shunmugasamy and coworkers (Shunmugasamy, Gupta et al. 2011) simulated different macro-geometries of unit implants and fixed the base of the outer sides of the cortical bone.

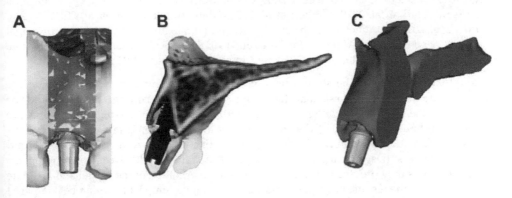

Fig. 9. (A) 3D CT-based model of an upper central incisor extraction socket and an implant model positioned inside of the alveolus. (B) Proximal view. (C) Final model: only the relevant segment were included.

7. Validation and interpretation of FE modeling

The validation of the results is the final and most important step in FEA, depending on the degree to which the system reflects its biologic influence reality. (Richmond, Wright et al. 2005) For instance, a model would be precise but inaccurate if the mesh is exceedingly dense but the loading and boundary conditions are unrealistic. (Richmond, Wright et al. 2005)

Validation of FEA entails comparing the behavior of the model with *in vivo* or *in vitro* data gathered from parts of the modeled structures. A combination of *in vitro* and *in vivo* experimentation potentially offers the best validation. *In vitro* validation allows one to carefully control the loads and boundary conditions in order to assess the validity of the model's geometry and elastic properties. (Richmond, Wright et al. 2005) Studies of the consistency of numeric models and their agreement with biologic data are scarce (Mellal et al. 2004; Ozan et al. 2010) and the level of agreement and consistency between different engineering methods, especially those regarding quantified stress/strain, remains a concern. (Iplikcioglu et al. 2003)

8. Analysis criteria involved in steps of the FEA

The analysis criteria widely used in implant prosthodontics are: (1) von Mises stress (σ_{vM}) (Yokoyama, Wakabayashi et al. 2005; Assuncao, Tabata et al. 2008; Huang, Hsu et al. 2008; Lim, Chew et al. 2008; Rubo and Souza 2008; Schrotenboer et al. 2008; Assuncao et al. 2009; Assuncao et al. 2009; Chaichanasiri, Nanakorn et al. 2009; Ding et al. 2009; Fazel, Aalai et al. 2009; Gomes et al. 2009; Kong et al. 2009; Li, Kong et al. 2009; Qian et al. 2009; Abreu et al. 2010; Ao et al. 2010; Caglar et al. 2010; Eraslan and Inan 2010; Eser et al. 2010; Faegh and Muftu 2010; Gomes et al. 2010; Miyamoto et al. 2010; Okumura, Stegaroiu et al. 2010; Pessoa, Muraru et al. 2010; Sagat, Yalcin et al. 2010; Takahashi et al. 2010; Teixeira et al. 2010; Winter et al. 2010; Bevilacqua et al. 2011; Burak Ozcelik, Ersoy et al. 2011; de Almeida, Rocha et al. 2011; Okumura et al. 2011), (2) maximum and minimum principal elastic strain (ε_{max} and ε_{min}) (Saab et al. 2007; Qian, Todo et al. 2009; Chou, Muftu et al. 2010; Danza, Quaranta et al. 2010; Eser, Tonuk et al. 2010; Limbert et al. 2010; Okumura, Stegaroiu et al. 2010; Pessoa, Muraru et al. 2010; de Almeida, Rocha et al. 2011) and (3) maximum and minimum principal stress (σ_{max} and σ_{min}). (de Almeida, Rocha et al. 2010; Degerliyurt et al. 2010; Hudieb et al. 2010; Jofre et al. 2010; Wu, Liao et al. 2010; de Almeida, Rocha et al. 2011)

Although the majority of the studies have used the σ_{vM} for evaluation of bone interface stress, several studies suggest that the magnitudes of the concentrations should be presented in the σ_{max} for evaluation of stress distribution in a brittle structure as bone, (de Almeida, Rocha et al. 2010; Degerliyurt, Simsek et al. 2010; Wu, Liao et al. 2010) as this criterion offers the possibility of making a distinction between tensile stress and compressive stress. (Degerliyurt, Simsek et al. 2010) In addition, displacement components of specific points may provide information about the deformation of the model and facilitate interpretation of the results. Principal stress values of fragile compact bone can be compared with its ultimate compressive strength and ultimate tensile strength values. (Ciftci and Canay 2000; Furmanski et al. 2009)

In fact, interfacial failure and bone resorption under different stress types are attributed to different mechanisms. Accordingly, it may be erroneous to emphasize the peak compressive or σ_{vM} without considering the risks of the tensile and the shear stresses at the interface. (Hudieb, Wakabayashi et al. 2010) However, when the titanium component is available as abutment, screw and implant, which are ductile structures, σ_{vM} is a recommended analysis criterion. (Cattaneo et al. 2005; Degerliyurt, Simsek et al. 2010; Wu, Liao et al. 2010)

Considering the FEA characteristics described above, the main advantage is the virtual simulation of real structures that are difficult to be clinically evaluated, i.e. stress and strain distribution on periimplant bone. Moreover, it is a low cost alternative in comparison to other *in vitro* methods since only a virtual model is used. Limitations of the FEA models include mainly the patient specific anatomy, and parameters such as the wet environment and damage accumulation under repetitive loading.

9. Summary and final remarks

The success of a dental implant depends on a variety of biomechanical factors including the design and position of the implant, implant-abutment connection, cantilever length, surface roughness, bone quality and type, depth of insertion, arch configuration, the nature of bone-implant interface, and occlusal conditions. (Randow, Glantz et al. 1986; Adell, Eriksson et al. 1990; van Steenberghe, Lekholm et al. 1990; Bozkaya and Muftu 2004; Bozkaya, Muftu et al. 2004; Misch et al. 2005; Yokoyama, Wakabayashi et al. 2005; De Smet et al. 2007; Abreu, Spazzin et al. 2010; Turkyilmaz 2011) All these biomechanical factors have been simulated by FEA in previous studies. While an increased number and quality of investigations has been published over the last decade, the results are often contradictory due to differences in model construction and meshing.

The biomechanical behavior of all components used in implant prosthodontics has been regarded as an important factor in determining the life expectancy of the restoration. Although substantial improvement has been made concerning implant/restorative component design, further improvement is achievable through better understanding the different dental implant treatment modalities' biomechanical behavior. While the field has gained key knowledge in model fabrication (experimental designing) and improvements in all steps here described concerning the basic FEA experimental design, it is expected that all steps will be further refined based on future gains in computer power and correlations made between modeling results and clinical observation, ultimately providing improved care for patients in need of oral rehabilitation.

10. References

Abreu, R. T., A. O. Spazzin, P. Y. Noritomi, et al. (2010). Influence of material of overdenture-retaining bar with vertical misfit on three-dimensional stress distribution, *Journal of prosthodontics: official journal of the American College of Prosthodontists*, Vol.19, No.6, pp. 425-431, 1532-849X (Electronic) 1059-941X (Linking).

Adell, R., B. Eriksson, U. Lekholm, et al. (1990). Long-term follow-up study of osseointegrated implants in the treatment of totally edentulous jaws, *The International journal of oral & maxillofacial implants*, Vol.5, No.4, pp. 347-359, 0882-2786 (Print) 0882-2786 (Linking).

Ao, J., T. Li, Y. Liu, et al. (2010). Optimal design of thread height and width on an immediately loaded cylinder implant: a finite element analysis, *Computers in biology and medicine*, Vol.40, No.8, pp. 681-686, 1879-0534 (Electronic) 0010-4825 (Linking).

Assuncao, W. G., V. A. Barao, L. F. Tabata, et al. (2009). Comparison between complete denture and implant-retained overdenture: effect of different mucosa thickness and resiliency on stress distribution, *Gerodontology*, Vol.26, No.4, pp. 273-281, 1741-2358 (Electronic) 0734-0664 (Linking).

Assuncao, W. G., V. A. Barao, L. F. Tabata, et al. (2009). Biomechanics studies in dentistry: bioengineering applied in oral implantology, *The Journal of craniofacial surgery*, Vol.20, No.4, pp. 1173-1177, 1536-3732 (Electronic) 1049-2275 (Linking).

Assuncao, W. G., E. A. Gomes, V. A. Barao, et al. (2009). Stress analysis in simulation models with or without implant threads representation, *The International journal of oral & maxillofacial implants*, Vol.24, No.6, pp. 1040-1044, 0882-2786 (Print) 0882-2786 (Linking).

Assuncao, W. G., L. F. Tabata, V. A. Barao, et al. (2008). Comparison of stress distribution between complete denture and implant-retained overdenture-2D FEA, *J Oral Rehabil*, Vol.35, No.10, pp. 766-774, 1365-2842 (Electronic) 0305-182X (Linking).

Assuncao, W. G., L. F. Tabata, V. A. Barao, et al. (2008). Comparison of stress distribution between complete denture and implant-retained overdenture-2D FEA, *Journal of oral rehabilitation*, Vol.35, No.10, pp. 766-774, 1365-2842 (Electronic) 0305-182X (Linking).

Bevilacqua, M., T. Tealdo, M. Menini, et al. (2011). The influence of cantilever length and implant inclination on stress distribution in maxillary implant-supported fixed dentures, *The Journal of prosthetic dentistry*, Vol.105, No.1, pp. 5-13, 1097-6841 (Electronic) 0022-3913 (Linking).

Bozini, T., H. Petridis&K. Garefis (2011). A meta-analysis of prosthodontic complication rates of implant-supported fixed dental prostheses in edentulous patients after an observation period of at least 5 years, *The International journal of oral & maxillofacial implants*, Vol.26, No.2, pp. 304-318, 0882-2786 (Print) 0882-2786 (Linking).

Bozkaya, D.&S. Muftu (2004). Efficiency considerations for the purely tapered interference fit (TIF) abutments used in dental implants, *Journal of biomechanical engineering*, Vol.126, No.4, pp. 393-401, 0148-0731 (Print) 0148-0731 (Linking).

Bozkaya, D., S. Muftu&A. Muftu (2004). Evaluation of load transfer characteristics of five different implants in compact bone at different load levels by finite elements analysis, *The Journal of prosthetic dentistry*, Vol.92, No.6, pp. 523-530, 0022-3913 (Print) 0022-3913 (Linking).

Branemark, P. I., R. Adell, U. Breine, et al. (1969). Intra-osseous anchorage of dental prostheses. I. Experimental studies, *Scand J Plast Reconstr Surg*, Vol.3, No.2, pp. 81-100, 0036-5556 (Print) 0036-5556 (Linking).

Branemark, P. I., B. O. Hansson, R. Adell, et al. (1977). Osseointegrated implants in the treatment of the edentulous jaw. Experience from a 10-year period, *Scand J Plast Reconstr Surg Suppl*, Vol.16, pp. 1-132, 0581-9474 (Print) 0581-9474 (Linking).

Brunski, J. B. (1992). Biomechanical factors affecting the bone-dental implant interface, *Clinical materials*, Vol.10, No.3, pp. 153-201, 0267-6605 (Print) 0267-6605 (Linking).

Burak Ozcelik, T., E. Ersoy&B. Yilmaz Biomechanical evaluation of tooth- and implant-supported fixed dental prostheses with various nonrigid connector positions: a finite element analysis, *J Prosthodont*, Vol.20, No.1, pp. 16-28, 1532-849X (Electronic) 1059-941X (Linking).

Burak Ozcelik, T., E. Ersoy&B. Yilmaz (2011). Biomechanical evaluation of tooth- and implant-supported fixed dental prostheses with various nonrigid connector positions: a finite element analysis, *Journal of prosthodontics : official journal of the American College of Prosthodontists*, Vol.20, No.1, pp. 16-28, 1532-849X (Electronic) 1059-941X (Linking).

Caglar, A., B. T. Bal, C. Aydin, et al. (2010). Evaluation of stresses occurring on three different zirconia dental implants: three-dimensional finite element analysis, *The International journal of oral & maxillofacial implants*, Vol.25, No.1, pp. 95-103, 0882-2786 (Print) 0882-2786 (Linking).

Canay, S.&K. Akca (2009). Biomechanical aspects of bone-level diameter shifting at implant-abutment interface, *Implant dentistry*, Vol.18, No.3, pp. 239-248, 1538-2982 (Electronic) 1056-6163 (Linking).

Cattaneo, P. M., M. Dalstra&B. Melsen (2005). The finite element method: a tool to study orthodontic tooth movement, *J Dent Res*, Vol.84, No.5, pp. 428-433, 0022-0345 (Print) 0022-0345 (Linking).

Chaichanasiri, E., P. Nanakorn, W. Tharanon, et al. (2009). Finite element analysis of bone around a dental implant supporting a crown with a premature contact, *Journal of the Medical Association of Thailand = Chotmaihet thangphaet*, Vol.92, No.10, pp. 1336-1344, 0125-2208 (Print) 0125-2208 (Linking).

Chang, C. L., C. S. Chen&M. L. Hsu (2010). Biomechanical effect of platform switching in implant dentistry: a three-dimensional finite element analysis, *The International journal of oral & maxillofacial implants*, Vol.25, No.2, pp. 295-304, 0882-2786 (Print) 0882-2786 (Linking).

Chen, J. H., C. Liu, L. You, et al. (2010). Boning up on Wolff's Law: mechanical regulation of the cells that make and maintain bone, *Journal of biomechanics*, Vol.43, No.1, pp. 108-118, 1873-2380 (Electronic) 0021-9290 (Linking).

Chou, H. Y., S. Muftu&D. Bozkaya (2010). Combined effects of implant insertion depth and alveolar bone quality on periimplant bone strain induced by a wide-diameter, short implant and a narrow-diameter, long implant, *The Journal of prosthetic dentistry*, Vol.104, No.5, pp. 293-300, 1097-6841 (Electronic) 0022-3913 (Linking).

Chun, H. J., S. Y. Cheong, J. H. Han, et al. (2002). Evaluation of design parameters of osseointegrated dental implants using finite element analysis, *Journal of oral rehabilitation*, Vol.29, No.6, pp. 565-574, 0305-182X (Print) 0305-182X (Linking).

Ciftci, Y.&S. Canay (2000). The effect of veneering materials on stress distribution in implant-supported fixed prosthetic restorations, *The International journal of oral & maxillofacial implants*, Vol.15, No.4, pp. 571-582, 0882-2786 (Print) 0882-2786 (Linking).

Cohen, Z. A., D. M. McCarthy, S. D. Kwak, et al. (1999). Knee cartilage topography, thickness, and contact areas from MRI: in-vitro calibration and in-vivo measurements, *Osteoarthritis and cartilage / OARS, Osteoarthritis Research Society*, Vol.7, No.1, pp. 95-109, 1063-4584 (Print) 1063-4584 (Linking).

Danza, M., A. Quaranta, F. Carinci, et al. (2010). Biomechanical evaluation of dental implants in D1 and D4 bone by Finite Element Analysis, *Minerva stomatologica*, Vol.59, No.6, pp. 305-313, 0026-4970 (Print) 0026-4970 (Linking).

de Almeida, E. O., E. P. Rocha, W. G. Assuncao, et al. (2011). Cortical bone stress distribution in mandibles with different configurations restored with prefabricated bar-prosthesis protocol: a three-dimensional finite-element analysis, *Journal of prosthodontics : official journal of the American College of Prosthodontists*, Vol.20, No.1, pp. 29-34, 1532-849X (Electronic) 1059-941X (Linking).

de Almeida, E. O., E. P. Rocha, A. C. Freitas, Jr., et al. (2010). Finite element stress analysis of edentulous mandibles with different bone types supporting multiple-implant superstructures, *The International journal of oral & maxillofacial implants*, Vol.25, No.6, pp. 1108-1114, 0882-2786 (Print) 0882-2786 (Linking).

De Smet, E., S. V. Jaecques, J. J. Jansen, et al. (2007). Effect of constant strain rate, composed of varying amplitude and frequency, of early loading on peri-implant bone (re)modelling, *Journal of clinical periodontology*, Vol.34, No.7, pp. 618-624, 0303-6979 (Print) 0303-6979 (Linking).

Degerliyurt, K., B. Simsek, E. Erkmen, et al. (2010). Effects of different fixture geometries on the stress distribution in mandibular peri-implant structures: a 3-dimensional finite element analysis, *Oral surgery, oral medicine, oral pathology, oral radiology, and endodontics*, Vol.110, No.2, pp. e1-11, 1528-395X (Electronic) 1079-2104 (Linking).

Ding, X., X. H. Zhu, S. H. Liao, et al. (2009). Implant-bone interface stress distribution in immediately loaded implants of different diameters: a three-dimensional finite element analysis, *Journal of prosthodontics : official journal of the American College of Prosthodontists*, Vol.18, No.5, pp. 393-402, 1532-849X (Electronic) 1059-941X (Linking).

Duyck, J., H. Van Oosterwyck, J. Vander Sloten, et al. (2000). Magnitude and distribution of occlusal forces on oral implants supporting fixed prostheses: an in vivo study,

Clinical oral implants research, Vol.11, No.5, pp. 465-475, 0905-7161 (Print) 0905-7161 (Linking).

Eraslan, O.&O. Inan (2010). The effect of thread design on stress distribution in a solid screw implant: a 3D finite element analysis, *Clinical oral investigations*, Vol.14, No.4, pp. 411-416, 1436-3771 (Electronic) 1432-6981 (Linking).

Eser, A., K. Akca, S. Eckert, et al. (2009). Nonlinear finite element analysis versus ex vivo strain gauge measurements on immediately loaded implants, *The International journal of oral & maxillofacial implants*, Vol.24, No.3, pp. 439-446, 0882-2786 (Print) 0882-2786 (Linking).

Eser, A., E. Tonuk, K. Akca, et al. (2010). Predicting time-dependent remodeling of bone around immediately loaded dental implants with different designs, *Medical engineering & physics*, Vol.32, No.1, pp. 22-31, 1873-4030 (Electronic) 1350-4533 (Linking).

Eskitascioglu, G., A. Usumez, M. Sevimay, et al. (2004). The influence of occlusal loading location on stresses transferred to implant-supported prostheses and supporting bone: A three-dimensional finite element study, *The Journal of prosthetic dentistry*, Vol.91, No.2, pp. 144-150, 0022-3913 (Print) 0022-3913 (Linking).

Esposito, M., J. M. Hirsch, U. Lekholm, et al. (1998). Biological factors contributing to failures of osseointegrated oral implants. (II). Etiopathogenesis, *Eur J Oral Sci*, Vol.106, No.3, pp. 721-764, 0909-8836 (Print) 0909-8836 (Linking).

Faegh, S.&S. Muftu (2010). Load transfer along the bone-dental implant interface, *Journal of biomechanics*, Vol.43, No.9, pp. 1761-1770, 1873-2380 (Electronic) 0021-9290 (Linking).

Fajardo, R. J., T. M. Ryan&J. Kappelman (2002). Assessing the accuracy of high-resolution X-ray computed tomography of primate trabecular bone by comparisons with histological sections, *Am J Phys Anthropol*, Vol.118, No.1, pp. 1-10, 0002-9483 (Print) 0002-9483 (Linking).

Fazel, A., S. Aalai&M. Rismanchian (2009). Effect of macro-design of immediately loaded implants on micromotion and stress distribution in surrounding bone using finite element analysis, *Implant dentistry*, Vol.18, No.4, pp. 345-352, 1538-2982 (Electronic) 1056-6163 (Linking).

Fontijn-Tekamp, F. A., A. P. Slagter, M. A. van't Hof, et al. (1998). Bite forces with mandibular implant-retained overdentures, *Journal of dental research*, Vol.77, No.10, pp. 1832-1839, 0022-0345 (Print) 0022-0345 (Linking).

Freitas, A. C., Jr., E. P. Rocha, P. H. dos Santos, et al. (2010). All-ceramic crowns over single implant zircon abutment. Influence of young's modulus on mechanics, *Implant Dent*, Vol.19, No.6, pp. 539-548, 1538-2982 (Electronic) 1056-6163 (Linking).

Freitas Junior, A. C., E. P. Rocha, P. H. Santos, et al. (2010). Mechanics of the maxillary central incisor. Influence of the periodontal ligament represented by beam elements, *Comput Methods Biomech Biomed Engin*, Vol.13, No.5, pp. 515-521, 1476-8259 (Electronic) 1025-5842 (Linking).

Freitas Junior, A. C., E. P. Rocha, P. H. Santos, et al. (2010). Mechanics of the maxillary central incisor. Influence of the periodontal ligament represented by beam elements, *Computer methods in biomechanics and biomedical engineering*, Vol.13, No.5, pp. 515-521, 1476-8259 (Electronic) 1025-5842 (Linking).

Furmanski, J., M. Anderson, S. Bal, et al. (2009). Clinical fracture of cross-linked UHMWPE acetabular liners, *Biomaterials*, Vol.30, No.29, pp. 5572-5582, 1878-5905 (Electronic) 0142-9612 (Linking).

Galantucci, L. M., G. Percoco, G. Angelelli, et al. (2006). Reverse engineering techniques applied to a human skull, for CAD 3D reconstruction and physical replication by rapid prototyping, *J Med Eng Technol*, Vol.30, No.2, pp. 102-111, 0309-1902 (Print) 0309-1902 (Linking).

Geng, J. P., K. B. Tan&G. R. Liu (2001). Application of finite element analysis in implant dentistry: a review of the literature, *The Journal of prosthetic dentistry*, Vol.85, No.6, pp. 585-598, 0022-3913 (Print) 0022-3913 (Linking).

Genna, F. (2003). On the effects of cyclic transversal forces on osseointegrated dental implants: experimental and finite element shakedown analyses, *Computer methods in biomechanics and biomedical engineering*, Vol.6, No.2, pp. 141-152, 1025-5842 (Print) 1025-5842 (Linking).

Gomes, E. A., W. G. Assuncao, V. A. Barao, et al. (2010). Passivity versus unilateral angular misfit: evaluation of stress distribution on implant-supported single crowns: three-dimensional finite element analysis, *The Journal of craniofacial surgery*, Vol.21, No.6, pp. 1683-1687, 1536-3732 (Electronic) 1049-2275 (Linking).

Gomes, E. A., W. G. Assuncao, L. F. Tabata, et al. (2009). Effect of passive fit absence in the prosthesis/implant/retaining screw system: a two-dimensional finite element analysis, *The Journal of craniofacial surgery*, Vol.20, No.6, pp. 2000-2005, 1536-3732 (Electronic) 1049-2275 (Linking).

Gotfredsen, K., T. Berglundh&J. Lindhe (2000). Anchorage of titanium implants with different surface characteristics: an experimental study in rabbits, *Clinical implant dentistry and related research*, Vol.2, No.3, pp. 120-128, 1523-0899 (Print) 1523-0899 (Linking).

Greco, G. D., W. C. Jansen, J. Landre Junior, et al. (2009). Stress analysis on the free-end distal extension of an implant-supported mandibular complete denture, *Brazilian oral research*, Vol.23, No.2, pp. 182-189, 1807-3107 (Electronic) 1806-8324 (Linking).

Guan, H., R. van Staden, Y. C. Loo, et al. (2009). Influence of bone and dental implant parameters on stress distribution in the mandible: a finite element study, *The International journal of oral & maxillofacial implants*, Vol.24, No.5, pp. 866-876, 0882-2786 (Print) 0882-2786 (Linking).

Hansson, S.&A. Halldin (2009). Re: effect of microthreads and platform switching on crestal bone stress levels: a finite element analysis, *Journal of periodontology*, Vol.80, No.7, pp. 1033-1035; authors response 1035-1036, 0022-3492 (Print) 0022-3492 (Linking).

Hasan, I., A. Rahimi, L. Keilig, et al. (2011). Computational simulation of internal bone remodelling around dental implants: a sensitivity analysis, *Computer methods in biomechanics and biomedical engineering*, pp. 1, 1476-8259 (Electronic) 1025-5842 (Linking).

Hsu, J. T., L. J. Fuh, D. J. Lin, et al. (2009). Bone strain and interfacial sliding analyses of platform switching and implant diameter on an immediately loaded implant: experimental and three-dimensional finite element analyses, *Journal of periodontology*, Vol.80, No.7, pp. 1125-1132, 0022-3492 (Print) 0022-3492 (Linking).

Huang, H. L., C. H. Chang, J. T. Hsu, et al. (2007). Comparison of implant body designs and threaded designs of dental implants: a 3-dimensional finite element analysis, *The*

International journal of oral & maxillofacial implants, Vol.22, No.4, pp. 551-562, 0882-2786 (Print) 0882-2786 (Linking).

Huang, H. L., L. J. Fuh, C. C. Ko, et al. (2009). Biomechanical effects of a maxillary implant in the augmented sinus: a three-dimensional finite element analysis, *The International journal of oral & maxillofacial implants*, Vol.24, No.3, pp. 455-462, 0882-2786 (Print) 0882-2786 (Linking).

Huang, H. L., J. T. Hsu, L. J. Fuh, et al. (2008). Bone stress and interfacial sliding analysis of implant designs on an immediately loaded maxillary implant: a non-linear finite element study, *Journal of dentistry*, Vol.36, No.6, pp. 409-417, 0300-5712 (Print) 0300-5712 (Linking).

Hudieb, M. I., N. Wakabayashi&S. Kasugai (2010). Magnitude and Direction of Mechanical Stress at the Osseointegrated Interface of the Microthread Implant, *Journal of periodontology*, pp. 1943-3670 (Electronic) 0022-3492 (Linking).

Hudieb, M. I., N. Wakabayashi&S. Kasugai (2011). Magnitude and direction of mechanical stress at the osseointegrated interface of the microthread implant, *Journal of periodontology*, Vol.82, No.7, pp. 1061-1070, 1943-3670 (Electronic) 0022-3492 (Linking).

Iplikcioglu, H., K. Akca, M. C. Cehreli, et al. (2003). Comparison of non-linear finite element stress analysis with in vitro strain gauge measurements on a Morse taper implant, *The International journal of oral & maxillofacial implants*, Vol.18, No.2, pp. 258-265, 0882-2786 (Print) 0882-2786 (Linking).

Ishigaki, S., T. Nakano, S. Yamada, et al. (2003). Biomechanical stress in bone surrounding an implant under simulated chewing, *Clinical oral implants research*, Vol.14, No.1, pp. 97-102, 0905-7161 (Print) 0905-7161 (Linking).

Jofre, J., P. Cendoya&P. Munoz (2010). Effect of splinting mini-implants on marginal bone loss: a biomechanical model and clinical randomized study with mandibular overdentures, *The International journal of oral & maxillofacial implants*, Vol.25, No.6, pp. 1137-1144, 0882-2786 (Print) 0882-2786 (Linking).

Kappelman, J. (1998). Advances in three-dimensional data acquisition and analysis. Primate locomotion: recent advances. E. F. Strasser, J.G.; Rosenberger, A.; McHenry, H.M. New York, Plenum Press: 205–222.

Kohal, R. J., G. Papavasiliou, P. Kamposiora, et al. (2002). Three-dimensional computerized stress analysis of commercially pure titanium and yttrium-partially stabilized zirconia implants, *The International journal of prosthodontics*, Vol.15, No.2, pp. 189-194, 0893-2174 (Print) 0893-2174 (Linking).

Kong, L., Z. Gu, T. Li, et al. (2009). Biomechanical optimization of implant diameter and length for immediate loading: a nonlinear finite element analysis, *The International journal of prosthodontics*, Vol.22, No.6, pp. 607-615, 0893-2174 (Print) 0893-2174 (Linking).

Kozlovsky, A., H. Tal, B. Z. Laufer, et al. (2007). Impact of implant overloading on the peri-implant bone in inflamed and non-inflamed peri-implant mucosa, *Clinical oral implants research*, Vol.18, No.5, pp. 601-610, 0905-7161 (Print) 0905-7161 (Linking).

Levine, R. A., D. S. Clem, 3rd, T. G. Wilson, Jr., et al. (1999). Multicenter retrospective analysis of the ITI implant system used for single-tooth replacements: results of loading for 2 or more years, *The International journal of oral & maxillofacial implants*, Vol.14, No.4, pp. 516-520, 0882-2786 (Print) 0882-2786 (Linking).

Li, T., L. Kong, Y. Wang, et al. (2009). Selection of optimal dental implant diameter and length in type IV bone: a three-dimensional finite element analysis, *International journal of oral and maxillofacial surgery*, Vol.38, No.10, pp. 1077-1083, 1399-0020 (Electronic) 0901-5027 (Linking).

Lim, K. H., C. M. Chew, P. C. Chen, et al. (2008). New extensometer to measure in vivo uniaxial mechanical properties of human skin, *Journal of biomechanics*, Vol.41, No.5, pp. 931-936, 0021-9290 (Print) 0021-9290 (Linking).

Limbert, G., C. van Lierde, O. L. Muraru, et al. (2010). Trabecular bone strains around a dental implant and associated micromotions--a micro-CT-based three-dimensional finite element study, *Journal of biomechanics*, Vol.43, No.7, pp. 1251-1261, 1873-2380 (Electronic) 0021-9290 (Linking).

Lin, C. L., S. H. Chang, W. J. Chang, et al. (2007). Factorial analysis of variables influencing mechanical characteristics of a single tooth implant placed in the maxilla using finite element analysis and the statistics-based Taguchi method, *European journal of oral sciences*, Vol.115, No.5, pp. 408-416, 0909-8836 (Print) 0909-8836 (Linking).

Lin, C. L., S. H. Chang, J. C. Wang, et al. (2006). Mechanical interactions of an implant/tooth-supported system under different periodontal supports and number of splinted teeth with rigid and non-rigid connections, *Journal of dentistry*, Vol.34, No.9, pp. 682-691, 0300-5712 (Print) 0300-5712 (Linking).

Lin, C. L., Y. H. Lin&S. H. Chang (2010). Multi-factorial analysis of variables influencing the bone loss of an implant placed in the maxilla: prediction using FEA and SED bone remodeling algorithm, *Journal of biomechanics*, Vol.43, No.4, pp. 644-651, 1873-2380 (Electronic) 0021-9290 (Linking).

Lin, C. L., J. C. Wang&W. J. Chang (2008). Biomechanical interactions in tooth-implant-supported fixed partial dentures with variations in the number of splinted teeth and connector type: a finite element analysis, *Clinical oral implants research*, Vol.19, No.1, pp. 107-117, 0905-7161 (Print) 0905-7161 (Linking).

Lin, D., Q. Li, W. Li, et al. (2010). Mandibular bone remodeling induced by dental implant, *Journal of biomechanics*, Vol.43, No.2, pp. 287-293, 1873-2380 (Electronic) 0021-9290 (Linking).

Lin, D., Q. Li, W. Li, et al. (2009). Dental implant induced bone remodeling and associated algorithms, *J Mech Behav Biomed Mater*, Vol.2, No.5, pp. 410-432, 1878-0180 (Electronic) 1878-0180 (Linking).

Maeda, Y., J. Miura, I. Taki, et al. (2007). Biomechanical analysis on platform switching: is there any biomechanical rationale?, *Clinical oral implants research*, Vol.18, No.5, pp. 581-584, 0905-7161 (Print) 0905-7161 (Linking).

Magne, P. (2007). Efficient 3D finite element analysis of dental restorative procedures using micro-CT data, *Dental materials : official publication of the Academy of Dental Materials*, Vol.23, No.5, pp. 539-548, 0109-5641 (Print) 0109-5641 (Linking).

Mellal, A., H. W. Wiskott, J. Botsis, et al. (2004). Stimulating effect of implant loading on surrounding bone. Comparison of three numerical models and validation by in vivo data, *Clinical oral implants research*, Vol.15, No.2, pp. 239-248, 0905-7161 (Print) 0905-7161 (Linking).

Menicucci, G., A. Mossolov, M. Mozzati, et al. (2002). Tooth-implant connection: some biomechanical aspects based on finite element analyses, *Clinical oral implants research*, Vol.13, No.3, pp. 334-341, 0905-7161 (Print) 0905-7161 (Linking).

Mericske-Stern, R., M. Piotti&G. Sirtes (1996). 3-D in vivo force measurements on mandibular implants supporting overdentures. A comparative study, *Clinical oral implants research*, Vol.7, No.4, pp. 387-396, 0905-7161 (Print) 0905-7161 (Linking).

Merz, B. R., S. Hunenbart&U. C. Belser (2000). Mechanics of the implant-abutment connection: an 8-degree taper compared to a butt joint connection, *The International journal of oral & maxillofacial implants*, Vol.15, No.4, pp. 519-526, 0882-2786 (Print) 0882-2786 (Linking).

Misch, C. E., J. B. Suzuki, F. M. Misch-Dietsh, et al. (2005). A positive correlation between occlusal trauma and peri-implant bone loss: literature support, *Implant dentistry*, Vol.14, No.2, pp. 108-116, 1056-6163 (Print) 1056-6163 (Linking).

Miyamoto, S., K. Ujigawa, Y. Kizu, et al. (2010). Biomechanical three-dimensional finite-element analysis of maxillary prostheses with implants. Design of number and position of implants for maxillary prostheses after hemimaxillectomy, *International journal of oral and maxillofacial surgery*, Vol.39, No.11, pp. 1120-1126, 1399-0020 (Electronic) 0901-5027 (Linking).

Morneburg, T. R.&P. A. Proschel (2002). Measurement of masticatory forces and implant loads: a methodologic clinical study, *The International journal of prosthodontics*, Vol.15, No.1, pp. 20-27, 0893-2174 (Print) 0893-2174 (Linking).

Natali, A. N., E. L. Carniel&P. G. Pavan (2009). Investigation of viscoelastoplastic response of bone tissue in oral implants press fit process, *J Biomed Mater Res B Appl Biomater*, Vol.91, No.2, pp. 868-875, 1552-4981 (Electronic) 1552-4973 (Linking).

Natali, A. N., E. L. Carniel&P. G. Pavan (2010). Modelling of mandible bone properties in the numerical analysis of oral implant biomechanics, *Computer methods and programs in biomedicine*, Vol.100, No.2, pp. 158-165, 1872-7565 (Electronic) 0169-2607 (Linking).

O'Mahony, A. M., J. L. Williams&P. Spencer (2001). Anisotropic elasticity of cortical and cancellous bone in the posterior mandible increases peri-implant stress and strain under oblique loading, *Clinical oral implants research*, Vol.12, No.6, pp. 648-657, 0905-7161 (Print) 0905-7161 (Linking).

Okumura, N., R. Stegaroiu, E. Kitamura, et al. (2010). Influence of maxillary cortical bone thickness, implant design and implant diameter on stress around implants: a three-dimensional finite element analysis, *Journal of prosthodontic research*, Vol.54, No.3, pp. 133-142, 1883-1958 (Print) 1883-1958 (Linking).

Okumura, N., R. Stegaroiu, H. Nishiyama, et al. (2011). Finite element analysis of implant-embedded maxilla model from CT data: comparison with the conventional model, *Journal of prosthodontic research*, Vol.55, No.1, pp. 24-31, 1883-1958 (Print) 1883-1958 (Linking).

Ozan, F., H. Yildiz, O. A. Bora, et al. (2010). The effect of head trauma on fracture healing: biomechanical testing and finite element analysis, *Acta Orthop Traumatol Turc*, Vol.44, No.4, pp. 313-321, 1017-995X (Print) 1017-995X (Linking).

Pessoa, R. S., L. Muraru, E. M. Junior, et al. (2010). Influence of implant connection type on the biomechanical environment of immediately placed implants - CT-based nonlinear, three-dimensional finite element analysis, *Clin Implant Dent Relat Res*, Vol.12, No.3, pp. 219-234, 1708-8208 (Electronic) 1523-0899 (Linking).

Pessoa, R. S., L. Muraru, E. M. Junior, et al. (2010). Influence of implant connection type on the biomechanical environment of immediately placed implants - CT-based nonlinear, three-dimensional finite element analysis, *Clinical implant dentistry and related research*, Vol.12, No.3, pp. 219-234, 1708-8208 (Electronic) 1523-0899 (Linking).

Pessoa, R. S., L. G. Vaz, E. Marcantonio, Jr., et al. (2010). Biomechanical evaluation of platform switching in different implant protocols: computed tomography-based three-dimensional finite element analysis, *The International journal of oral & maxillofacial implants*, Vol.25, No.5, pp. 911-919, 0882-2786 (Print) 0882-2786 (Linking).

Peterson, J.&P. C. Dechow (2003). Material properties of the human cranial vault and zygoma, *Anat Rec A Discov Mol Cell Evol Biol*, Vol.274, No.1, pp. 785-797, 1552-4884 (Print) 1552-4884 (Linking).

Pieri, F., N. N. Aldini, C. Marchetti, et al. (2011). Influence of implant-abutment interface design on bone and soft tissue levels around immediately placed and restored single-tooth implants: a randomized controlled clinical trial, *The International journal of oral & maxillofacial implants*, Vol.26, No.1, pp. 169-178, 0882-2786 (Print) 0882-2786 (Linking).

Pierrisnard, L., G. Hure, M. Barquins, et al. (2002). Two dental implants designed for immediate loading: a finite element analysis, *The International journal of oral & maxillofacial implants*, Vol.17, No.3, pp. 353-362, 0882-2786 (Print) 0882-2786 (Linking).

Pierrisnard, L., F. Renouard, P. Renault, et al. (2003). Influence of implant length and bicortical anchorage on implant stress distribution, *Clin Implant Dent Relat Res*, Vol.5, No.4, pp. 254-262, 1523-0899 (Print) 1523-0899 (Linking).

Qian, L., M. Todo, Y. Matsushita, et al. (2009). Effects of implant diameter, insertion depth, and loading angle on stress/strain fields in implant/jawbone systems: finite element analysis, *The International journal of oral & maxillofacial implants*, Vol.24, No.5, pp. 877-886, 0882-2786 (Print) 0882-2786 (Linking).

Rancourt, D., A. Shirazi-Adl, G. Drouin, et al. (1990). Friction properties of the interface between porous-surfaced metals and tibial cancellous bone, *Journal of biomedical materials research*, Vol.24, No.11, pp. 1503-1519, 0021-9304 (Print) 0021-9304 (Linking).

Randow, K., P. O. Glantz&B. Zoger (1986). Technical failures and some related clinical complications in extensive fixed prosthodontics. An epidemiological study of long-term clinical quality, *Acta Odontol Scand*, Vol.44, No.4, pp. 241-255, 0001-6357 (Print) 0001-6357 (Linking).

Rayfield, E. J. (2004). Cranial mechanics and feeding in Tyrannosaurus rex, *Proc Biol Sci*, Vol.271, No.1547, pp. 1451-1459, 0962-8452 (Print) 0962-8452 (Linking).

Richmond, B. G., B. W. Wright, I. Grosse, et al. (2005). Finite element analysis in functional morphology, *Anat Rec A Discov Mol Cell Evol Biol*, Vol.283, No.2, pp. 259-274, 1552-4884 (Print) 1552-4884 (Linking).

Roberts, D. H. (1970). The failure of retainers in bridge prostheses. An analysis of 2,000 retainers, *Br Dent J*, Vol.128, No.3, pp. 117-124, 0007-0610 (Print) 0007-0610 (Linking).

Rodriguez-Ciurana, X., X. Vela-Nebot, M. Segala-Torres, et al. (2009). Biomechanical repercussions of bone resorption related to biologic width: a finite element analysis of three implant-abutment configurations, *The International journal of periodontics & restorative dentistry*, Vol.29, No.5, pp. 479-487, 0198-7569 (Print) 0198-7569 (Linking).

Romanos, G. E. (2004). Present status of immediate loading of oral implants, *The Journal of oral implantology*, Vol.30, No.3, pp. 189-197, 0160-6972 (Print) 0160-6972 (Linking).

Romeed, S. A., S. L. Fok&N. H. Wilson (2006). A comparison of 2D and 3D finite element analysis of a restored tooth, *J Oral Rehabil*, Vol.33, No.3, pp. 209-215, 0305-182X (Print) 0305-182X (Linking).

Ross, C. F. (2005). Finite element analysis in vertebrate biomechanics, *Anat Rec A Discov Mol Cell Evol Biol*, Vol.283, No.2, pp. 253-258, 1552-4884 (Print) 1552-4884 (Linking).

Rubo, J. H.&E. A. Souza (2008). Finite element analysis of stress in bone adjacent to dental implants, *The Journal of oral implantology*, Vol.34, No.5, pp. 248-255, 0160-6972 (Print) 0160-6972 (Linking).

Ryan, T. M.&B. van Rietbergen (2005). Mechanical significance of femoral head trabecular bone structure in Loris and Galago evaluated using micromechanical finite element models, *Am J Phys Anthropol*, Vol.126, No.1, pp. 82-96, 0002-9483 (Print) 0002-9483 (Linking).

Saab, X. E., J. A. Griggs, J. M. Powers, et al. (2007). Effect of abutment angulation on the strain on the bone around an implant in the anterior maxilla: a finite element study, *The Journal of prosthetic dentistry*, Vol.97, No.2, pp. 85-92, 0022-3913 (Print) 0022-3913 (Linking).

Sagat, G., S. Yalcin, B. A. Gultekin, et al. (2010). Influence of arch shape and implant position on stress distribution around implants supporting fixed full-arch prosthesis in edentulous maxilla, *Implant dentistry*, Vol.19, No.6, pp. 498-508, 1538-2982 (Electronic) 1056-6163 (Linking).

Schrotenboer, J., Y. P. Tsao, V. Kinariwala, et al. (2008). Effect of microthreads and platform switching on crestal bone stress levels: a finite element analysis, *Journal of periodontology*, Vol.79, No.11, pp. 2166-2172, 0022-3492 (Print) 0022-3492 (Linking).

Sevimay, M., F. Turhan, M. A. Kilicarslan, et al. (2005). Three-dimensional finite element analysis of the effect of different bone quality on stress distribution in an implant-supported crown, *The Journal of prosthetic dentistry*, Vol.93, No.3, pp. 227-234, 0022-3913 (Print) 0022-3913 (Linking).

Shunmugasamy, V. C., N. Gupta, R. S. Pessoa, et al. (2011). Influence of clinically relevant factors on the immediate biomechanical surrounding for a series of dental implant designs, *J Biomech Eng*, Vol.133, No.3, pp. 031005, 1528-8951 (Electronic) 0148-0731 (Linking).

Silva, G. C., J. A. Mendonca, L. R. Lopes, et al. (2010). Stress patterns on implants in prostheses supported by four or six implants: a three-dimensional finite element analysis, *The International journal of oral & maxillofacial implants*, Vol.25, No.2, pp. 239-246, 0882-2786 (Print) 0882-2786 (Linking).

Sotto-Maior, B. S., E. P. Rocha, E. O. Almeida, et al. (2010). Influence of high insertion torque on implant placement: an anisotropic bone stress analysis, *Braz Dent J*, Vol.21, No.6, pp. 508-514, 1806-4760 (Electronic) 0103-6440 (Linking).

Stanford, C. M.&R. A. Brand (1999). Toward an understanding of implant occlusion and strain adaptive bone modeling and remodeling, *The Journal of prosthetic dentistry*, Vol.81, No.5, pp. 553-561, 0022-3913 (Print) 0022-3913 (Linking).

Tada, S., R. Stegaroiu, E. Kitamura, et al. (2003). Influence of implant design and bone quality on stress/strain distribution in bone around implants: a 3-dimensional finite element analysis, *The International journal of oral & maxillofacial implants*, Vol.18, No.3, pp. 357-368, 0882-2786 (Print) 0882-2786 (Linking).

Takahashi, T., I. Shimamura&K. Sakurai (2010). Influence of number and inclination angle of implants on stress distribution in mandibular cortical bone with All-on-4 Concept, *Journal of prosthodontic research*, Vol.54, No.4, pp. 179-184, 1883-1958 (Print) 1883-1958 (Linking).

Teixeira, E. R., Y. Sato, Y. Akagawa, et al. (1998). A comparative evaluation of mandibular finite element models with different lengths and elements for implant biomechanics, *Journal of oral rehabilitation*, Vol.25, No.4, pp. 299-303, 0305-182X (Print) 0305-182X (Linking).

Teixeira, M. F., S. A. Ramalho, I. A. de Mattias Sartori, et al. (2010). Finite element analysis of 2 immediate loading systems in edentulous mandible: rigid and semirigid splinting of implants, *Implant dentistry*, Vol.19, No.1, pp. 39-49, 1538-2982 (Electronic) 1056-6163 (Linking).

Turkyilmaz, I. (2011). 26-year follow-up of screw-retained fixed dental prostheses supported by machined-surface Branemark implants: a case report, *Texas dental journal*, Vol.128, No.1, pp. 15-19, 0040-4284 (Print) 0040-4284 (Linking).

Ugural, A. C. F. S. (2003). *Advanced strength and applied elasticity*. New Jersey, Prentice-Hall.

Van Oosterwyck, H., J. Duyck, J. Vander Sloten, et al. (1998). The influence of bone mechanical properties and implant fixation upon bone loading around oral implants, *Clinical oral implants research*, Vol.9, No.6, pp. 407-418, 0905-7161 (Print) 0905-7161 (Linking).

Van Staden, R. C., H. Guan&Y. C. Loo (2006). Application of the finite element method in dental implant research, *Computer methods in biomechanics and biomedical engineering*, Vol.9, No.4, pp. 257-270, 1025-5842 (Print) 1025-5842 (Linking).

van Steenberghe, D., U. Lekholm, C. Bolender, et al. (1990). Applicability of osseointegrated oral implants in the rehabilitation of partial edentulism: a prospective multicenter study on 558 fixtures, *The International journal of oral & maxillofacial implants*, Vol.5, No.3, pp. 272-281, 0882-2786 (Print) 0882-2786 (Linking).

Wakabayashi, N., M. Ona, T. Suzuki, et al. (2008). Nonlinear finite element analyses: advances and challenges in dental applications, *Journal of dentistry*, Vol.36, No.7, pp. 463-471, 0300-5712 (Print) 0300-5712 (Linking).

Walton, J. N., F. M. Gardner&J. R. Agar (1986). A survey of crown and fixed partial denture failures: length of service and reasons for replacement, *The Journal of prosthetic dentistry*, Vol.56, No.4, pp. 416-421, 0022-3913 (Print) 0022-3913 (Linking).

Wang, Z. L., J. C. Teo, C. K. Chui, et al. (2005). Computational biomechanical modelling of the lumbar spine using marching-cubes surface smoothened finite element voxel meshing, *Computer methods and programs in biomedicine*, Vol.80, No.1, pp. 25-35, 0169-2607 (Print) 0169-2607 (Linking).

Williams, J. A. (2000). *Engineering Tribology*. Oxford, Oxford University Press.

Winter, W., S. M. Heckmann&H. P. Weber (2004). A time-dependent healing function for immediate loaded implants, *Journal of biomechanics*, Vol.37, No.12, pp. 1861-1867, 0021-9290 (Print) 0021-9290 (Linking).

Winter, W., S. Mohrle, S. Holst, et al. (2010). Bone loading caused by different types of misfits of implant-supported fixed dental prostheses: a three-dimensional finite element analysis based on experimental results, *The International journal of oral & maxillofacial implants*, Vol.25, No.5, pp. 947-952, 0882-2786 (Print) 0882-2786 (Linking).

Wu, T., W. Liao, N. Dai, et al. (2010). Design of a custom angled abutment for dental implants using computer-aided design and nonlinear finite element analysis, *Journal of biomechanics*, Vol.43, No.10, pp. 1941-1946, 1873-2380 (Electronic) 0021-9290 (Linking).

Yang, H. S., L. A. Lang&D. A. Felton (1999). Finite element stress analysis on the effect of splinting in fixed partial dentures, *J Prosthet Dent*, Vol.81, No.6, pp. 721-728, 0022-3913 (Print) 0022-3913 (Linking).

Yang, H. S., L. A. Lang, A. Molina, et al. (2001). The effects of dowel design and load direction on dowel-and-core restorations, *J Prosthet Dent*, Vol.85, No.6, pp. 558-567, 0022-3913 (Print) 0022-3913 (Linking).

Yokoyama, S., N. Wakabayashi, M. Shiota, et al. (2005). Stress analysis in edentulous mandibular bone supporting implant-retained 1-piece or multiple superstructures, *The International journal of oral & maxillofacial implants*, Vol.20, No.4, pp. 578-583, 0882-2786 (Print) 0882-2786 (Linking).

Zarone, F., A. Apicella, L. Nicolais, et al. (2003). Mandibular flexure and stress build-up in mandibular full-arch fixed prostheses supported by osseointegrated implants, *Clinical oral implants research*, Vol.14, No.1, pp. 103-114, 0905-7161 (Print) 0905-7161 (Linking).

Zhou, X., Z. Zhao, M. Zhao, et al. (1999). [The boundary design of mandibular model by means of the three-dimensional finite element method], *Hua xi kou qiang yi xue za zhi = Huaxi kouqiang yixue zazhi = West China journal of stomatology*, Vol.17, No.1, pp. 29-32, 1000-1182 (Print) 1000-1182 (Linking).

4

Evaluation of Stress Distribution in Implant-Supported Restoration Under Different Simulated Loads

Paulo Roberto R. Ventura[1], Isis Andréa V. P. Poiate[1],
Edgard Poiate Junior[2] and Adalberto Bastos de Vasconcellos[1]
[1]Federal Fluminense University,
[2]Pontifical Catholic University,
Brazil

1. Introduction

Technical and scientific developments, in the form of osseointegrated implants, have made Restorative Dentistry evolve to become a highly successful form of treatment (Adell *et al.*, 1981; Binon, 2000), considered as an additional safe tool oral surgeons have at their disposal to rehabilitate partially and totally edentulous patients. However, many negative factors can interfere in the treatment, causing clinical and functional complications that may culminate in the loss of osseointegration (Adell *et al.*, 1981; Jemt *et al.*, 1989; Naert *et al.*, 2001a, 2001b; Isidor, 2006). The attachment of the crown to the abutment and of the latter to the implant, the loosening of the fixation screw, and even its fracture may arise from a poor distribution of occlusal loads in the prosthesis-implant set (Binon, 1994).

The finite element method (FEM) has been applied to the prognosis of stress distribution in both the implant and its interface with the adjacent bone, for a comparison not only of several geometries and applied loads (Hansson, 1999; Bozkaya, Muftu, Muftu, 2004), but also of different clinical situations (Van Oostewyck *et al.*, 2002) and prosthesis designs (Papavasiliou *et al.*, 1996). The study of stresses using the FEM is basically a virtual simulation of two- or three-dimensional mathematical models, in which all biological and material structures involved can be discretized, that is, subdivided into smaller structures that preserve individual anatomical and mechanical features.

Considering the physical alterations in components and the peri-implant tissue alterations to which they are subject under an unbalanced distribution of occlusal loads, it is fundamental to evaluate the distribution of the stresses generated by functional and parafunctional masticatory loads on implant-supported restorations to reach a better understanding of their possible biomechanical etiologic agents. The purpose of this study was to analyze the stress distribution in single implant-supported restorations, as well as in the peri-implant bone tissue, by means of a three-dimensional model, using the finite element method.

2. Material and methods

The finite element method is a technique to solve a complex problem subdividing it into a set of smaller and simpler problems (elements), which can be solved using numerical techniques. In other words, it is a method in which a formulation of equations for each finite element combines elements to get a solution for the whole problem, rather than solving it in just one operation (Holmgren *et al.*, 1998).

To divide the region of interest into elements, it is necessary to generate a mesh. The process to generate the mesh, the elements, their respective nodes, and the boundary conditions is referred to as the problem's own "discretization" (Geng et al., 2001).

The method of analysis with three-dimensional finite elements (3D FEA) was used. It is composed of three sequential and well-defined phases: pre-processing, processing, and post-processing, all described in detail below.

2.1 Pre-processing

In the pre-processing phase, a mathematical model of the object or structure under study is developed using computer-aided design (CAD). In the generation of the FEM, this geometry is discretized. Next, a physical phase is considered, in which properties equal to those of the materials or structures of the real model they represent are attributed to the elements of the mesh, and some hypotheses are formulated to make the analysis (linear elasticity) viable or just to address the lack of knowledge about the behavior of the material or structure represented (homogeneous and isotropic). Finally, the conditions of model fixation and the characteristics of force (load) are also applied.

2.1.1 The development of a geometric model

A model simulating a single implant-supported prosthesis was developed in the region of the second upper premolar. It is composed of a metallo-ceramic crown on a machined-surface external hexagon implant of regular diameter (Master Screw, Conexão Sistemas de Prótese Ltda – SP, Brasil). The development of this prosthesis model and of the peri-implant support bone was carried out based on the sound tooth model constructed by Poiate (2007). The model was generated with the MSC/PATRAN 2005r2 software and the simulation was carried out with the MSC/NASTRAN 2005r1 software (The MacNeal-Schwendler Corporation - USA). To make the support structures, the anatomical dimensions of the cortical and cancelous bone were based on an image of a vestibulo-lingual cross-section of the upper premolar region presented by Berkovitz, Holland and Moxham (2004). This image was digitized with a high-resolution scanner. The vestibulo-lingual dimensions of the cortical and cancelous bone were taken by the DigXY 1.2 software, developed to digitize bitmap data, largely employed to digitize x,y coordinates from graphs. The model includes the geometries of cortical and cancelous bone, implant, screw, abutment, infrastructure, and of the crown's ceramic layer.

The dental structure of the original model was removed and models of abutment, fixation screw, and implant were imported, supplied by Conexão Sistemas de Prótese Ltda. The modeling carried out included components equivalent to one external hexagon implant Master Screw of 3.75 mm in diameter and 13 mm in length; one titanium straight abutment for cemented prosthesis with a 2 mm cervical collar (sectioned at the occlusal end to obtain a final length of 3.55 mm); and one titanium fixation screw with a torque of 20 Ncm.

The implant was installed with the platform cervical boundary coinciding with the boundary of the bone crest, and the axial, mesio-distal, and labio-palatine positions equivalent to the root portion of the original sound tooth. Thus, the prosthetic metallo-ceramic crown could be constructed from a cervical prolongation of the sound tooth crown, also favoring the same spatial positioning.

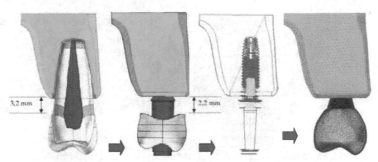

Fig. 1. Development of the model based on a sound tooth (Poiate, 2007).

The model geometries were used to generate volumetric meshes using the tetrahedral element topology (Tet4), that is, a pyramidal element of four faces with six edges and one node on each edge. Elements with edges of 0.05 mm were used in regions of high curvature, small size, or regions of transition between structures of up to 0.3 mm.

To create the volumetric meshes, it was necessary to proceed from the smaller or most internal structure to the larger or most external, that is, in this case the volumetric mesh was created first in the screw, and then the sequence adopted continued to the extremities, following the procedures in Poiate et al. (2008, 2009a, 2009b, 2011). This procedure assures perfect congruence in the FEM. The degree of discretization in the model derives from studies on the convergence of results and from the capacity of the computer used in the analyses, in order to assure adequate density in the finite element mesh for each model, describing the geometry of different components in a rather realistic way. The discretization detailed above corresponded to the maximum discretization established on a Pentium Core Duo 1.6 MHz computer, with 3.0 GB of RAM and a 160 GB hard disk. Thus, the model discretization generated 164,848 node points and 1,011,727 elements.

2.1.2 Load conditions

The occlusal pattern adopted in this study was the cusp-marginal ridge (one tooth to two teeth), thus justifying the positioning of the simulated occlusal loads. Therefore, four load conditions were applied, with different inclinations and points of application for a total static load of 291.36 N (Ferrario et al., 2004):

a. load distributed among 38 node points, 19 of which on an area of 0.85 mm^2 of the vestibular cusp, and 19 on an area of 0.75 mm^2 of the lingual cusp (Kumagai et al., 1999), with an inclination of 45°, but with the resultant (291.36 N) parallel to the tooth long axis, aiming at evaluating the effect of the axial force.

b.   oblique load, with an inclination of 45°, distributed among 19 node points on an area of 0.85 mm² of the transverse ridge of the vestibular cusp, aiming at evaluating the vestibular lever effect.

c.   load with 0° of inclination distributed among 19 node points on an area of 0.80 mm² of the mesial marginal ridge, aiming at evaluating the proximal lever effect.

d.   oblique load, with an inclination of 45° in the vestibular direction, distributed among 19 node points on an area of 0.80 mm² of the mesial marginal ridge, aiming at evaluating the effect of torsion on the long axis.

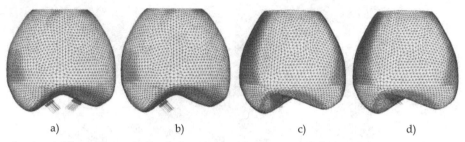

a)   b)   c)   d)

Fig. 2. a) axial load, b) vestibular lever, c) proximal lever and d) torsion.

2.1.3 Fixation conditions

Contour conditions, also called fixation or bonding conditions, are those determined for the edges or extremities of modeled structures, so that they have some spatial support, with displacement and/or rotation constraint, to allow for the analysis under the applied loads (Bathe, 1996).

For the simulation, the following fixation conditions were applied: in the maxillary sinus, translation restricted to the directions x, y and z, and rotations restricted to the axes x, y and z; at the mesial and distal ends of the cortical and cancelous bone, translation restricted to the direction x and rotations restricted to the axes y and z.

2.1.4 Definition of the mechanical properties of anatomic structures

Mechanical properties were attributed to each element in the discretized model, such as the elastic modulus (E) and the Poisson's ratio (v), considering the particularities of all anatomical structures and materials represented by the elements in the composition of the three-dimensional model. Table 1 shows the values applied to each of these properties, as well as their respective bibliographical references.

Structure/Material	Elastic Modulus E (GPa)	Poisson's Ratio (v)	Bibliographical Reference
Cortical bone	13.70	0.30	Ko *et al.* (1992)
Cancelous bone	1.37	0.30	Ko *et al.* (1992)
Ni-Cr	188.00	0.33	Vasconcellos (1999)
Titanium (implant, abutment and screw)	110.00	0.35	Iplikçioglu and Akça (2002)
Feldspathic Ceramic	82.20	0.35	Peyton and Craig (1963)

Table 1. Mechanical properties of structures and bibliographical references.

2.1.5 Formulated hypotheses

As in any type of numerical analysis, some hypotheses needed to be formulated to make modeling and problem solving processes viable. It must be emphasized that the numerical tools available to analyze stresses are much more advanced than the knowledge about the mechanical properties of the structures involved. As in Poiate's studies (2005, 2006), all constant structures in the model behaved isotropically (mechanical properties did not vary according to the direction), homogeneously (properties were constant independent of location), and were linearly elastic (strains were directly proportional to the force applied), characterized by two material constants, the Elastic Modulus (E) and the Poisson's Ratio ($\square$), with interfaces between structures presumed to be perfectly bonded. Therefore, the interface implant-bone was considered fully osseointegrated, and the cement layer regarded as negligible, since this study did not intend to analyze stresses on that structure nor its interrelation with the others (Çiftçi, Canay, 2000; Silva, 2005).

2.2 Processing

The software used in the analysis was *MSC/NASTRAN* (The MacNeal-Schwendler Corporation – USA), version 2005r1, on a Pentium Dual-Core computer, with a 1.7 GHz processor, a 160 Gb hard disk, and 2 Gb of RAM. This software takes information from the previous phase (pre-processing) into account. Based on the contact relationship among mesh elements, it makes a series of mathematical calculations organized in an algorithm, that is, a sequence of instructions logically ordered to solve a problem in a finite number of phases (Silva, 2005). What is mathematically analyzed, in general, is the displacement of element nodes according to the load applied (Holmgren *et al.*, 1998).

2.3 Post-processing

The software *MSC/PATRAN* 2005r2 employed in the pre-processing was also used in the post-processing to visualize and evaluate results.

2.3.1 Analysis of results

The principal stresses peak values were compared to the values of tensile and compressive strength in the model structures to analyze whether these loads were potentially harmful to the structures.

Structure/Material	Tensile Strength (MPa)	Compressive Strength (MPa)	Bibliographical Reference
Bone	121	167	Tanaka *et al.* 2003 / O'Brien, 2005
Cortical Bone	-	173	Reilly and Burstein, 1975
Cancelous Bone	-	167	Çiftçi and Canay, 2000
Ni-Cr	790	-	O'Brien, 2005
Titanium (implant, abutment and screw)	930	-	O'Brien, 2005
Feldspathic Ceramic	37.2	150	O'Brien, 2005

Table 2. Tensile and compressive strength and bibliographical references.

The von Mises criterion, or theory of the maximum distortion energy, was also used in this study to analyze the results from the implant and its components. It is a rupture criterion to evaluate ductile materials, based on the determination of the maximum distortion energy of a structure, that is, of the energy related to changes in form (as opposed to the energy related to changes in the volume of material) (Beer and Johnston, 1995). The structure failure occurs when, at any point of the material, the distortion energy per unit of volume is higher than the yield strength value obtained for the material in a tensile test.

The scale of stresses (which appear in different colors in the figures) does not have equal intervals. This is a result of the stresses in action on each group of models (different types of load). Thus, a single scale was defined for all models (except for the axially loaded models) to make the comparison easier.

3. Results and discussion

3.1 Axial load – Load with resultant parallel to the long axis

In Figure 3 (perspective views), compressive stresses can be seen in the region of the abutment and in the cervical region of the implant, but not intense enough to cause harm (between 1 and 5 MPa).

Again in Figure 3 (internal views), it can be observed the result of the internal maximum principal stresses. The concentration of tensile stresses on the region of the central groove of the occlusal surface (40 to 60 MPa), seen in Figure 4, shows a narrow strip on the external surface, which indicates a stronger tendency for rupture or formation of small cracks in the ceramic surface only, since the decreasing gradient of stresses reach a peak value of only 20 MPa on the ceramic close to the infrastructure. These tensile stresses reach the infrastructure (Ni-Cr), but they are not harmful, given their low intensity, with a peak value of 20 MPa, well below the tensile strength of the material.

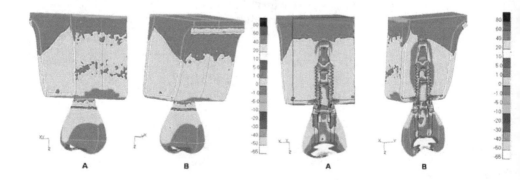

Fig. 3. Perspective and internal views of all structures. Maximum principal stresses in model under longitudinal load.

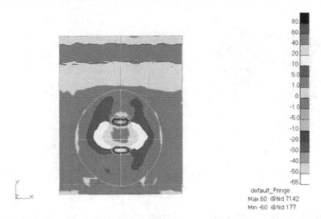

Fig. 4. Occlusal view. Maximum principal stresses in model under longitudinal load.

In Figure 4, it can be observed that the higher compressive stress values are located at points under the area where the load was applied (10 to 55 MPa, well below the tensile strength value of 150 MPa). On the other hand, the increasing tensile stress gradient, between 40 and 60 MPa, concentrated on the region of the central groove, is higher than the tensile strength value (37.2 MPa), suggesting a stronger tendency for micro-cracks in the ceramic.

To compare and discuss the results obtained by Poiate (2007), the stress values found in the cortical and cancelous bone of a sound tooth model are twice lower than the values found in the implant model (between 5 and 10 MPa), in consequence of the absence of periodontal ligament and tooth in the latter.

In Figure 5, very low compressive stresses can be seen in the region of the cortical bone in contact with the implant platform, with a peak value of 8 MPa.

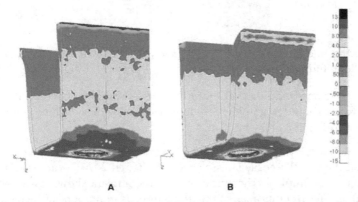

Fig. 5. Perspective view of the cortical bone. Maximum principal stresses in model under longitudinal load. In A, disto-vestibular view; in B, mesio-palatine view.

Figure 6 shows that the concentration of compressive stresses on the region in contact with the implant is on a narrow strip on the external surface of the cortical bone. Next to the cancelous bone, it shows a concentration of tensile stresses of 10 MPa.

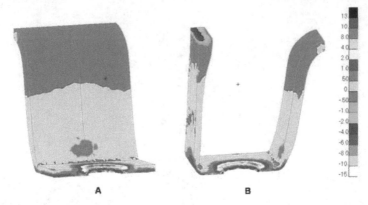

A   **B**

Fig. 6. Internal view of the cortical bone. Maximum principal stresses in model under longitudinal load. In A, mesio-distal section; in B, vestibulo-palatine section.

In Figure 7, there is a concentration of bulb-shaped tensile stresses on the cancelous bone in contact with the implant, with values ranging from 0.5 to 2 MPa approximately. It is also possible to see that the compressive stresses in contact with the implant thread decrease towards the implant apex, even though the tensile stresses in contact with the implant thread are constant (2 to 4 MPa).

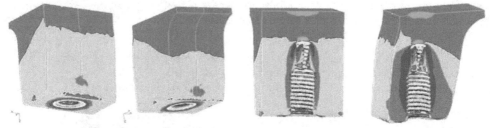

Fig. 7. Perspective and internal view of the cancelous bone. Maximum principal stresses in model under longitudinal load.

According to Lehmann and Elias (2008), to minimize this higher concentration of stresses, new implant forms should be employed, among them platform switching and micro-threads in the most cervical region of the implant.

The implant and its components withstood the stress and did not reach the tensile strength of the material (Figure 8). Tensile stresses between 10 and 20 MPa can be seen in the most cervical region of the implant. The apical third of the implant shows lower tensile stresses (up to 1 MPa) as a result of its anatomy, with no threads in portions of each quadrant, thus minimizing stress concentration. Again in Figure 8, there are compressive stresses on the apex (between 1 and 10 MPa), as well as on the medium third of the implant, a region corresponding to the screw apex (between 1 and 10 MPa). Low tensile stresses are found in the region in contact with the apical third of the screw, in the medium third of the implant (between 10 and 60 MPa), which might result in the loss of stability of the fixation screw.

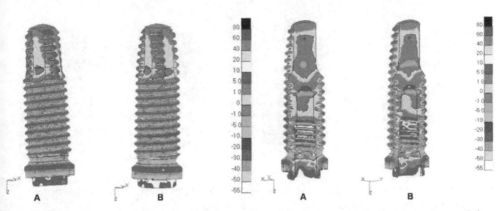

Fig. 8. Perspective and internal view of the implant. Maximum principal stresses in model under longitudinal load.

According to Dinato (2001), occlusal forces with axial resultant produce a vertical load and do not exert force on the screw nor cause screw loosening. In Figure 9 (perspectives A and B), it can be seen a prevalence of low-intensity tensile stresses on the screw, with higher values concentrated on the apex, on the distal surface (74 MPa). Again in Figure 9 (sections A and B), compressive stresses between 1 and 10 MPa are seen on the coronal third, in the region without spindle. Tensile stresses from 20 to 60 MPa are seen on the screw apex.

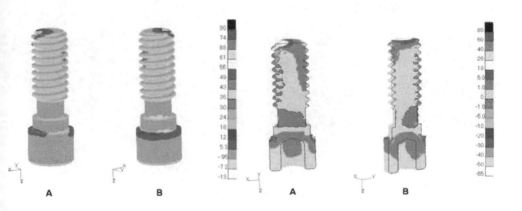

Fig. 9. Perspective and internal view of the screw. Maximum principal stresses in model under longitudinal load.

To obtain uniform stress distribution in implants, it is necessary a precise adjustment of the abutment, since the unit closest to the load will be subject to the greatest stresses (Rangert, Jemt and Jörneus, 1989). Contrary to this statement, when under axial load the highest stress values are below the cervical third of the abutment (in contact with the screw, see Figure 10), in the medium third and in the platform of the implant (Figure 8).

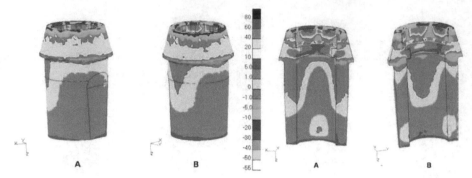

Fig. 10. Perspective and internal view of the abutment. Maximum principal stresses in model under longitudinal load.

The results of the longitudinal load revealed that there is a concentration of von Mises stresses, with a peak value of 80 MPa, on the implant head in contact with the cortical bone and on the abutment collar in contact with the implant, as well as on the screw apex, with a peak value of 130 MPa.

3.2 Vestibular lever effect – Load of 45° on the vestibular cusp

Mastication produces mainly vertical forces, but also transverse forces originating from the horizontal movement of the jaw and from the inclination of the cusps. These forces are transferred to the implant through the prosthesis, transforming occlusal forces into bone stresses (Rangert, Jemt e Jörneus, 1989).

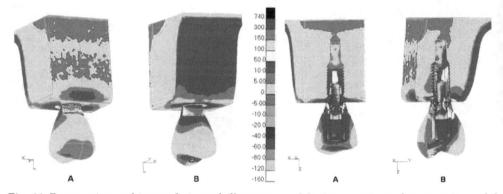

Fig. 11. Perspective and internal view of all structures. Maximum principal stresses in model with a load of 45° on the vestibular cusp.

Applying load on the vestibular cusp (Figure 11) produces compression on the vestibular cusp, more concentrated on the region of contact with the cortical bone, which acts as a fulcrum. The tensile stress concentrates and reaches its maximum on the opposite side, the palatine. In Figure 11 (perspective B), tensile stresses between 300 and 740 MPa are seen in the abutment neck, but they are lower than the titanium tensile strength of 930 MPa. When analyzing the ceramic, a concentration of tensile stresses with peak value of 740 MPa can be seen on the implant and on the abutment (Figure 11, section B). It was observed that stress concentration under the place where the load is applied is punctual and well located. In Figure 11 (section A), there is an accumulation of tensile stresses ranging from 50 to 100 MPa on the screw apex, in the medium third of the implant.

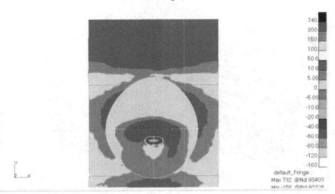

Fig. 12. Occlusal view. Maximum principal stresses in model with a load of 45° on the vestibular cusp.

In Figure 12, it can be seen that the highest compressive stress values are under the area of application of a load of approximately 156 MPa, rather higher than the values found in the model under axial load (between 10 and 55 MPa, see Figure 4), higher than the strength value (150 MPa). Also, as it occurred under axial load, the tensile stress values adjacent to the area of load application exceeded its strength value, suggesting the possibility of micro-cracks in the ceramic.

A study by Faulkner *et al.* (1998) showed, as this one does, that implants loaded with forces that are distant from their axis bear considerable stresses on the bone crest. In Figure 13, there are tensile stresses concentrated in the region of contact with the implant on the palatine surface, with a peak value of 300 MPa, higher than the bone tensile strength, suggesting that there will be a fracture in the cortical bone.

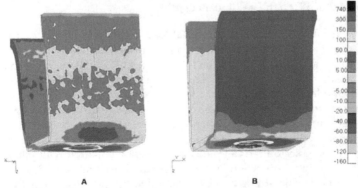

Fig. 13. Perspective view of the cortical bone. Maximum principal stresses in model with a load of 45° on the vestibular cusp. In A, disto-vestibular view; in B, mesio-palatine view.

Figure 14 shows that only a narrow strip, close to the implant, exceeds the cortical bone tensile strength. However, the incidence of excess stresses on the cortical bone produces micro-fractures and consequent resorption (Burr *et al.*, 1985; Papavasiliou *et al.*, 1996; Holmgren *et al.*, 1998).

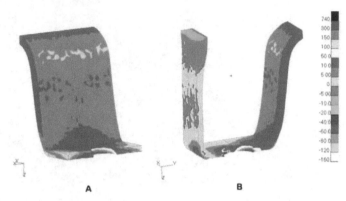

A   B

Fig. 14. Internal view of the cortical bone. Maximum principal stresses in model with a load of 45° on the vestibular cusp. In A, mesio-distal section; in B, vestibulo-palatine section.

The maximum principal stresses on the cancelous bone are shown in Figure 15.

Fig. 15. Perspective and internal view of the cancelous bone. Maximum principal stresses in model with a load of 45° on the vestibular cusp.

Comparing the results from models with axial load and vestibular lever effect, it could be observed that the result from the internal maximum principal stresses on the cancelous bone differs in intensity and in the pattern of stress distribution. There is no concentration of bulb-shaped tensile stresses on the cancelous bone in contact with the implant, as occurred in the previous model. Moreover, the tensile stress peak value in the model with vestibular lever effect is 56.9 times higher. It is also possible to see that the compressive stresses in contact with the implant thread appear only on the vestibular surface, which was already expected, since the load was applied to the vestibular cusp, where the implant is being bent. It can be seen that stresses on the cancelous bone in contact with thread pitches or between them reach high values, especially in the coronal third (between 300 and 740 MPa), exceeding the bone tensile strength (121 MPa, Table 2) and suggesting the formation of micro-fractures and consequent resorption (Papavasiliou *et al.*, 1996; Holmgren *et al.*, 1998; De Tolla *et al.*, 2000).

In Figure 16 (perspective B), there are tensile stresses on the medium and cervical third of the implant, which was expected, since the load was applied to the vestibular cusp. The implant is being bent, with a fulcrum in the region close to the cortical bone, where there is a decreasing dissipation of tensile stresses on the palatine surface. Again in Figure 16 (internal view), there are tensile stresses on the cervical third of the implant, on the lingual surface, which was expected as well, since the load was applied to the vestibular cusp, with a peak value of 300 MPa, rather lower than the tensile strength of the material (930 MPa).

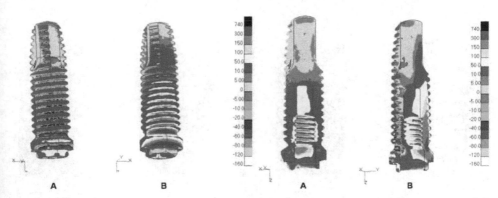

Fig. 16. Perspective and internal view of the implant. Maximum principal stresses in model with a load of 45° on the vestibular cusp.

In Figure 17, there is a prevalence of tensile stresses on the cervical third of the screw, between the beginning of the thread and the head, on the palatine surface, which was already expected, since the load was applied to the vestibular cusp and the screw was being bent, with a fulcrum in the cervical region in contact with the implant (above the cortical bone).

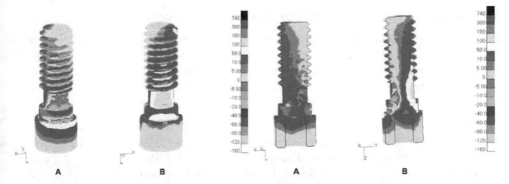

Fig. 17. Perspective and internal view of the screw. Maximum principal stresses in model with a load of 45° on the vestibular cusp.

Understanding this pattern of stress distribution in the screw is important to explain the problems and complications encountered, such as screw fractures, as described by Zarb and

Schmitt (1990), although they seem slightly probable when we compare the stress values (150 MPa) with the amount necessary for a rupture (930 MPa).

The results revealed in Figure 18 showed a narrow strip of tensile stresses between 300 and 740 MPa on the external surface of the abutment neck, with dissipation to the palatine and internal surface, but not exceeding its tensile strength. There are tensile stress values (650 MPa in the abutment collar in contact with the implant) rather higher than those in the model under longitudinal load (peak value of 80 MPa).

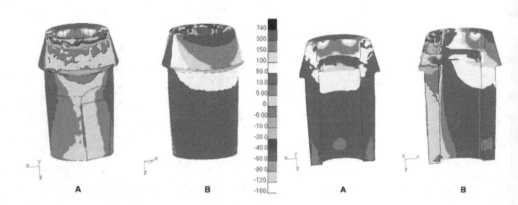

Fig. 18. Perspective and internal view of the abutment. Maximum principal stresses in model with a load of 45° on the vestibular cusp.

The results from the distribution of von Mises stresses in the implant and its components showed that higher stress values are found also in the implant head, when compared with the axial model.

3.3 Proximal lever effect – Load of 0° on the mesial marginal ridge

In Figure 19, there are tensile stresses between 150 and 300 MPa on the cortical bone in contact with the implant. They were already expected, given the direction of the force on the crown, which pulled the implant neck in this area. It is also possible to see tensile stresses (5 MPa) on the line between the abutment and the ceramic, which could be explained by compression on materials of different rigidity.

In Figure 19, it can also be seen that the highest compressive stress values are under the application area of a load ranging from approximately 60 to 120 MPa, slightly lower than the values found in the model with vestibular lever effect (150 MPa), and below the ceramic compressive strength (150 MPa, Table 2). It shows the place of load application with a well-located and punctual stress concentration; however, this stress concentration does not transfer to the vestibular surface (Figure 19, perspective A) and transfers only slightly to the palatine surface (Figure 19, perspective B). On the other hand, there are tensile stresses close to the load application point, with values between 50 and 100 MPa, higher than the ceramic tensile strength, suggesting a possible formation of micro-cracks.

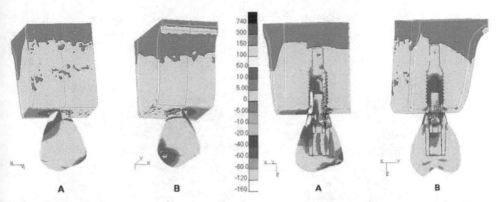

Fig. 19. Perspective and internal view of all structures. Maximum principal stresses in model with a load of 0° on the mesial marginal ridge.

According to Gross (2001), neither axial nor non-axial load generate stress concentration on the implant apex. Under the load that generates the proximal lever effect, these results are confirmed in this study. We could observe that the region of the apex does not concentrate compressive stresses, which are dissipated throughout the apical third of the model (with a peak value of 5 MPa).

However, there is a concentration of tensile stresses on the screw apex, with a peak value of 50 MPa (medium third of implant, Figure 19, section A), and tensile stresses on the distal surface of the implant and on the apical third of the abutment, with values between 50 and 100 MPa. Although the tensile stress values found in the implant and its components are lower than their tensile strength value (Table 2), its presence is important, so that we understand complications, such as screw loss and/or loosening, caused by the micro-movements generated between these two surfaces. Again in Figure 19, it is possible to see a concentration of low-intensity tensile stresses only on the cortical bone, in a mesio-distal section, in the occlusal distal region (5 to 10 MPa).

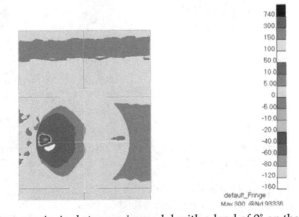

Fig. 20. Occlusal view. Maximum principal stresses in model with a load of 0° on the mesial marginal ridge.

In Figure 21, low-intensity tensile stresses can be seen in the distal cervical region. However, the internal surface of the cortical bone in contact with the implant shows tensile stresses between 50 and 150 MPa, exceeding the bone tensile strength (121 MPa, Table 2), which suggests the formation of micro-fractures and consequent resorption.

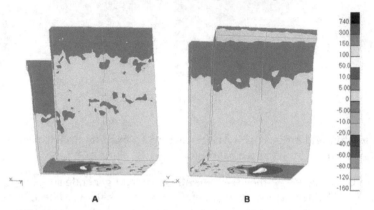

A   **B**

Fig. 21. View of the cortical bone. Maximum principal stresses in model with a load of 0° on the mesial marginal ridge. In A, disto-vestibular view; in B, mesio-palatine view.

In Figure 22, it is clear the presence of tensile stresses with intensity between 50 and 150 MPa on the cortical bone in contact with the implant, on the distal surface (Figure 22B), exceeding the bone tensile strength (Table 2). In implant-supported restorations, stresses are close to the bone crest (Misch *et al.*, 2001), altering the existing process of bone crest remodeling.

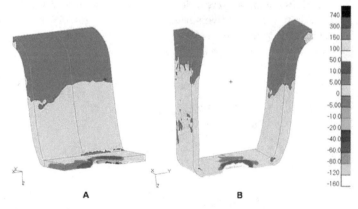

A   **B**

Fig. 22. Internal view of the cortical bone. Maximum principal stresses in model with a load of 0° on the mesial marginal ridge. In A, mesio-distal section; in B, vestibulo-palatine section.

In Figure 23, it can be observed that there is a dissipation of low-intensity tensile stresses in the distal cervical region from the cortical to the cancelous bone, as well as a concentration of stresses in contact with the implant. The bone anatomy itself leads to the concentration of local forces, given the existence of an external layer of rigid cortical bone and of an internal layer of the more elastic cancelous bone. Natural teeth themselves under load generate higher stresses next to the cortical bone (Caputo and Standlee, 1987).

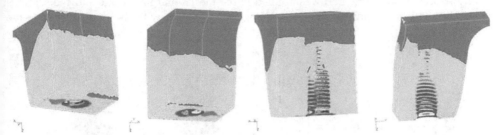

Fig. 23. Perspective and internal view of the cancelous bone. Maximum principal stresses in model with a load of 0° on the mesial marginal ridge.

Also in Figure 23, it can be observed that, with a force with inclination of 0° to the axial axis of the set, there was a higher concentration of tensile stresses on the side opposite to the applied force, but with stresses limited to the screw threads of the component, with a peak value of 300 MPa, exceeding the bone tensile strength (121 MPa, Table 2), which suggests the formation of micro-fractures and consequent resorption.

In Figure 24, there is a concentration of tensile stresses only on the cervical third of the implant (with a peak value of 300 MPa). Comparing the models with vestibular and proximal lever effect, it can be considered that the area of tensile stress concentration is larger than in the model that produces a vestibular lever effect, suggesting a stronger possibility of osseointegration failure. The clinical success derived from osseointegration proves that implants resist firmly to the masticatory load; however, the concentration of stresses can result in the loss of osseointegration (Adell et al., 1981).

Again in Figure 24 (internal view), there are tensile stresses located on the cervical third of the implant, on the distal surface, which was expected, since the load was applied to the mesial marginal ridge, with a peak value of 150 MPa, rather lower than the tensile strength of the material (930 MPa) and 50% lower than the peak value of tensile stresses found in the implant of the model that produces a vestibular lever effect.

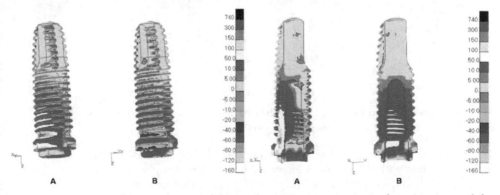

Fig. 24. Perspective and internal view of the implant. Maximum principal stresses in model with a load of 0° on the mesial marginal ridge.

In Figure 25, there is a prevalence of tensile stresses on the cervical third of the screw, between the beginning of the thread and its head or platform, on the distal side, which was already expected, since the load was applied to the mesial marginal ridge and the screw is being bent, with a fulcrum in the cervical region, in contact with the implant (above the cortical bone).

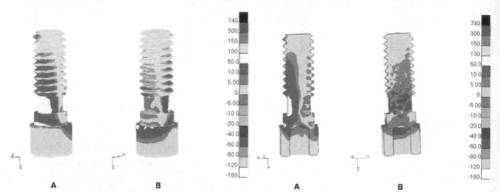

Fig. 25. Perspective and internal view of the screw. Maximum principal stresses in model with a load of 0° on the mesial marginal ridge.

Sealing, obtained by the precise adjustment of the abutment surface to the implant, would prevent problems of a peri-implant nature and minimize the development of tangential forces harmful to the interface implant-bone tissue, which could lead to osseointegration failure (Adell *et al.*, 1981). The maximum principal stresses on the abutment are shown in Figure 26.

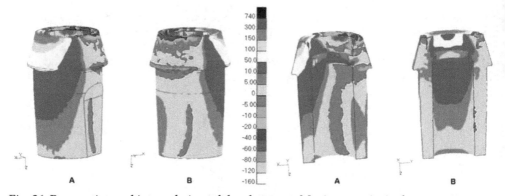

Fig. 26. Perspective and internal view of the abutment. Maximum principal stresses in model with a load of 0° on the mesial marginal ridge.

The results revealed values of von Mises stresses (peak value of 300 MPa in the abutment collar in contact with the implant) lower than those found in the model that produces the vestibular lever effect (peak value of 650 MPa), but higher than those found in the model under longitudinal load (peak value of 80 MPa). Higher values of von Mises stresses are also found in the implant platform, when compared to the axial model.

3.4 Torsion effect – Load of 45° on the mesial marginal ridge

Single implant-supported restorations can also be subject to rotational or torsion forces, whenever they are clinically demanded through the functional contacts cusp/marginal ridge, that is, in a one tooth to two teeth relation (Cohen *et al.*, 1995).

Figure 27 shows tensile stresses between 50 and 300 MPa on the abutment, which was already expected, in view of the direction of the force on the crown, which pulls the implant in this area. Again in Figure 27 (section B), it can be seen that there is a concentration of tensile stresses on the cervical and medium third of the implant (peak value of 300 MPa) on the palatine surface, lower than the mechanical strength.

In Figure 28, it can be seen that the highest compressive stress values are under the area where the load was applied (10 to 80 MPa, lower than the strength value of 150 MPa), which is a stress concentration point with no transmission to the vestibular surface. However, the existence of tensile stresses close to the application of load (50 to 100 MPa) exceeds the ceramic strength (37.2 MPa) and suggests a localized micro-crack.

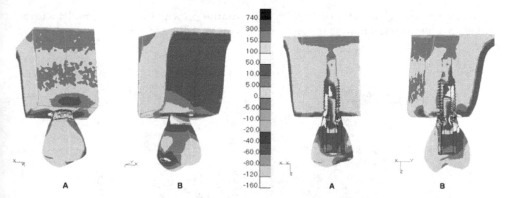

Fig. 27. Perspective and internal view of all structures. Maximum principal stresses in model with a load of 45° on the mesial marginal ridge.

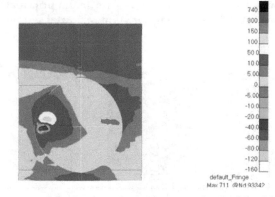

Fig. 28. Occlusal view of all structures. Maximum principal stresses in model with a load of 45° on the mesial marginal ridge.

In Figure 29, a dissipation of low-intensity compressive stresses can be seen in the vestibular surface of the cortical bone (5 to 20 MPa), and there are tensile stresses on the palatine surface (peak value of 10 MPa). However, high-intensity tensile stresses are present on the cortical bone in contact with the implant (palatine). This is better shown in Figure 30.

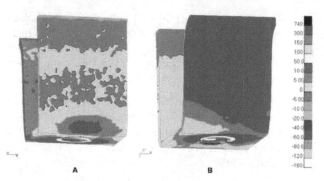

Fig. 29. Perspective view of the cortical bone. Maximum principal stresses in model with a load of 45° on the mesial marginal ridge. In A, disto-vestibular view; in B, mesio-palatine view.

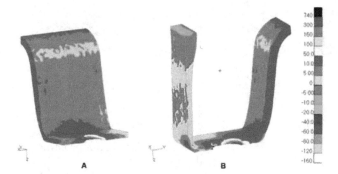

Fig. 30. Internal view of the cortical bone. Maximum principal stresses in model with a load of 45° on the mesial marginal ridge. In A, mesio-distal section; in B, vestibulo-lingual section.

Fig. 31. Perspective and internal view of the cancelous bone. Maximum principal stresses in model with a load of 45° on the mesial marginal ridge.

In Figure 31, there are tensile stresses on the cortical bone (on the palatine surface) in contact with the implant that exceed the tensile strength value and overload the implant. According to Rangert *et al.* (1995), in a retrospective analysis, this overload induces bone resorption, which seems to precede and contribute to the fracture of implant components.

Also in Figure 31, there are distribution of low-intensity tensile stresses (5 to 10 MPa) and stress concentration on the palatine surface and around the implant.

The concentration of tensile stresses on the cancelous bone in contact with the implant (especially on the threads) reaches the cervical (higher intensity) and medium (lower intensity) thirds on their palatine surface. This was expected, as a result of the direction of the load application. According to Rangert *et al.* (1989), threads reduce the shear stress on the implant-bone interface when it is under axial load.

Tensile stresses between 50 and 100 MPa (Figure 32) on the implant platform are five times higher than the stresses found in the axial model (10 to 20 MPa). In Figure 32, it can also be observed that there is a concentration of tensile stresses on the medium and cervical third of the implant, as a consequence of load application, in addition to the torsion effect. Moreover, the implant is being bent, with a fulcrum in the region close to the cortical bone, where there is dissipation of decreasing tensile stresses on the palatine surface. Tensile stresses are also seen in the medium and cervical third of the implant, on the distal and lingual surface, a region corresponding to the fulcrum in the cortical bone and to the screw thread (between 50 and 150 MPa). Tensile stresses are found in the region in contact with the apical third of the screw, medium third of the implant. Although the tensile stress value at the end of the screw thread is lower than the mechanical strength of the material, it might contribute to the loss of stability in the fixation screw.

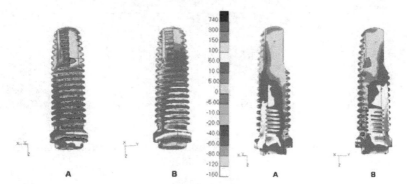

Fig. 32. Perspective view of the implant. Maximum principal stresses in model with a load of 45° on the mesial marginal ridge.

In a disto-vestibular view, Figure 33 shows compressive stresses between approximately 5 and 20 MPa, concentrated on the coronal third of the screw, in the region without spindle. Tensile stresses can also be seen in the apex (between 50 and 150 MPa) and in the screw thread (5 to 10 MPa), in a mesio-lingual view, suggesting the possibility of screw loosening in this region. In Figure 33 (sections A and B), there is a prevalence of tensile stresses between 50 and 150 MPa, but concentrated on the medium-coronal third of the screw, a region without thread, and not exceeding the tensile strength of the material.

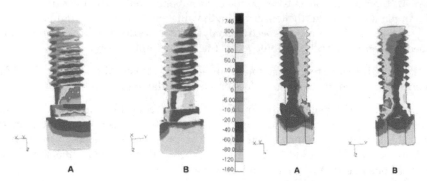

Fig. 33. Perspective view of the screw. Maximum principal stresses in model with a load of 45° on the mesial marginal ridge.

In Figure 34, the prevalence of tensile stresses can be seen on the abutment external surface (it shows only a narrow strip of low-intensity compressive stresses, of up to 5 MPa, on the vestibular surface). This prevalence of tensile stresses on the abutment external surface is important to explain complications such as the loss and/or loosening of screws, since these stresses are directly related to the abutment-implant interface, as a result of the creation of micro-movements between the two surfaces when a non-axial load is applied to them.

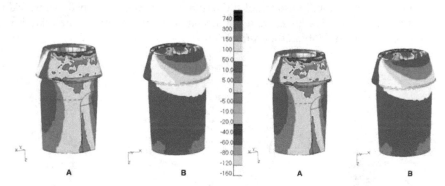

Fig. 34. Perspective and internal view of the abutment. Maximum principal stresses in model with a load of 45° on the mesial marginal ridge.

The result of von Mises stresses revealed tensile stresses values in the abutment collar in contact with the implant (peak value of 600 MPa) that are higher than those found in the proximal lever model (peak value of 300 MPa) and close to those found in the model that produces a vestibular lever effect (peak value of 650 MPa), and considerably higher than those found in the model under longitudinal load (peak value of 80 MPa).

4. Conclusions

Taking into account that the results produced in this study, for all models, revealed a higher concentration of forces on the cervical region and higher stress values in the models under load with a lever effect, no load was able to fracture the components of the implant system simulated here. However, they may suggest loosening of the screw, micro-cracks in the ceramic, and bone micro-fractures (resorptions), except for the model under axial load, which proved to be the least harmful to the stability of the rehabilitating system under consideration. It can be stated that a careful observation of the criteria for rehabilitation using implant-supported restorations, as regards the direction of occlusal loads, is crucially important to achieve success in this therapy.

5. Acknowledgements

This study was partially based on a thesis submitted to The Fluminense Federal University, in fulfillment of the requirements for the degree of Master of Science. The authors are grateful to the State of Rio de Janeiro Research Foundation (FAPERJ), the Conexão Sistemas de Prótese Ltda and to the Finite Element Laboratory in Dentistry (LEFO), Fluminense Federal University, for their support.

6. References

Adell, R; Lekholm, U; Rockler, B & Brånemark PI. (1981). A 15-year study of osseointegrated implants in the treatment of edentulous jaw. *Int J Oral Surg*, Vol. 10, pp. 387-416.

Bathe, KJ. (1996). *Finite Elements Procedures* (1st ed). Prentice Hall Inc, New Jersey.

Beer, FP; Johnston Jr, ER. (1995). *Resistência dos Materiais*. Makron Books. São Paulo.

Berkovitz, BKB; Holland, CR & Moxham, BJ. (2004). *Anatomia, embriologia e histologia buccal* (3rd ed.). Artmed Editora S.A., Porto Alegre.

Binon, PP. (1994). The role of screws in implant systems. *Int J Oral Maxillofac Implants*, Vol.9, pp. 48-63.

Binon, PP. (2000). Implants and components: entering the new millenium. *Int J Oral Maxillofac Implants*, Vol. 15, No. 1, pp. 76-94.

Bozkaya, D; Muftu, S & Muftu A. (2004). Evaluation of load transfer characteristics of five different implants in compact bone at different load levels by finite elements analysis. *J Prosthet Dent*, Vol. 92, No. 6, pp. 523-30.

Burr, DB; Martin, RB; Shaffler, B & Radin, E. (1985). Bone remodeling in response to in vivo fatigue microdamage. *J Biomech*, Vol. 18, pp. 189-200.

Caputo, AA; Standlee, JP. (1987). Force transmission during function, In: *Biomechanics in Clinical Dentistry*, Quintessence, pp. 1-29, Chicago.

Cohen, BI; Pagnillo, M; Condos, S & Deutsch, AS. (1995). Comparison of torsional forces at failure for seven endodontic post systems. *J Prosthet Dent*, Vol. 74, No. 4, pp. 350-57.

Çiftçi, Y; Canay, Ş. (2000). The effects of veneering materials on stress distribution in implant-supported fixed prosthetic restoration. *Int J Oral Maxillofac Implants*, Vol. 15, No. 4, pp. 5715-82.

Dinato, JC; Polido, WD. (2000). Adaptação passiva: ficção ou realidade? In: *Implantes osseointegrados: cirurgia e prótese*, Artes Médicas, pp. 1-283, São Paulo.

De Tolla, DH; Andreana, S; Patra, A; Buhite, R & Comella B. (2000). The role of the finite element model in dental implants. *J Oral Implant*, Vol. 24, No. 2, pp. 77-81.

Ferrario, VF; Sforza, C; Serrao, G; Dellavia, C & Tartaglia, GM. (2004). Single tooth bite forces in healthy young adults. *J Oral Rehabil*, Vol. 31, No. 1, pp.

Faulkner, G; Wolfaardt, J & Valle, V. (1998).Console abutment loading in craniofacial osseointegration. *Int. J. Oral Maxillofac Implants*, Vol. 13, No. 2, pp. 245-52.

Geng, JP; Tan, KB & Liu GR. (2001). Application of finite element analysis in implant dentistry; A review of the literature. *J Prosthet Dent*, Vol. 85, No. 6, pp. 585-98.

Gross, MD; Nissan, Samuel R. (2001). Stress distribution around maxillary implants in anatomic photoelastic models of varying geometry. Part II. *J Prosthetic Dent*, Vol. 85, No. 5, pp. 450-454.

Hansson, S. (1999). The implant neck: smooth or provided with retention elements. A biomechanical approach. *Clin Oral Impl Res*, Vol. 10, No. 5, pp. 394-405.

Holmgren, EP; Seckinger, RJ; Kilgren, LM & Mante F. (1988). Evaluating parameters of osseointegrated dental implants using finite element analysis – A two-dimensional comparative study examining the effects of implant diameter, implant shape, and load direction. *J Oral Implantol*, Vol. 24, No. 2, pp. 80-88.

Iplikçioğlu, H; Akça, K. (2002). Comparative evaluation of the effect of diameter, length and number of implant supporting three-unit fixed partial prostheses on stress distribution in the bone. *J Dent*, Vol. 30, No. 1, pp. 41-6.

Isidor F. (2006). Influence of forces on peri-implant bone. *Clin Oral Impl Res*, Vol. 17, No. 2, pp. 8-18.

Jemt, T; Lekholm, U & Adell, R. (1989). Osseointegrated implants in the treatment of partially edentulous patients: a preliminary study on 876 consecutively placed fixtures. *Int J Oral Maxillofac Implants*, Vol. 4, pp. 211-217.

Ko, CC; Chu, CS; Chung,,KH & Lee, MC. (1992). Effects of posts on dentin stress distribution in the bone in pulpless teeth. *J Prosthet Dent*, Vol. 68, No. 2, pp. 421-27.

Kumagai, H; Suzuki, T; Hamada, T; Sondang, P; Fugitani, M & Nikawa H. (1999). Occlusal force distribution on the dental arch during various levels of clenching. *J Oral Rehabil*, Vol. 26, No. 12, pp. 932-35.

Lehmann, RB; Elias, CN. (2008). Tensões em implantes cônicos com hexágono externo e com hexágono interno. *Rev Dental Press Periodontia Implantol*, Vol. 2, pp.

Naert, IE; Duyck, JA; Hosny, MM; Quirynen, M & Van Steenbherg, D. (2001a). Freestanding in tooth-implants connected prostheses in the treatment of partially edentulous patients: Part I: an up to 15-years radiographic evaluation. *Clin Oral Impl Res*, Vol. 12, pp. 237-244.

Naert, IE; Duyck, JA; Hosny, MM; Quirynen, M & Van Steenbherg, D. (2001b). Freestanding in tooth-implants connected prostheses in the treatment of partially edentulous patients: Part II: an up to 15-years radiographic evaluation. *Clin Oral Impl Res*, Vol. 12, pp. 245-251.

O' Brien, WJ. (April 1996). Biomaterials Properties Database, In: *University of Michigan*, 25 december 2006. Available from:
    http://www.lib.umich.edu/dentlib/Dental_tables/toc.html

Papavasiliou, G; Kamposiora, P; Bayne, SC & Felton, DA. (1996). Three-dimensional finite element analysis of stress-distribution around single tooth implants as a function of bony support, prosthesis type, and loading during function. *J Prostet Dent*, Vol. 76, No. 6, pp. 633-40.

Peyton, FA; Craig, RG. (1963). Current evaluation of plastics in crow and bridge prostesis. *J Prosthet Dent*, Vol. 13, pp. 743-53.

Poiate IAVP. (2005). Análise da distribuição de tensões em modelos tridimensionais de um incisivo central superior, gerados a partir de tomografia computadorizada, sob diferentes magnitudes de carga: método dos elementos finitos, In: *Universidde federal Fluminense*. Master's dissertation, Niterói.

Poiate IAVP. (2007). Análise biomecânica de dentes restaurados com retentor intra-radicular fundido, com e sem ferula, In: Universidade de São Paulo. Doctoral Thesis, São Paulo.

Poiate, IAVP; Vasconcellos, AB; Andueza, A; Pola, IRV & Poiate Jr, E. (2008). Three dimensional finite element analyses of oral structures by computerized tomograghy. *J Biosc Bioeng*, Vol. 106, No. 6, pp. 906-9.

Poiate, IAVP; Vasconcellos, AB; Santana, RB &, Poiate Jr, E. (2009a). Three-Dimensional Stress Distribution in the Human Periodontal Ligament in Masticatory, Parafunctional, and Trauma Loads: Finite Element Analysis. *J Periodontology*, Vol. 80, pp. 1859-1867.

Poiate, IAVP; Vasconcellos, AB; Poiate Jr, E & Dias KRC. (2009b). Stress distribution in the cervical region in a 3D FE. *Brazilian Oral Research*, Vol. 23, pp. 161-168.

Poiate, IAVP; Vasconcellos, AB; Mori, M &, Poiate Jr, E. (2011). 2D and 3D finite element analysis of central incisor generated by computerized tomography. *Comput Methods Programs Biomed*, Vol 104, No. 2, pp. 292-9.

Rangert B, Jemt, T, Jörneus L. Forces and moments on Brånemark Implants. Int J Oral Maxillofac Implants, 4(3):241-47, 1989.

Reilly, DT; Burstein, AH. (1975). The elastic and ultimate properties of compact bone tissue. *J Biomech*, Vol. 8, pp. 393-405.

Silva MG. (2005). Influência da esplintagem de restaurações protéticas fixas e do número de implantes na distribuição de tensões em mandíbula edentada posterior – análise em elementos finitos. In: Universidade de São Paulo. Master's dissertation, São Paulo.

Tanaka, M; Naito, T & Yokota M. (2003). Finte element analysis of the possible mechanism of cervical lesion formation by occlusal forces. *J Oral Rehabil*, Vol. 30, pp. 60-67.

Van Oostewyck, H; Duyck, J; Vander Sloten, J; Van Der Perri, G & Naert, I. (2002). Peri-implant bone tissue strains in cases of dehiscence: a finite element study. *Clin Oral Impl Res*, Vol. 13, No. 3, pp. 327-33.

Vasconcellos, AB; Mori, M; Andueza, A; Silva EM (1999). Tensões internas em prótese parcial fixa com dois sistemas de retenção corono-radicular: método dos elementos finitos. *Rev Bras Odontol*, Vol. 59, pp. 206-10.

Zarb, GA; Schmitt, A. (1990). The longitudinal clinical effectiveness of osseointegrated dental implants: Toronto study. Part III: Problems and complications encountered. *J Posthet Dent*, Vol. 64, No. 2, pp. 185-94.

5

Biomechanical Analysis of Restored Teeth with Cast Intra-Radicular Retainer with and Without Ferrule

Isis Andréa Venturini Pola Poiate[1],
Edgard Poiate Junior[2] and Rafael Yagüe Ballester[3]
[1]Federal Fluminense University,
[2]Pontifical Catholic University,
[3]University of São Paulo,
Brazil

1. Introduction

For the prosthetic reconstruction of endodontically treated teeth with large loss of tooth structure a lot of times become indispensable to obtain retention by the use of post and core systems. The retention loss and the dental fractures are the two failures more commonly described in this restoration type (Ferrari & Mannocci, 2000).

The dental fracture tends to happen longitudinally with the end below the alveolar bone crest, what constitutes a no restorative failure and leads to the tooth's loss. This failure type is attributed mainly to the use of posts with length and/or diameter incorrect and deficiencies in dental structure preservation[1].

The intra-radicular post is used to provide retention to a core, but would also have the function of distributing the functional load in a larger area of the remaining coronary structure and dental root. However, many authors (Assif & Gofil, 1994; Martinez-Insua et al., 1999; Rundquist & Versluis, 2006; Whitworth et al., 2002) have been demonstrating that this reconstruction method doesn't restore the original strength of a vital tooth, what is attributed to the wedge effect that this restoration type causes and to the stiffness difference between the post and the tooth.

This effect appears when the load induces the post intrusion inside the dental root. When being pushed, the wedge tends to increase the perimeter of the remnant transversal section due to the tensile stress guided parallel to the circular outlines of the tooth's transversal section. These tangent stresses can cause vertical fracture (Rundquist & Versluis, 2006).

An attempt to increase the root strength front to the physiologic load is the ferrule making that is a metallic necklace of 360° that surrounds the axial walls of the remnant dentine and can be propitiated by the core or by the crown, and tends to produce the encirclement of the root. This "ferrule effect" would protect the pulpless tooth against fracture by counteracting spreading forces generated by the post. The core should extend apical to the shoulder of the preparation to provide a 1.5 to 3.0 mm ferrule of the intact tooth structure (Tan et al., 2005; Pereira et al., 2006; Morgano, 1996; Morgano & Bracket, 1999).

According to Loney et al (1990), the ferrule supplied by the core contributes for the most balanced stress distribution in the dental root and it could compress the remnant structure.

The understanding of the biomechanical principles and restorations applicability is important to design restorations to provide larger strength and retention. The aim of this study was to evaluate the ferrule geometry formed by core on the stress developed in the dental root when a second superior premolar is submitted to four different load conditions.

2. Materials and methods

A natural healthy tooth model (H) was generated in the MSC/PATRAN 2005 (MSC Software Corporation, Santa Ana, CA, USA). On that model modifications were accomplished for the creation of five new models with retainer intra-radicular.

The 3D model of the maxillary second premolar was built based on data presented by Shillinburg & Grace (1973), regarding the measures of the dentin thickness in the mesio-distal and vestibular-lingual axis in four horizontal slices (with 3.5 mm interval) accomplished along the main axis of the dental root starting from the cementoenamel junction. The dentin's thickness in the remnant of the root was built by the interpolation with a spline curve starting from the existent data, following the root anatomy. The pulp dimensions were built based on the data presented by Green & Brooklyn (1960), that it supplies the average diameter of the apical foramen and Shillinburg & Grace (1973), that it supplies the average dimensions in the cervical area.

For the making of the crown geometry the enamel thickness and coronary dentin were based on the data presented by Shillinburg & Grace (1973), through slices of interval of 1 mm, by Cantisano at al. (1987) and Ueti at al. (1997). The height of the vestibular cusp was considered larger than the lingual cusp in 0.9 mm (Shillinburg et al., 1972).

The lamina dura was represented with mechanical properties equal of the cortical bone and as a uniform layer of 0.25 mm thickness as well as the periodontal ligament (Lee et al., 2000).

The models that represented intra-radicular retainer were built from the natural healthy tooth model already. Specify modifications were accomplished in the area corresponding to the pulp and to the dentin, in the places corresponding to the core and porcelain-faced crown.

The dimension of the pulp was enlarged to reproduce the endodontic access stage (Cohen & Hargreaves 2005), the biomechanical prepare and the filling of the root canal.

The porcelain-faced crown was represented with the same outline of the enamel, being average thickness of 1.5 mm in the vestibular face, 1.2 mm in the lingual face, 1.0 mm in the proximal face and 2.0 mm in the occlusal face, with an end in chamfer around the core. The minimum thickness of the metal (NiCr) in the restoration porcelain-faced was 0.3 mm (Yamamoto, 1985).

The zinc phosphate cement was select because is usually used for cementation of the crown and of intra-radicular retainer with thickness from 50 to 100 μm (Anusavice, 2003) and the gutta-percha height was of 5 mm (Morgano, 1996).

In four models with intra-radicular retainer was varied the ending line of the core preparation (with ferrule geometry given by the core) and another model was simulated with simple core (without ferrule), staying the same coronary restoration.

The width and the height of the ferrule were determined as a proportion of the thickness of radicular dentine found in the cementoenamel junction region on the lingual side (1.85 mm), separated for three thirds. For other words, the ferrule with x height varying of 0.62 x 0.62 (T1H1), 0.62 x 1.34 (T1H2), 1.34 x 0.62 (T2H1) and 1.34 x 1.34 mm (T2H2), (Fig. 1).

All of the models with intra-radicular retainer presented the same characteristics in relation to the supporting structures, porcelain-faced crown, cement thickness and apical seal (Fig. 2).

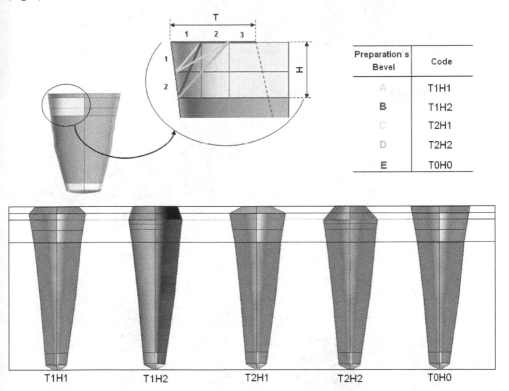

Preparation's Bevel	Code
A	T1H1
B	T1H2
C	T2H1
D	T2H2
E	T0H0

Fig. 1. Ferrule geometries of the radicular dentine in a premolar root.

With the measures of the thickness in each one of the structures (pulp, dentin, enamel) in each face (mesial, distal, lingual, vestibular) a coordinates system was generated with the origin in the pulpal apex and all the values of thickness of the structures were transformed in coordinates. As many coordinates existed to generate the points in the Finite Element (FE) program, a routine in PCL (Patran Command Language) was created to read the spreadsheet and to generate the points. After the points generation (Fig. 3a), the curves were generated and, soon afterwards, the surfaces. Starting from the dental structures surfaces built, the superficial meshes were generated with triangular linear elements (Tri3). After that, volumetric meshes (Fig. 3d) with tetrahedral linear elements (Tet4) were generated, following the procedures in Poiate et al. (2008, 2009a, 2009b, 2011).

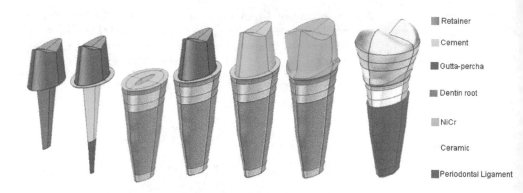

Fig. 2. Structures of the models with intra-radicular retainer.

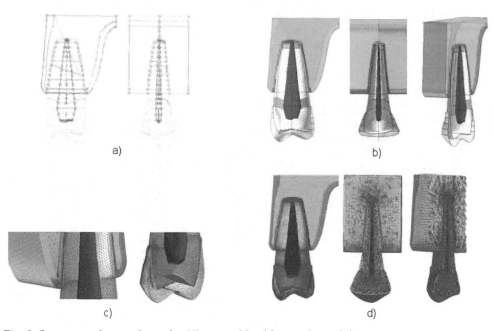

Fig. 3. Sequence of procedures for 3D natural healthy tooth model generation, points generation, dental structures surfaces, superficial and volumetric mesh, respectively.

The degree of discretization of the FE models (Table 1) was established from convergence studies of the results in computer modeling (Pentium 4 3.2 GHz computer with 2.0 Gb RAM memory) to ensure that a proper FE model mesh density was generated and in this way model a realistically anatomic geometry.

Code	Ferrule Thickness x Height	Number of nodes	Number of elements
H	Natural healthy tooth	179403	1109929
T1H1	1 x 1	193034	1226486
T1H2	1 x 2	179430	1142542
T2H1	2 x 1	189438	1200778
T2H2	2 x 2	180720	1148853
T0H0	0 x 0	184750	1175175

Table 1. Models and mesh size.

It was assumed that all the structures in the models were, homogeneous, isotropic and linearly elastic behavior as characterized by two physical properties: Young´s Modulus (E) and Poisson's Ratio (v), Table 2. The interfaces between the structures were presumed to be perfectly united, because the aim was to provide a comparison of our approach, that this simplification was justified.

Structure / Material	Young´s Modulus (GPa)	Poisson´s Ratio	Reference
Pulp	0.02	0.45	Farah & Craig (1974)
Dentin	18.60	0.31	Ko et al. (1992)
Enamel	41.00	0.30	Ko et al. (1992)
Periodontal ligament	0.0689	0.45	Weinstein et al. (1980)
Cortical bone	13.70	0.30	Ko et al. (1992)
Spongy bone	1.37	0.30	Ko et al. (1992)
Gutta-percha	0.00069	0.45	Friedman et al.(1975)
NiCr	188.00	0.33	Black & Hastings (1998)
Phosphate of Zinc Cement	13.00	0.35	Powers at al. (1976)
Feldspathic Ceramic	82.80	0.35	Peyton & Craig (1963)
Cast radicular retainer (ILOR56: gold-alloy post)	93.00	0.33	Pegoretti et al. (2002)

Table 2. Physical properties of the anatomical structures and materials.

Finally, boundary conditions or the model constraint and the loads are also applied. The constraint conditions applied were: in the maxillary sinus, translation in x, y and z directions and rotations in x, y and z axis, fully anchor; in the mesial and distal extremities of the cortical and spongy bone, translation in x direction (perpendicular to this faces) and rotations in y and z axis were anchor. Four load cases were built varying the site, inclination and the application area of a 291.36 N total static load, maximum masticatory force (Ferrario et al., 2004).

Initially, all of the models received the resultant load intensity of 291.36 N parallel to the tooth's long axis, with the aim to evaluate the wedge effect (Fig. 4). The load was applied on the lingual and vestibular cusp above the occlusal surface, at 45º in relation to the tooth's long axis (Holmes et al. 1996), distributed in 19 nodal points of 0.85 mm² area in vestibular cusp and in 19 nodal points of 0.75 mm² area of in lingual cusp (Kumugai et al., 1999).

After that, the model with ferrule that best minimized the wedge effect, as well as the models that represent natural healthy tooth and restored with simple core (T0H0, without ferrule, Fig. 1) were submitted to the following loads (Fig. 4):

a. Oblique Load with 45º in relation to the tooth's long axis distributed in 19 nodal points of 0.85 mm² area in the vestibular cusp, with the aim to evaluate the vestibular lever effect;

b. Load parallel to the tooth's long axis distributed in 19 nodal points of 0.80 mm² area in the mesial marginal ridge, with the aim to evaluate the proximal lever effect;

c. Oblique Load with 45º in relation to the tooth's long axis pointed to vestibular cusp, distributed in 19 nodal points of 0.80 mm² area in the mesial marginal ridge, with the aim to evaluate the torsion effect.

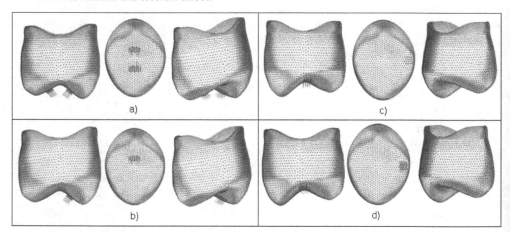

Fig. 4. Place, area and orientation of the load application, a)wedge effect, b)vestibular lever effect, c)proximal lever effect and d)torsion effect.

The processing stage or the solution analysis was performed using MSC/NASTRAN 2005 software (MSC Software Corporation, Santa Ana, CA, USA). The MSC/PATRAN 2005 software, used in the pre-processing, was also used for the post-processing, visualization, and evaluation of the results.

In this study, the maximum principal stress was used as the stress criterion to present the stress patterns distribution in the analyzed models. The cement and dentin's tensile strength of 8.3 MPa (Powers at al., 1976) and 103 MPa (Tanaka et al., 2003), respectively, as well as dentin's compressive strength of 282 MPa (Tanaka et al., 2003), respectively, will serve as reference for results comparison to evaluate if the loads would be potentially harmful to the studied structures.

3. Results

3.1 Wedge effect

Under load with parallel resultant to the long axis, all of the models answered equally. The differences were just in the dentin area in contact with the ferrule (Fig. 5). The results

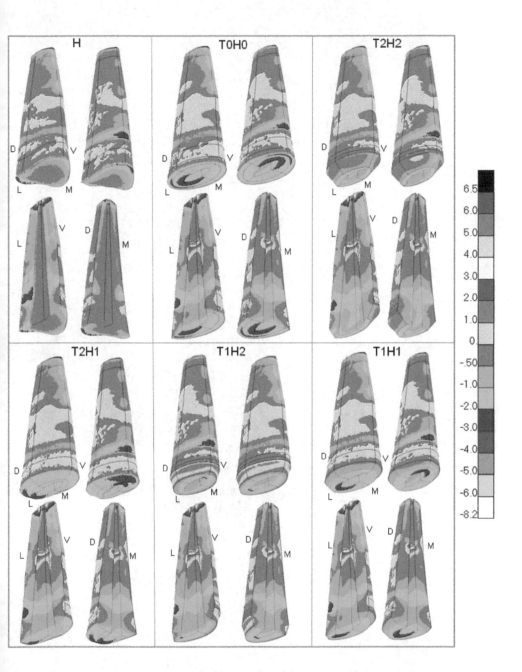

Fig. 5. MPS in all models loaded longitudinally with Vestibular (V)-Lingual (L) and Mesial (M)–Distal (D) slices.

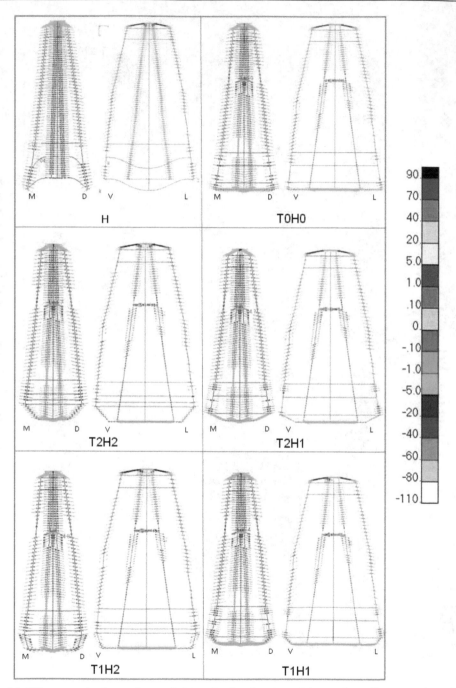

Fig. 6. MPS vectors in the surfaces of the radicular dentine in the wedge effect in M-D and V-L slices.

obtained with models T1H1 and T2H1 presented similar pattern of stress distribution in Maximum Principal Stress (MPS), but different from models T1H2 and T2H2 (that present low tensile stress under the ferrule). In the model H, the dentin area between the bone and the enamel presents some tensile stress, but the Fig. 6 show that the stress vector direction is not totally tangential to the transversal section and it wouldn't be deleterious, nor for the intensity nor for the direction. Similar stress appears in all the models with retainer immediately for apical of the ferrule and cannot are responsible for eventual root fracture.

The Fig. 7 represents a perspective view of the distribution of the MPS in the cement between the retainer and dentin. The results obtained with models T2H1, T0H0 and T1H1 presented a similar pattern of stress distribution in cement below the ferrule, but different from the model T1H2, which shows higher tensile stresses (up to 4 MPa), which by direction (Fig. 6) suggests greater tendency to loosen or break the cement layer in the region, because the stresses are significantly higher, but the failure of the cement seems unlikely when comparing the absolute values of stress with the stress required to the cement cohesive failure (8.3 MPa).

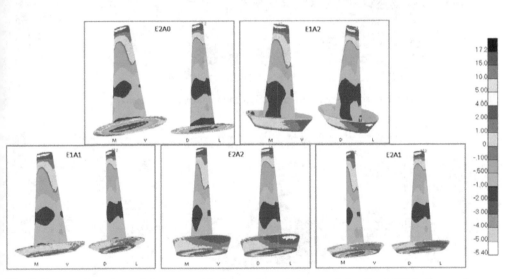

Fig. 7. Perspective view of the cement between the retainer and dentin, MPS in the wedge effect.

3.2 Vestibular lever effect

Under load with inclination of 45º in the vestibular cusp the T2H2, T0H0 and H models answered equally (Fig. 8), except in the radicular dentin that is being doubled and it is observed compressive stress in the vestibular, concentrated in the contact area with the cortical bone that acts as fulcrum (Fig. 9). The tensile stress concentrates and reaches the maximum on the opposite side, lingual, but, unexpectedly, it presents a larger extension in the model with ferrule, with maximum of 104, 110 and 109 MPa for H, T0H0 and T2H2 models, respectively.

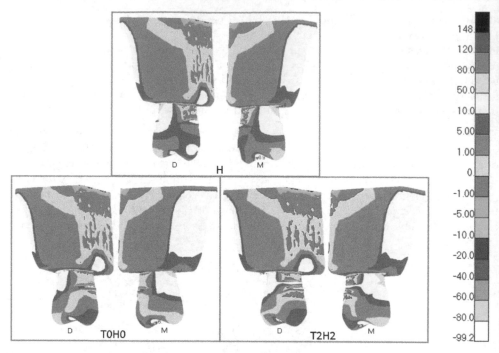

Fig. 8. Perspective view of all structures, MPS in the vestibular lever effect.

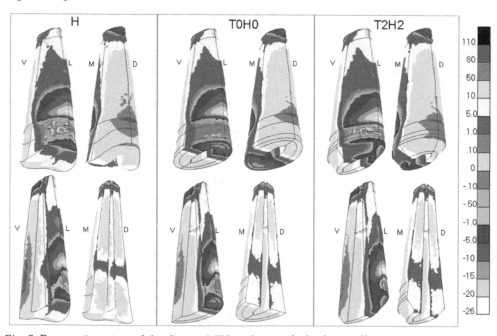

Fig. 9. Perspective view of the dentin, MPS in the vestibular lever effect.

The Fig. 11 represents a perspective view of the distribution of the MPS in the cement between the retainer and dentin and the Fig. 10 shows the MPS vectors in the internal and external surfaces of the radicular dentine at the intersections with in M-D and V-L planes.

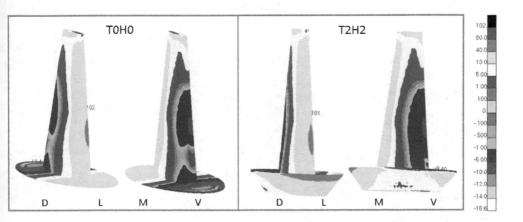

Fig. 10. Perspective view of the cement between the retainer and dentin, MPS in the vestibular lever effect.

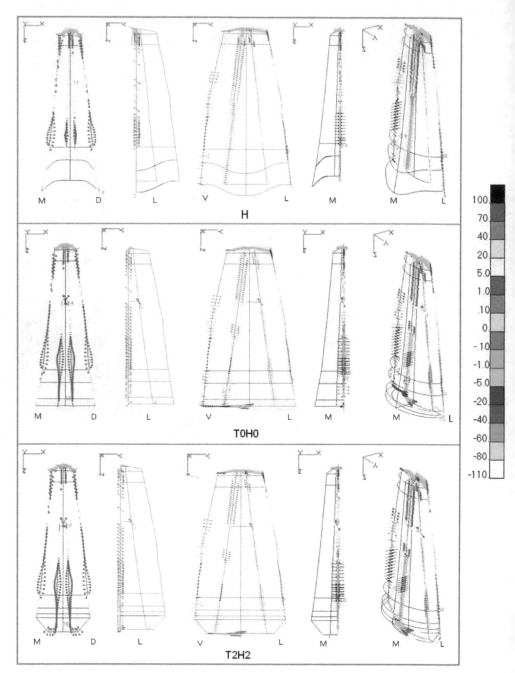

Fig. 11. MPS vectors in the surfaces of the radicular dentine in the vestibular lever effect in M-D and V-L slices.

3.3 Proximal lever effect

The Fig. 12 shows the results under load parallel to the tooth's long axis in the mesial marginal ridge. The maximum tensile stress in dentin (Fig. 13) at the distal edge are also much seemed (68, 68 and 67 MPa), smaller than in the vestibular lever effect.

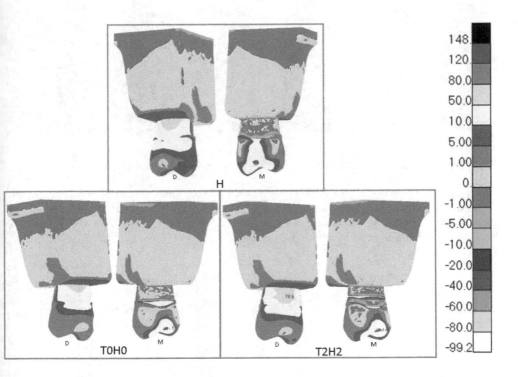

Fig. 12. Perspective view of all structures, MPS in the proximal lever effect.

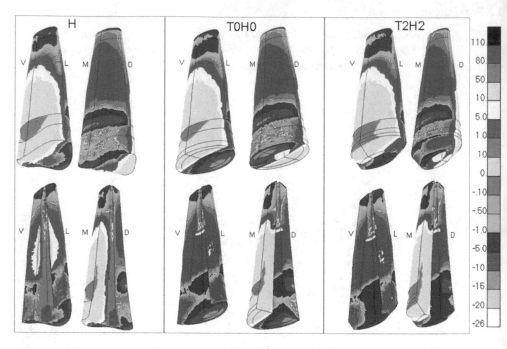

Fig. 13. Perspective view of the dentin, MPS in the proximal lever effect.

The Fig. 14 represents a perspective view of the distribution of the MPS in the cement between the retainer and dentin and the Fig.15 shows the MPS vectors in the internal and external surfaces of the radicular dentine at the intersections with in M-D and V-L planes. The stress orientation in all of the cases is similar.

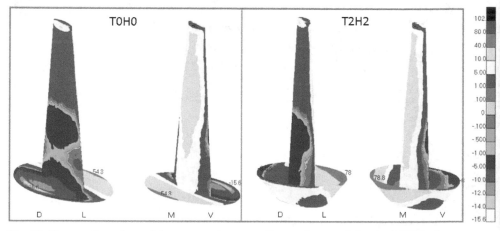

Fig. 14. Perspective view of the cement between the retainer and dentin, MPS in the proximal lever effect.

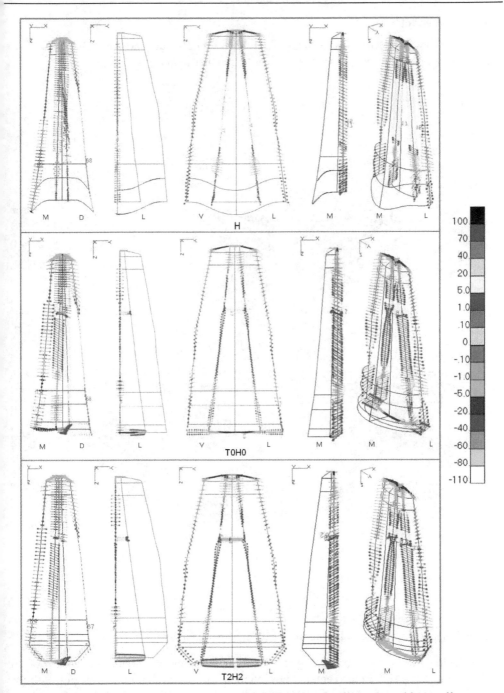

Fig. 15. MPS vectors in the surfaces of the radicular dentine in the proximal lever effect in M-D and V-L slices.

3.4 Torsion effect

The Fig. 16 shows the MPS results in all structures under oblique load with 45° in relation to the tooth's long axis pointed to vestibular cusp in the mesial marginal ridge.

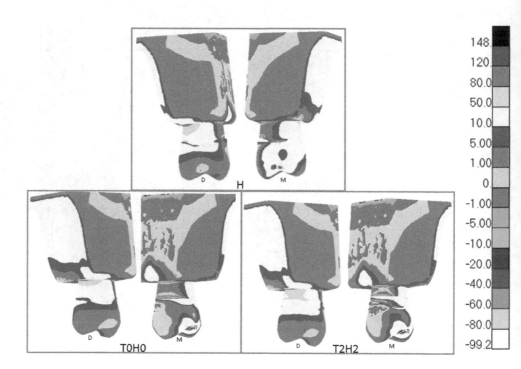

Fig. 16. Perspective view of all structures, MPS in the torsion effect.

The MPS in dentin are present in Fig. 17. The Fig. 18 represents a perspective view of the distribution of the MPS in the cement between the retainer and dentin and the Fig.19 shows the MPS vectors in the internal and external surfaces of the radicular dentine at the intersections with in M-D and V-L planes.

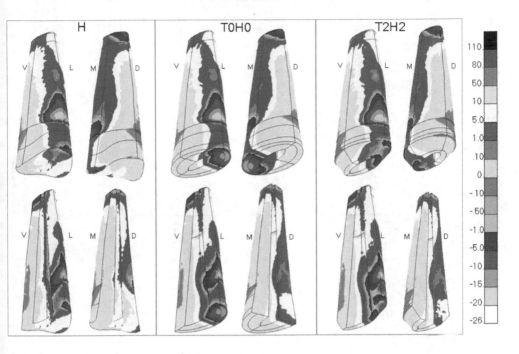

Fig. 17. Perspective view of the dentin, MPS in the torsion effect.

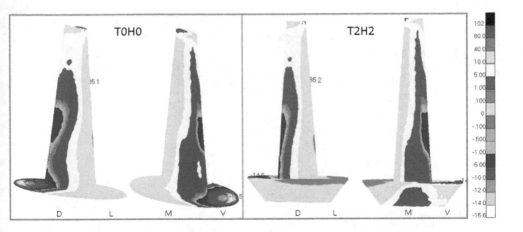

Fig. 18. Perspective view of the cement between the retainer and dentin, MPS in the torsion effect.

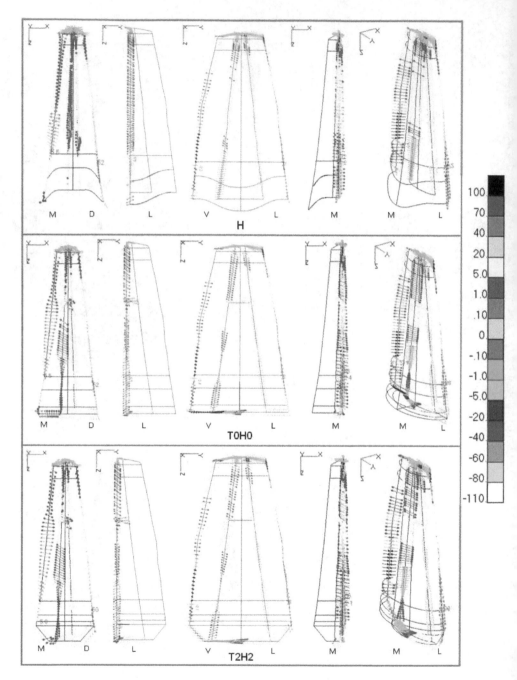

Fig. 19. MPS vectors in the surfaces of the radicular dentine in the torsion effect in M-D and V-L slices.

3.5 Stress plot in wedge, vestibular lever, proximal lever and torsion effect

The Figs 20 to 28 shows the MPS in a ridge A to B, indicated in the drawing corresponding to each graph, in the surface of root dentin (interface with periodontal ligament), depending on the relative position of the node under every load, due to the abcissa axis is adimensional (Distance/Total length),.

The Fig. 20 shows the outside edge of the lingual surface. It is observed that the models that generate the wedge or proximal lever present in the lingual edge, very low stresses, all of the same order of magnitude. Since the models that generate vestibular lever and torsion have high tensile stresses in this edge, but below the dentin´s strenght limit.

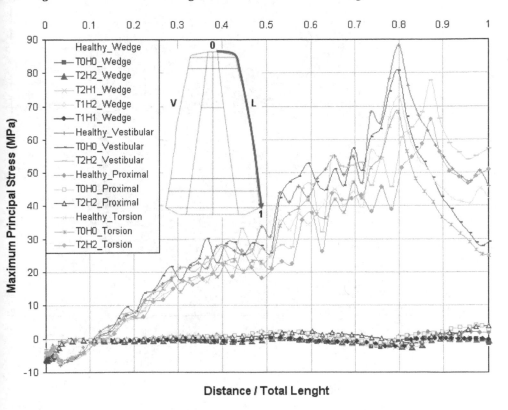

Fig. 20. MPS at the edge of the root dentin on the lingual surface, from the apex to the cervical direction.

The Fig. 21 shows the outside edge of the vestibular surface. It is observed that the models that generate proximal wedge and proximal lever have the same order of magnitude of stresses acting on the vestibular edge with inflection behavior (tensile to compressive) from the apex to the cervical direction. In models that generate proximal lever, the tensile stresses near the apex are close to the wedge effect, however, near the neck, the stress tensile increase up to 5 MPa.

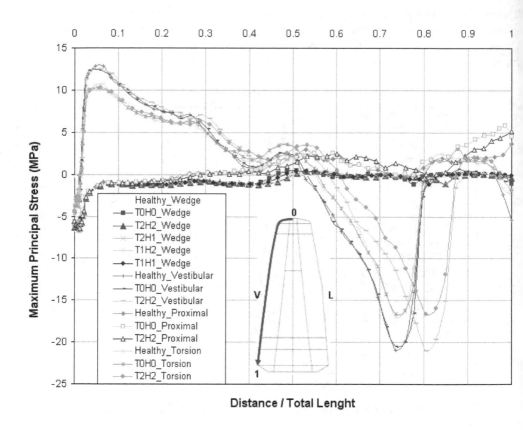

Fig. 21. MPS at the edge of the root dentin on the vestibular surface, from the apex to the cervical direction.

The Fig. 22 shows the outside edge of the distal surface. We observed lowest stresses in wedge models, followed by vestibular lever models, torsion models and reaching the maximum stresses in proximal lever models. In models of proximal lever and torsion effect are generated stress peaks on the same nodes of the same models; the lower stress in torsion models could be due to the lower longitudinal component of the load in this case.

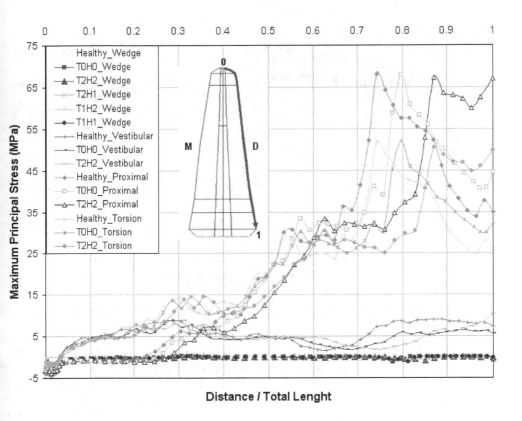

Fig. 22. MPS at the edge of the root dentin, the distal surface, from the apex to the cervical, under all loads

The Fig. 23 shows the outside edge of the mesial surface. We observed lowest stresses in wedge models. The inversion of tensile to compression stress occurs so evident in the proximal lever models and so milder in torsion models.

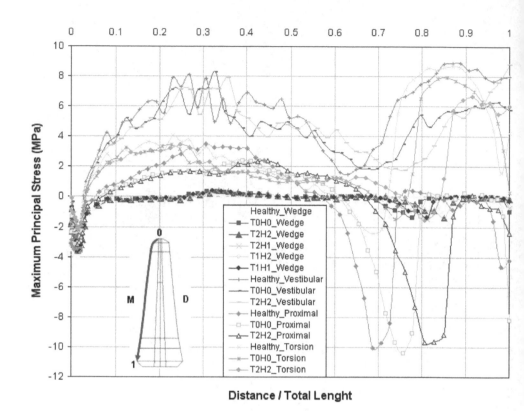

Distance / Total Lenght

Fig. 23. MPS at the edge of the root dentin in the mesial side, from the apex to the cervical, under all loads.

The Fig. 24 shows the inner edge of the lingual surface. We observed lowest stresses in the models of wedge and tensile stresses up to 45 MPa in the models of vestibular lever, followed by models of torsion effect, but in both types of load, in models of healthy tooth the stresses are smaller. It seems important to note that the order of values is much lower than those in Fig. 20, corresponding to the outer surface, which leads to think that the fracture is expected to start on the outside, apparently motivated by folding of the tooth as a whole. Discard the possibility of fracture driven by stress concentration on the inside face, which could be motivated by the wedging effect promoted by the retainer.

The Fig. 20 shows that the highest stress on the outside edge was achieved by the healthy tooth model. However, it is unlikely that the healthy tooth had a greater tendency to fracture: this is not what is observed in practice.

On the other hand Fig. 24 shows the stress almost doubles in inner edge, for the cases of vestibular lever with retainer, compared to the healthy tooth. This suggests that the failure criterion should not be simply the value of the maximum principal stress, but must be influenced by the stress gradient.

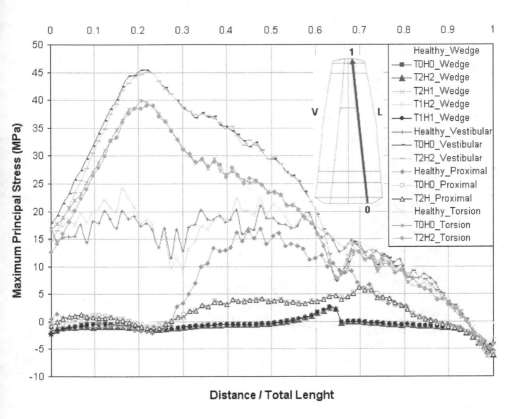

Fig. 24. MPS at the edge of the root dentin on the lingual surface, from the apex to the cervical, under all loads

The Fig. 25 shows the inner edge of the vestibular surface. Tensile stresses are observed up to 27 MPa in the region near the retainer apex in the models that generate vestibular lever, followed by models that generate torsion, but in both types of loading, the stresses are smaller in the healthy tooth models. Figure 11 shows that the direction of the peak tensile is radial and therefore not likely to cause longitudinal fracture but delamination in the dentin and detachment of the retainer apex. In fact, presents a direction parallel to the tensile stress on the edge of the external vestibular root, which appears to be motivated by the apical ligament region, to oppose the rotation of the tooth that is supported by the vestibular cortical bone.

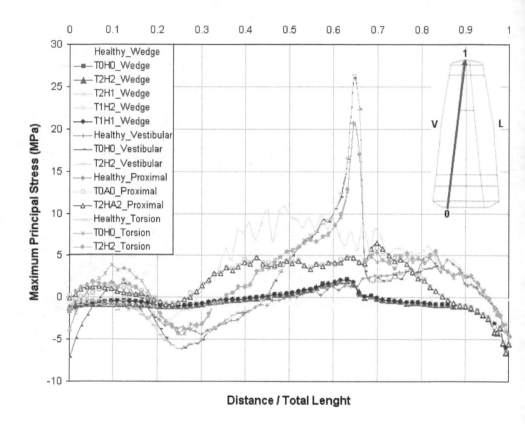

Fig. 25. MPS at the edge of the root dentin on the vestibular surface, from the apex to the cervical, under all loads

The Fig. 26 shows the inner edge of the distal surface. We observed lowest stresses in wedge models. Except the loading wedge models, all have tensile stresses, which shows that practically the entire thickness of the wall works in traction. Moreover, none of the models exceeded the stress developed by the model of healthy tooth with proximal lever load, which leads to the hypothesis that in no case has reached the stress level compatible with the fracture.

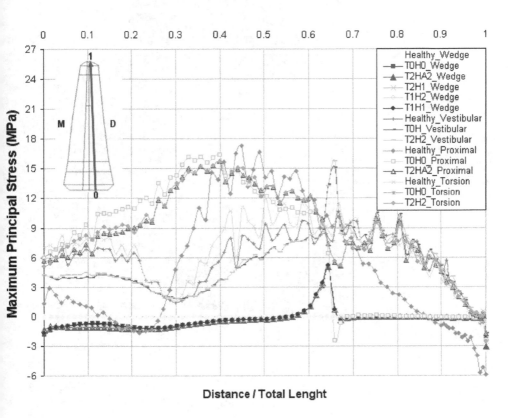

Fig. 26. MPS at the edge of the root dentin, the distal surface, from the apex to the cervical, under all loads.

The Fig. 27 shows the inner edge of the mesial surface. We observed lowest stresses in wedge models. The highest tensile stress occur to the torsion effect, regardless of the presence or not of the ferrula, which also does not help to reduce the stress on the lever vestibular.

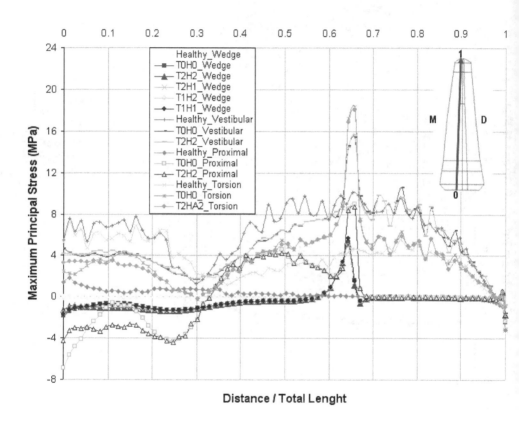

Fig. 27. MPS at the edge of the root dentin in the mesial side, from the apex to the cervical, under all loads.

In Figure 28 shows the circular edge of the inner dentin around the apex of the retainer. The abscissa axis is adimensional (Distance/Total length), where the 0 and 1 represents the vestibular, 0.25 the mesial, 0.5 the lingual and 0.75 the distal aspect. We observed lowest stresses in wedge models, followed by models of proximal lever, torsion and vestibular lever models. The region that remains the most tensioned is the vestibular.

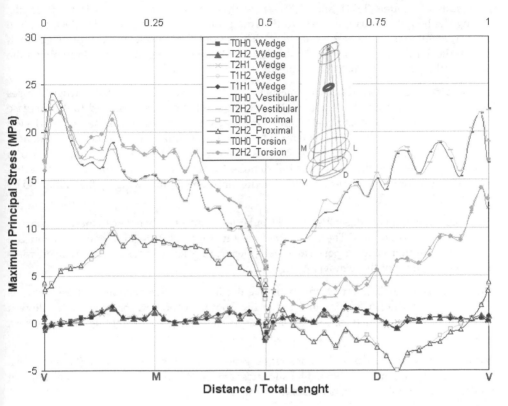

Fig. 28. MPS shown in the circular edge of the inner dentin (VMLDV) around the apex of the retainer under all loads.

4. Discussion

Several studies (Pierrisnard et al., 2002; Zhi-Yue & Yu-Xing, 2003) show that the ferrule creates a positive effect in the reduction of the stress concentration in the dentin-core junction and helps to maintain the integrity of the cement seal in the crown (Libman & Nicholls 1995).

However, in spite of some authors recommend a coronary minimum height of ferrule (Tan et al., 2005; Pereira et al., 2006; Morgano, 1996; Morgano & Bracket, 1999; Aykent, 2006) to increase the fracture strength values, in this study the ferrule didn't influence on the stress distribution. An important observation should be made: for the perfect adhesion among the structures, it was not possible to notice a wedge effect appreciable.

Under load with parallel resultant to the long axis, all of the models present compression stress in the root apex, but with very low intensity when compared with the compression strength of dentin. The light stress concentration found around of the post apex cannot are responsible by longitudinal fracture, because the orientation and the intensity would not justify the fracture.

The results of models T1H1 and T2H1 show that the increase in the width of ferrule, keeping the height, it was important to the root protection with the ferrule effect, because the compressive stress increased four times. The model T2H2 presented compressive stress in the radicular dentin on the ferrule (0.5 to 2.0 MPa) and in the radicular dentin above the periodontal ligament (0.5 to 20 MPa), mesial and distal face (Fig. 4).

On the other hand, the results of models T1H2 and T2H2 show that tensile stresses disappears in the dentine in contact with the ferrule, what can be attributed to the increase in the ferrule width, keeping the height. This characteristic confers to the ferrule design some superiority, because it can be inferred that it will be more difficult than the cement is unstuck in that area, since the tensile stress isn't submitted. The model T2H2 seems also to minimize the tensile stress in the cervical dentine (compressive stress from 0.5 to 2.0 MPa) and in the radicular dentine near to the periodontal ligament, following by the model T2H1, T1H1 and T1H2.

Posteriors teeth can be subject also to vestibular lever effects whenever requested eccentric efforts in lateral excursion. The contact oclusal on the work side in lateral excursion can reach the vestibular cusp in posteriors teeth, generating lever force on the involved roots that serve as guide for those movements.

Under load with inclination of 45° in the vestibular cusp, the maximum stress value is compatible with the fracture occurrence. If the fracture begins at that place can follow a perpendicular plan to the tensile vector and to spread tending the uprighting that suggests a vertical radicular fracture would pass exactly in the bone crest limit.

Fractures vertical appear as a result of stresses generated inside the root canal (Lertchirakarn et al., 2003), but the models showed tensile stress concentrated in the post apex, although of smaller magnitude that in the height of the bone crest.

In all of the models, the root canal stress in the mesial-distal slice presented tensile stress (except for the area near to the post apex) and near to the bone crest presents change in the vector orientation due to the rotation of the same ones, that it seems related with the presence of cortical bone. Therefore, the fracture area not just depends on the post and ferrule, but also of location of the tooth in the alveolus.

An interesting finding was the relatively small difference among the three models that could be attributed to the fact of the interfaces between the structures of the models was simulated with perfect adhesion. The perfect adhesion can losing in the course of time, what would explain that the failures didn't use to happen in retainer recently cemented. On the other hand, the models showed high tensile stresses concentration in more than half of the cement layer, which could be responsible for cohesive fracture.

Under the load area the tensile stresses on the ferrule generates 40 MPa, exceeding the cement's tensile strength of 8.3 MPa and associated with the stress orientation (Fig. 11) suggests the tendency to the cement layer failure in this area.

Under load parallel to the tooth's long axis in the mesial marginal ridge, the tooth restored with post would present a tendency to the rupture similar to the one of the natural, what is not supported by the clinical observations. The explanation for this discrepancy could be again in the fact that the cement layer presented stresses much larger than the necessary ones for cohesive fracture. After the cement failure the stress distribution change and could propitiate the root fracture.

The intra-radicular retainers can be subject also to rotation force or torsion, whenever requested clinically through contacts functional cusp in the ridge, in other words, tooth relationship to two teeth (Hemmings et al., 1991).

The occlusal contacts in the marginal ridge of the premolar could produce a torsion loads that tends to rotate the root along the tooth's axis. For this reason is extremely important have in mind the use of retainers that offer larger safety to the radicular remnant in any situation of mechanical effort what the root is submitted.

Under oblique load with 45º in relation to the tooth's long axis pointed to vestibular cusp in the mesial marginal ridge, the maximum stress and orientation evidences again that cement layer is prone to failure and that the stress concentration is more serious in the ferrule case. When the cement around of the most coronary portion deteriorates, the fulcrum migrates to apical, what increases the lever arm progressively (Cohen et al., 1993).

The modeling in this study considered all the components without setting out eventual contact problems, in other words, without cement failure, what couldn't happen in the clinic reality.

Regarding the clinical significance of the results, it seems that the best way to protect the radicular remnant when restored with post would be to guarantee the occlusal adjustment, that don't happen loads different from the longitudinal. The most vulnerable part of the whole restorative system is the cement layer, that doesn't resist to the tensile stresses result from the loads application different from the longitudinal load. The post perfectly adhered doesn't seem lead to the occurrence of longitudinal fracture (below the bone crest), but probably to fracture that would begin in the vestibular in the height of the bone crest.

5. Conclusions

Within the limitations of this study, the following conclusions were drawn:

1.   In spite of none of the simulated cases to evidence the wedge effect, the longitudinal load produced stresses that don't justify the rupture nor of the dentine nor of the cement layer and the stresses are very inferior to developed by all the other load types;
2.   The ferrule isn't necessary to improve the stresses distribution, except for longitudinal load, in that presents discreet beneficial effect;
3.   In the vestibular load were found, in the lingual face, tensile stresses guided parallel to the longitudinal axis and magnitude enough to induce vertical fractures above the bone crest. These stresses are associated to the tooth's bend, leaning in the cortical bone that it acts as a fulcrum;
4.   In the cement layer were found stresses that are prone to fracture in all of the loads, except for the case of longitudinal load;
5.   A concerted effort to develop other FE models to improve the results may provide more reliable data about the non-linearly of the structures and the cement failures in the clinic reality;
6.   The analyses help the field better understand the biomechanical analysis of restored teeth with cast intra-radicular retainer with and without ferrule.

6. Acknowledgements

This study was partially based on a thesis submitted to the University of São Paulo, in fulfillment of the requirements for PhD degree. The authors are grateful to the CAPES (Coordenação de Aperfeicoamento de Pessoal de Nível Superior, Coordination for the Improvement of Higher Education Personnel) that supported this research project.

7. References

Anusavice, JK. (2003). *Phillips' Science of Dental Materials* (11th ed.), Elsevier Science, Saunders.

Assif, D; Gofil, C. (1994). Biomechanical considerations in restoring endodontically treated teeth. *J Prosthet Dent*, Vol. 71, pp. 565-67.

Aykent, F; Kalkan, M; Yucel, MT & Ozyesil, AG. (2006). Effect of dentin bonding and ferrule preparation on the fracture strength of crowned teeth restored with dowels and amalgam cores. *J Prosthet Dent*, Vol. 95, pp. 297-301.

Black, J; Hastings, G. (1998). *Handbook of Biomaterial Properties* (1th ed.), Chapman & Hall, London.

Cantisano, W; Palhares, WR & Santos, HJ. (1987). *Anatomia dental e escultura* (3rd ed.), Guanabara Koogan, Rio de Janeiro.

Cohen, BI; Musikant, BL & Deutsch AS. (1993). Comparison of the retentive properties of two hollow-tube post systems to those of a solid post design. *J Prosthet Dent*, Vol. 70, pp. 234-8.

Cohen, S; Hargreaves, KM. (2005). *Pathways of the pulp* (9th ed.), Mosby, St. Louis.

Farah, J; Craig, R. (1974). Finite element stress analysis of a restored axisymmetric first molar. *J Dent Res*, Vol. 53, pp. 859-66.

Ferrari, M; Mannocci, F. (2000). Bonding of an esthetic fiber post into root canal with a 'one-bottle' system: a clinical case. *Int J Endodont*, Vol. 33, pp. 397-400.

Ferrario, VF; Sforza, C; Serrao, G; Dellavia, C & Tartaglia, GM. (2004). Single tooth bite forces in healthy young adults. *J Oral Rehabil*, Vol. 31, pp. 18-22.

Friedman, C; Sandrik, J; Heuer, M & Rapp, G. (1975). Composition and mechanical properties of gutta-percha endodontic points. *J Dent Res*, Vol. 54, pp. 921-25.

Green, D; Brooklyn, NY. (1960). Stereomicroscopic study of 700 root apices of maxillary and mandibular posterior teeth. *Oral Surg Oral Med Oral Pathol*, Vol. 13, pp. 728-33.

Hemmings, KW; King, PA & Setchell, DJ. (1991). Resistance to torsional forces of various posts and core designs. *J Prosthet Dent*, (1991), Vol. 66, pp. 325-9.

Holmes, DC; Diaz-Arnold, AM & Leary JM. (1996). Influence of post dimension on stress distribution in dentin. *J Prosthet Dent*, Vol. 75, pp. 140-7.

Ko, CC; Chu, CS; Chung, HK & Lee, MC. (1992). Effects of post on dentin stress distribution in pulpless teeth. *J Prosthet Dent*, Vol. 68, pp. 421-27.

Kumugai, H; Suzuki, T; Hamada, T; Sondang, P; Fujitani, M & Nikawa, H. (1999). Occlusal force distribution on the dental arch during various levels of clenching. *J Oral Rehabil*, Vol. 26, pp. 932-35.

Lee, SY; Huang, HM & Lin, CY. (2000). In vivo and in vitro natural frequency analysis of periodontal conditions, in innovative method. *J Periodontol*, *Vol.* 71, pp. 632-40.

Lertchirakarn, V; Palamara, JE & Messer HH. (2003). Finite element analysis and strain-gauge studies of vertical root fracture. *J Endod*, Vol. 29, pp. 529-34.

Libman, WJ; Nicholls JI. (1995). Load fatigue of teeth restored with cast posts and cores and complete crowns. *Int J Prosthodont*, Vol. 8, pp. 155-61.

Loney, RW; Kotowicz, WE & McDowel, GC. (1990). Three-dimensional photoelastic stress analysis of the ferrule effect in cast post and cores. *J Prosthet Dent*, Vol. 63, pp. 506-12.

Martinez-Insua, A; Da Silva, L; Rilo, B & Santana, U. (1999). Comparison of the fracture resistances of pulpless teeth restored with a cast post and core or fiber post with a composite core. *J Prosthet Dent*, Vol. 80, pp. 527-32.

Morgano S. (1996). Restoration of pulpless teeth: application of traditional principles in present and future contexts. *J Prosthet Dent*, Vol. 75, pp. 375-80.

Morgano, SM; Bracket SE. (1999). Foundation restoration in fixed prosthodontics: current knowledge and future needs. A literature review. *J Prosthet Dent, Vol.* 82, pp. 643-57.

Pegoretti, A; Fambri, L; Zappini, G & Bianchetti, M. (2002). Finite element analysis of a glass fibre reinforced composite endodontic post. *Biomaterials*, Vol. 23, pp. 2667-82.

Pereira, JR; Ornelas, F; Conti, PCR & Valle, AL. (2006). Effect of a crown ferrule on the fracture resistance of endodontically treated teeth restored with prefabricated posts. *J Prosthet Dent*, Vol. 95, pp. 50-4.

Peyton, FA; Craig, RG. (1963). Current evaluation of plastics in crown and bridge prosthesis. *J Prosthet Dent*, Vol. 13, pp. 743-53.

Pierrisnard, L; Bohin, F; Renault, P & Barquins M. (2002). Corono-radicular reconstruction of pulpless teeth: a mechanical study using finite element analysis. *J Prosthet Dent*, Vol. 88, pp. 442-8.

Poiate, IAVP; Vasconcellos, AB; Andueza, A; Pola, IRV & Poiate Jr, E. (2008). Three dimensional finite element analyses of oral structures by computerized tomograghy. *J Biosc Bioeng*, Vol. 106, No. 6, pp. 906-9.

Poiate, IAVP; Vasconcellos, AB; Santana, RB &, Poiate Jr, E. (2009a). Three-Dimensional Stress Distribution in the Human Periodontal Ligament in Masticatory, Parafunctional, and Trauma Loads: Finite Element Analysis. *J Periodontology*, Vol. 80, pp. 1859-1867.

Poiate, IAVP; Vasconcellos, AB; Poiate Jr, E & Dias KRC. (2009b). Stress distribution in the cervical region in a 3D FE. Brazilian Oral Research, Vol. 23, pp. 161-168.

Poiate, IAVP; Vasconcellos, AB; Mori, M &, Poiate Jr, E. (2011). 2D and 3D finite element analysis of central incisor generated by computerized tomography. *Comput Methods Programs Biomed*, Vol 104, No. 2, pp. 292-9.

Powers, M; Farah, JW & Craig RG (1976). Modulus of elasticity and strength properties of dental cements. *J Am Dent Assoc*, Vol. 92, no paginated.

Rundquist, BD; Versluis, A. (2006). How does canal taper affect root stresses? *Int Endod J*, Vol. 39, pp. 226-37.

Shillingburg, HT; Kaplan, MJ & Grace, CS. (1972). Tooth dimensions – A comparative study. *J South Calif Dent Assoc*, Vol. 40, pp. 830-9.

Shillingburg, HT; Grace, CS. Thickness of enamel and dentin. *J South Calif Dent Assoc*, Vol. 41, pp. 33-52.

Tan, PLB; Aquilino, SA; Gratton, DG; et al. (2005). In vitro fracture resistance of endodontically treated central incisors with varying ferrule heights and configurations. *J Prosthet Dent*; Vol . 93, pp. 331-6.

Tanaka, M; Naito, T; Yokota, M & Kohno, M. (2003). Finite element analysis of the possible mechanism of cervical lesion formation by occlusal force. *J Oral Rehabil*, Vol. 30, pp. 60-67.

Ueti, H; Todescan, R & Gil, C. (1997). Study of the thickness enamel/dentin in function of age, group of teeth and distance in relation to the external portion of the clinical crown. *Rev Pós-Grad da USP*, Vol 4, pp. 153-9.

Weinstein, AM; Klaawitter, JJ & Cook, SD. (1980). Implant-bone interface characteristics of bioglass dental implants. *J Biomed Mater Res*, Vol. 14, pp. 23-29.

Whitworth, JM; Walls, AWG & Wassell, RW. (2002). Crowns and extra-coronal restorations: Endodontic considerations: the pulp, the root-treated tooth and the crown. *Br Dental J*, 2002, Vol. 192, pp. 315-27.

Yamamoto, M. (1985). *Metal-ceramics: Principles and methods of Makoto Yamamoto* (1st ed.), Quintessence, Chicago.

Zhi-Yue, L; Yu-Xing, Z. (2003). Effects of post-core design and ferrule on fracture resistance of endodontically treated maxillary central incisors. *J Prosthet Dent*, Vol. 89, pp. 368-73.

6

Past, Present and Future of Finite Element Analysis in Dentistry

Ching-Chang Ko[1,2,*], Eduardo Passos Rocha[1,3] and Matt Larson[1]
[1]Department of Orthodontics,
University of North Carolina School of Dentistry,
[2]Department of Material Sciences and Engineering,
North Carolina State University Engineering School, Raleigh,
[3]Faculty of Dentistry of Araçatuba, UNESP,
Department of Dental Materials and Prosthodontics, Araçatuba, Saõ Pauló,
[1,2]USA
[3]Brazil

1. Introduction

Biomechanics is fundamental to any dental practice, including dental restorations, movement of misaligned teeth, implant design, dental trauma, surgical removal of impacted teeth, and craniofacial growth modification. Following functional load, stresses and strains are created inside the biological structures. Stress at any point in the construction is critical and governs failure of the prostheses, remodeling of bone, and type of tooth movement. However, *in vivo* methods that directly measure internal stresses without altering the tissues do not currently exist. The advances in computer modeling techniques provide another option to realistically estimate stress distribution. Finite element analysis (FEA), a computer simulation technique, was introduced in the 1950s using the mathematical matrix analysis of structures to continuum bodies (Zienkiewicz and Kelly 1982). Over the past 30 years, FEA has become widely used to predict the biomechanical performance of various medical devices and biological tissues due to the ease of assessing irregular-shaped objects composed of several different materials with mixed boundary conditions. Unlike other methods (e.g., strain gauge) which are limited to points on the surface, the finite element method (FEM) can quantify stresses and displacement throughout the anatomy of a three dimensional structure.

The FEM is a numerical approximation to solve partial differential equations (PDE) and integral equations (Hughes 1987, Segerlind 1984) that are formulated to describe physics of complex structures (like teeth and jaw joints). Weak formulations (virtual work principle) (Lanczos 1962) have been implemented in FEM to solve the PDE to provide stress-strain solutions at any location in the geometry. Visual display of solutions in graphic format adds attractive features to the method. In the first 30 years (1960-1990), the development of FEM

---

programs focused on stability of the solution including minimization of numerical errors and improvement of computational speed. During the past 20 years, 3D technologies and non-linear solutions have evolved. These developments have directly affected automobile and aerospace evolutions, and gradually impacted bio-medicine. Built upon engineering achievement, dentistry shall take advantage of FEA approaches with emphasis on mechanotherapy. The following text will review history of dental FEA and validation of models, and show two examples.

2. History of dental FEA

2.1 1970-1990: Enlightenment stage -2D modeling

Since Farah's early work in restorative dentistry in 1973, the popularity of FEA has grown. Early dental models were two dimensional (2D) and often limited by the high number of calculations necessary to provide useful analysis (Farah and Craig 1975, Peters et al., 1983, Reinhardt et al., 1983, Thresher and Saito 1973). During 1980-1990, the plane-stress and plane-strain assumptions were typically used to construct 2D tooth models that did not contain the hoop structures of dentin because typically either pulp or restorative material occupied the central axis of the tooth (Anusavice et al., 1980). Additional constraints (e.g., side plate and axisymmetric) were occasionally used to patch these physical deficiencies (hoop structures) to prevent the separation of dentin associated with the 2D models (Ko, 1989). As such a reasonable biomechanical prediction was derived to aid designs of the endodontic post (Ko et al., 1992). Axisymmetric models were also used to estimate stress distribution of the dental implants with various thread designs (Rieger et al., 1990). Validation of the FE models was important in this era because assumptions and constraints were added to overcome geometric discontinuity in the models, leading to potential mathematical errors.

2.2 "1990-2000" beginnings stage of 3D modeling

As advancements have been made in imaging technologies, 3D FEA was introduced to dentistry. Computer tomography (CT) data provide stacks of sectional geometries of human jaws that could be digitized and reconstructed into the 3D models. Manual and semi-automatic meshing was gradually evolved during this time. The 3D jaw models and tooth models with coarse meshes were analyzed to study chewing forces (Korioth 1992, Korioth and Versluis 1997, Jones et al., 2001) and designs of restorations (Lin et al., 2001). In general, the element size was relatively large due to the immature meshing techniques at that time, which made models time consuming to build. Validation was required to check accuracy of the stress-strain estimates associated with the coarse-meshed models. In addition to the detail of 3D reconstruction, specific solvers (e.g., poroelasticity, homogenization theory, dynamic response) were adapted from the engineering field to study dental problems that involved heterogeneous microstructures and time-dependent properties of tissues. Interfacial micromechanics and bone adaptation around implants were found to be highly non-uniform, which may dictate osseointegration patterns of dental implants (Hollister et al., 1993; Ko 1994). The Monte Carlo model (probability prediction), with incorporation of the finite element method for handling irregular tooth surface, was developed by Wang and

Ko et al (1999) to stimulate optical scattering of the incipient caries (e.g., white spot lesion). The simulated image of the lesion surface was consistent with the true image captured *in clinic* (Figure 1). Linear fit of the image brightness between the FE and clinical images was 85% matched, indicating the feasibility of using numerical model to interpret clinical white spot lesions. The similar probability method was recently used to predict healing bone adaptation in tibia (Byrne et al., 2011). Recognition of the importance of 3D models and specific solutions were the major contributions in this era.

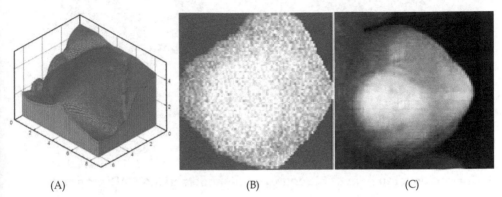

(A)   (B)   (C)

Fig. 1. **A.** Finite element mesh of *in vivo* carious tooth used for Monte Carlo simulation; **B.** Image rendered from Monte Carlo 3D simulation, **C.** True image of carious tooth obtained from a patient's premolar using an intra-oral camera.

2.3 "2000-2010" age of proliferation, 3D with CAD

As advancements have been made in computer and software capability, more complex 3D structures (e.g., occlusal surfaces, pulp, dentin, enamel) have been simulated in greater detail. Many recent FE studies have demonstrated accurate 3D anatomic structures of a sectioned jaw-teeth complex using μCT images. Increased mathematical functions in 3D computer-aid-design (CAD) have allowed accurate rendition of dental anatomy and prosthetic components such as implant configuration and veneer crowns (Figure 2). Fine meshing and high CPU computing power appeared to allow calculation of mechanical fields (e.g., stress, strain, energy) accounting for anatomic details and hierarchy interfaces between different tissues (e.g., dentin, PDL, enamel) that were offered by the CAD program. It was also recognized that inclusion of complete dentition is necessary to accurately predict stress-strain fields for functional treatment and jaw function (Field et al. 2009). Simplified models containing only a single tooth overlooked the effect of tooth-tooth contacts that is important in specified biomechanical problems such as orthodontic tooth movement and traumatic tooth injury. CAD software such as SolidWorks© (Waltham, MA, USA), Pro Engineering© (Needham, MA), and Geomagic (Triangle Park, NC, USA) have been adapted to construct dentofacial compartments and prostheses. These CAD programs output solid models that are then converted to FE programs (e.g., Abaqus, Ansys, Marc, Mimics) for meshing and solving. The automeshing capability of FE programs significantly improved during this era.

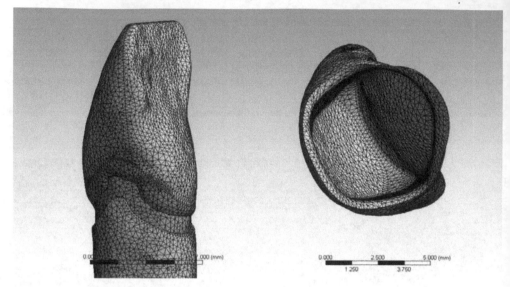

Fig. 2. Fine finite element mesh generated for ceramics veneer simulation.

3. Current development of 3D dental solid models using CAD programs

Currently, solid models have been created from datasets of computer tomography (CT) images, microCT images, or magnetic resonance images (MRI). To create a solid model from an imaging database, objects first need to be segregated by identifying interfaces. This is performed through the creation of non-manifold assemblies either through sequential 2D sliced or through segmentation of 3D objects. For this type of model reconstruction, the interfaces between different bodies are precisely specified, ensuring the existence of common nodes between different objects of the contact area. This provides a realistic simulation of load distribution within the object. For complex interactions, such as bone-implant interfaces or modeling the periodontal ligament (PDL), creation of these coincident nodes is essential.

When direct engineering (forward engineering) cannot be applied, reverse engineering is useful for converting stereolithographic (STL) objects into CAD objects (.iges). Despite minor loss of detail, this was the only option for creation of 3D organic CAD objects until the development of 3D segmentation tools and remains a common method even today. The creation of STL layer-by-layer objects requires segmentation tools, such as ITK-SNAP (Yushkevich et al., 2006) to segment structures in 3D medical images. SNAP provides semi-automatic segmentation using active contour methods, as well as manual delineation and image navigation.

Following segmentation, additional steps are required to prepare a model to be imported into CAD programs. FEA requires closed solid bodies – in other words, each part of the model should be able to hold water. Typical CT segmentations yield polygon surfaces with irregularities and possible holes. A program capable of manipulating these polygons and creating solid CAD bodies is required, such as Geomagic (Triangle Park, NC, USA).

Although segmentations may initially appear very accurate (Figure 3A), there are often many small irregularities that must be addressed (Figure 3B). Obviously, organic objects will have natural irregularities that may be important to model, but defects from the scanning and segmentation process must be removed. Automated processes in Geomagic such as mesh doctor can identify problematic areas (Figure 3B) and fix many minor problems. For larger defects, defeaturing may be required. Once the gaps in the surface have been filled, some amount of smoothing is typically beneficial. Excess surface detail that will not affect results only increases the file size, meshing times, mesh density, and solution times. To improve surfacing, a surface mesh on the order of 200,000 polygons is recommended. Geomagic has a tool ("optimize for surfacing") that redistributes the polygons nodes on the surface to create a more ideal distribution for surfacing (Figure 3C). Following these optimization steps, it is important to compare the final surface to the initial surface to verify that no significant changes were made.

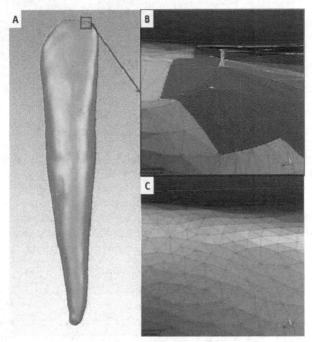

Fig. 3. Although initial geometry following segmentation can appear smooth (A), many small defects are present that Geomagic will highlight in red using "mesh doctor" as potentially problematic (B). Following closing gaps, smoothing, minor defeaturing, and optimization for surfacing, the polygon mesh is greatly improved (C).

With the optimized surfaces prepared using the previous steps, closed solid bodies can be created. Although the actual final bodies with the interior and exterior surfaces can be created at this stage, we have observed that closing each surface independently and using Booleans in the CAD program typically improves results. For example, this forces the interior surface of the enamel to be the identical surface as the exterior of the dentin. If the Boolean operations are done prior to surfacing, minor differences in creating NURB surfaces may affect the connectivity of the objects. Some research labs (Bright and Rayfield 2011) will simply transfer the polygon surfaces over to a FEA program for analysis without using a CAD program. This can be very effective for relatively simple models, but when multiple solid bodies are included and various mesh densities are required this process becomes cumbersome.

To use a CAD program with organic structures, the surface cannot be a polygon mesh, but rather needs to have a mathematical approximation of the surface. This is typically done with NURB surfaces, so the solid can be saved as an .iges or .step file. This process involves multiple steps – laying out patches, creating grids within these patches, optimizing the surface detail, and finally creating the NURB surface (Figure 4). Surfacing must be done carefully as incorrectly laying out the patches on the surface or not allowing sufficient detail may severely distort the surface. In the end, the surfaced body should not have problematic geometry, such as sliver faces, small faces, or small edges.

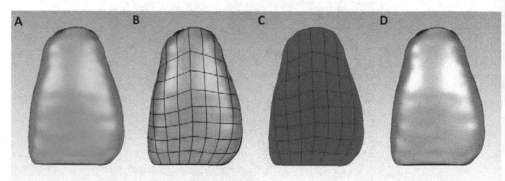

Fig. 4. Process of NURB surface generation using Geomagic. (A) Contour lines are defined that follow the natural geometry - in this case, line angles were used. (B) Patches are constructed and shuffled to create a clean grid pattern. (C) Grids are created within each patch. (D) NURB surfaces are created by placing control points along the created grids.

CAD programs allow the incorporation of high definition materials or parts from geometry files (e.g. .iges, .step), such as dentures, prosthesis, orthodontics appliances, dental restorative materials, surgical plates and dental implants. They even allow partial modification of the solid model obtained by CT or µCT to more closely reproduce accurate organic geometry. Organic modeling (biomodeling) extensively uses splines and curves to model the complex geometry. FE software or other platforms with limited CAD tools typically do not provide the full range of features required to manipulate these complicated organic models. Therefore, the use of a genuine CAD program is typically preferred for detailed characterization of the material and its contact correlation with surrounding structures. This is especially true for models that demand strong modification of parts or incorporation of multiple different bodies.

When strong modification is required, the basic parts of the model such as bone, skin or basic structures can be obtained in .stl format. They are then converted to a CAD file allowing modification and/or incorporation of new parts before the FE analysis. It is also possible to use the CT or microCT dataset to directly create a solid in the CAD program.

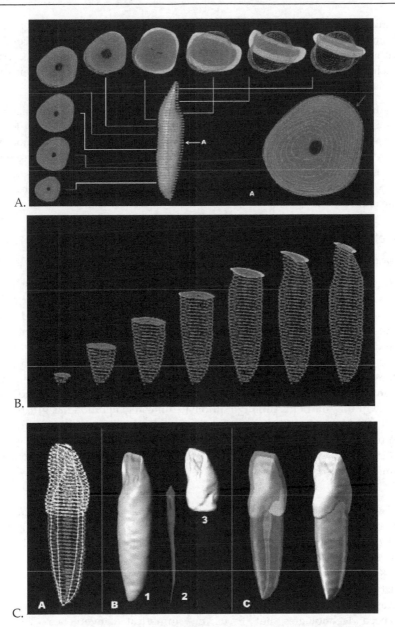

Fig. 5. The solid model of a maxillary central incisor was created through the following steps. (A) Multiple sketches were created in various slices of the microCT data. The sketch defined the contour of the root. (B) Sequential contours were used to reconstruct outer surface of dentin and other parts (e.g., enamel and pulp - not shown). (C) All parts (enamel, dentin and pulp) were combined to form the solid model of the central incisor. All procedures were performed using SolidWorks software.

Initially, this procedure might be time-consuming. However, it is useful for quickly and efficiently making changes in parts, resizing multiple parts that are already combined, and incorporating new parts. This also allows for serial reproduction of unaltered parts of the model, such as loading areas and unaltered support structures, keeping their dimensions and Cartesian coordinates.

This procedure involves the partial or full use of the dataset, serially organized, to create different parts. (Figures 5 A & B) In models with multiple parts, additional tools such as lofts, sweeps, surfaces, splines, reference planes, and lines can be used to modify existing solids or create new solids (Figures 5C). Different parts may be combined through Boolean operations to generate a larger part, to create spaces or voids, or to modify parts. The parts can be also copied, moved, or mirrored in order to reproduce different scenarios without creating an entirely new model.

4. Finite element analysis of the current dental models

4.1 Meshing

For descritization of the solid model, most FE software has automated mesh generating features that produce rather dense meshes. However, it is important to enhance the controller that configures the elements including types, dimensions, and relations to better fit the analysis to a particular case and its applications. Most of current FE software is capable of assessing the quality of the mesh according to element aspect ratio and the adaptive method. The ability of the adaptive method to automatically evaluate and modify the contact area between two objects overlapping the same region and to refine the mesh locally in areas of greater importance and complexity has profoundly improved the accuracy of the solution. Although automated mesh generation has greatly improved, note that it still requires careful oversight based on the specific analysis being performed. For example, when examining stresses produced in the periodontal ligament with orthodontic appliances, the mesh will be greatly refined in the small geometry of the orthodontic bracket, but may be too coarse in the periodontal ligament – the area of interest.

The validation that was concerned with meshing errors and morphological inaccuracy during 1970 − 2000 is no longer a major concern as the CAD and meshing technology evolves. However, numerical convergence (Huang et al., 2007) is still required, which is frequently neglected in dental simulations (Tanne et al., 1987; Jones et al., 2001; Liang et al., 2009; Kim et al., 2010). Some biologists ignore all results from FEA, requesting an unreasonable level of validation for each model, but overlooking the valuable contributions of engineering principles. A rational request should recognize evolution of the advanced technologies but focus on numerical convergence. The numerical convergence is governed by two factors, continuity and approximation methods, and can be classified to strong convergence $||X_n|| \rightarrow ||X||$ as $n \rightarrow \infty$ and weak convergence $\int(X_n) \rightarrow \int(X)$ as $n \rightarrow \infty$ where X_n represents physical valuables such as displacement, temperature, and velocity. X represents the exact solution and $\int$ indicates the potential energy. It is recommended that all dental FE models should test meshing convergence prior to analyses.

4.2 Validity of the models

The validity of the dental FEA has been a concern for decades. Two review articles (Korioth and Versluis, 1997; Geng et al., 2001) in dentistry provided thorough discussions about effects of geometry, element type and size, material properties, and boundary conditions on the accuracy of solutions. In general these discussions echoed an earlier review by Huiskes and Cao (1983). The severity of these effects has decreased as the technologies and knowledge evolved in the field. In the present CAD-FEA era, the consideration of FEA accuracy in relation to loading, boundary (constraint) conditions, and validity of material properties are described as follows:

4.2.1 Loading

The static loading such as bite forces is usually applied as point forces to study prosthetic designs and dental restorations. The bite force, however, presents huge variations (both magnitude and direction) based on previous experimental measures (Proffit et al., 1983; Proffit and Field 1983). Fortunately, FEA allows for easy changes in force magnitudes and directions to approximate experimental data, which can serve as a reasonable parametric study to assess different loading effects. On the other hand, loading exerted by devices such as orthodontic wires is unknown or never measured experimentally, and should be simulated with caution (see the section 5.2)

4.2.2 Boundary Condition (BC)

The boundary condition is a constraint applied to the model, from which potential energy and solutions are derived. False solutions can be associated at the areas next to the constraints. As a result, most dental models set constraints far away from the areas of interest. Based on the Saint-Venant's principle, the effects of constraints at sufficiently large distances become negligible. However, some modeling applies specific constraints to study particular physical phenomenon. For example, the homogenization theory was derived to resolve microstructural effects in composite by applying periodic constraints (Ko et al., 1996). It was reported that using homogenization theory to estimate bone-implant interfacial stresses by accounting for microstructural effects might introduce up to 20% error (Ko 1994).

4.2.3 Material properties

Mechanical properties of biological tissues remain a major concern for the FE approach because of the viscoelastic nature of biological tissues that prevents full characterization of its time-dependent behaviors. Little technology is available to measure oral tissue properties. Most FE studies in dentistry use the linear elastic assumption. Data based on density from CT images can be used to assign heterogeneous properties. Few researches attempting to predict non-linear behaviors using bilinear elastic constants aroused risks for a biased result (Cattaneo et al., 2009). Laboratory tests excluding tissues (e.g., PDL) were also found to result in less accurate data than computer predictions (Chi et al., 2011). Caution must be used when laboratory data is applied to validate the model. To our knowledge, the most valuable data for validation resides on clinical assessments such as measuring tooth movement (Yoshida et al., 1998; Brosh et al., 2002).

4.3 Solution/principle

The weak form of the equilibrium equation for classic mechanics is given below:

$\int_{\Omega^\varepsilon} c_{ijkl}\varepsilon_{ij}(v)\ \varepsilon_{kl}(u)\ d\Omega^\varepsilon = \int_\Gamma t_i v_i d\Gamma$, where Ω^ε represents the total domain of the object, and

t_i represents tractions. ε is obtained by applying the small strain-displacement relationship

$\varepsilon_{ij}(u) = \frac{1}{2}(\frac{\partial u_i}{\partial x_j} + \frac{\partial u_j}{\partial x_i})$. Stresses will be obtained by the constitutive law $\sigma_{ij} = E_{ijkl}\varepsilon_{kl}$. Using

the variational formulation and mesh descritization, this equilibrium equation can be assembled by the individual element $\int_{\Omega e}[B]^t[D][B]d\Omega e$. plus the boundary integral where B is the shape function and D is element stiffness matrix. The element stiffness matrix represents material property of either a linear or non-linear function. As mentioned above, mechanical properties of oral tissues are poorly characterized. The most controversial oral tissue is the PDL due to its importance in supporting teeth and regulating alveolar bone remodeling. To date, studies conducted to characterize non-linear behaviors of the PDL are not yet conclusive. One approximation of PDL properties assumes zero stiffness under low compression resulting in very low stress under compression (Cattaneo et al., 2009). Interpretation of such non-linear models must be approached with cautious. Consequently, linear elastic constants are frequently used for dental simulations to investigate initial responses under static loading.

In addition to the commonly used point forces, the tractions (t_i) in dental simulations should consider preconditions (e.g., residual stress, polymerization shrinkage and unloading of orthodontic archwire). Previously, investigation of composite shrinkage yielded valuable contributions to restorative dentistry (Magne et al., 1999). In the following section, we will demonstrate two applications using submodels from a full dentition CAD model: one with static point loading and the other with deactivated orthodontic archwire bending.

5. Examples of dental FEA

As described in Section 3, a master CAD model with full dentition was developed. The model separates detailed anatomic structures such as PDL, pulp, dentin, enamel, lamina dura, cortical bone, and trabecular bone. This state-of-the-art model contains high order NURB surfaces that allow for fine meshing, with excellent connectivity so the model can be conformally meshed with concurrent nodes at all interfaces. Many submodels can be isolated from this master model to study specified biomechanical questions. Two examples presented here are the first series of applications: orthodontic miniscrews and orthodontic archwires for tooth movement.

5.1 Orthodontic miniscrews

5.1.1 Introduction

The placement of miniscrews has become common in orthodontic treatment to enhance tooth movement and to prevent unwanted anchorage loss. Unfortunately, the FE biomechanical miniscrew models reported to date have been oversimplified or show incomplete reflections of normal human anatomy. The purpose of this study was to construct a more anatomically accurate FE model to evaluate miniscrew biomechanics. Variations of miniscrew insertion angulations and implant materials were analyzed.

5.1.2 Methods

A posterior segment was sectioned from the full maxillary model. Borders of the model were established as follows: the mesial boundary was at the interproximal region between the maxillary right canine and first premolar; the distal boundary used the distal aspect of the maxillary tuberosity; the inferior boundary was the coronal anatomy of all teeth ; and the superior boundary was all maxillary structures (including sinus and zygoma) up to 15mm superior to tooth apices (Figure 6A).

An orthodontic miniscrew (TOMAS®, 8mm long, 1.6mm diameter) was created using Solidworks CAD software. The miniscrew outline was created using the Solidworks sketch function and revolved into three dimensions. The helical sweep function was used to create a continuous, spiral thread. Subtraction cuts were used to create the appropriate head configuration after hexagon ring placement. The miniscrew was inserted into the maxillary model from the buccal surface between the second premolar and first molar using Solidworks. The miniscrew was inserted sequentially at angles of 90°, 60° and 45° vertically relative to the surface of the cortical bone (Figure 6B), and was placed so that the miniscrew neck/thread interface was coincident with the external contour of the cortical bone. For each angulation, the point of intersection between the cortical bone surface and the central axis of the miniscrew was maintained constant to ensure consistency between models. Boolean operations were performed and a completed model assembly was created at each angulation.

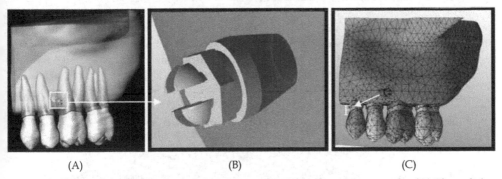

(A)   (B)   (C)

Fig. 6. The FE model of the orthodontic miniscrew used in the present study. (A) The solid model of four maxillary teeth plus the miniscrew was created using SolidWorks. (B) Close look of the miniscrew inserted to the bone. (C) FE mesh was generated by Ansys Workbench 10.0. F indicates the force (1.47 N = 150gm) applied to the miniscrew.

The IGES format file of each finished 3D model was exported to ANSYS 10.0 Workbench (Swanson Analysis Inc., Huston, PA, USA), and FE models with 10-node tetrahedral h-elements were generated for each assembly. The final FE mesh generated for each model contained approximately 91,500 elements, which was sufficient to obtain solution convergence. Following FE mesh generation, the model was fixed at the palatal, mesial, and superior boundaries. A 150 grams loading force to the mesial was then applied to the miniscrew to simulate distalization of anterior teeth (Figure 6C). All materials were linear and isotropic (Table 1), and the miniscrew/bone interface was assumed to be rigidly bonded. Three material properties (stainless steel, titanium, and composite) were used for the miniscrew. Each model was solved under the small displacement assumption. Two-way ANOVA was used to compare effects of angulation and material.

	Cortical	Trabecular	PDL	Dentin	Enamel	Pulp	Stainless Steel (SS)	Titanium (Ti)	Composite
Young's Modulus (MPa)	13,700	1370	175	18,000	77,900	175	190,000	113,000	20,000
Poisson's Ratio	0.3	0.3	0.4	0.3	0.3	0.4	0.3	0.3	0.3

Table 1. Computer model component material properties (O'Brien, 1997)

5.1.3 Results

Angle effect

Stress patterns in both cortical bone and the miniscrew from each simulation were concentrated in the second premolar/first molar area immediately around the implant/bone interface (Figure 7). Peak stress values for each model simulation are listed in Table 2. Peak maximum principal stress (MaxPS) within the miniscrew was greatest when angle placement was 45°. Peak MaxPS was lowest at the 60° placement angle. Peak MaxPS in cortical bone was greatest at 45° angulation, except for the stainless steel implant. In each angulation, the location of greatest maximum principle stress was located at the distal aspect of the miniscrew/cortical bone interface. Similarly, peak minimum principal stress (MinPS) was lowest on the miniscrew at 60° and greatest at 45°. Figure 8 shows mean stress plots for all angulations and materials. Angulation difference was statistically significant for miniscrews at 45° compared to 60° and 90° for all stress types analyzed (MaxPS p=0.01, MinPS p=0.01, von Mises stress (vonMS) p=0.01). There is no significant difference in cortical bone stress at any angulation.

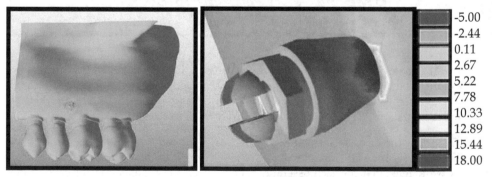

Fig. 7. Stress distributions of the orthodontic miniscrew showed that stresses concentrated in the neck region of the miniscrew at the interface between bone and the screw.

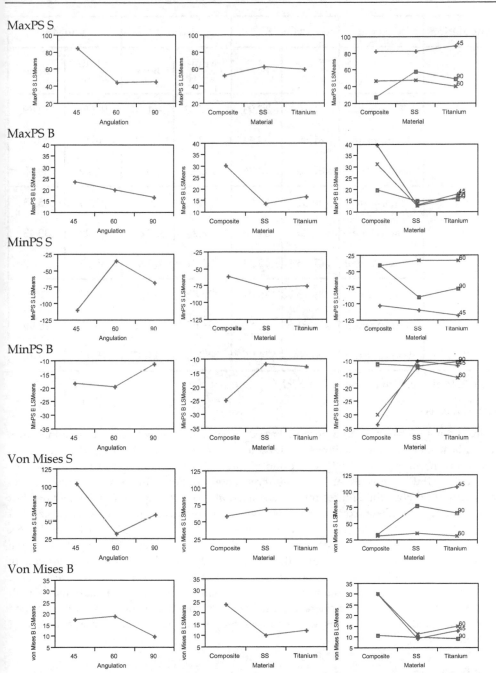

Fig. 8. Plots of mean stress (MPa) averaged over angulations (left column), materials (middle), and cross-action between angulation and materials (right). Symbols - MaxPS: maximum principal stress; MinPS: minimum principal stress; S: screw; B: Bone.

Angulation	Material	MaxPS		MinPS		vonMS	
		Mini-screw	Bone	Mini-screw	Bone	Mini-screw	Bone
45	Titanium	89.3	17.93	-117.63	-11.68	107.54	12.89
	Composite	82.99	39.94	-102.34	-33.26	110.09	30.33
	Stainless Steel	82.75	13.26	-109.24	-10.05	94.49	9.47
60	Titanium	40.31	16.55	-32.18	-16.23	31.56	15.13
	Composite	46.79	31.26	-40.29	-29.83	32.02	30.43
	Stainless Steel	47.53	12.74	-32.24	-12.62	35.35	11.42
90	Titanium	49.73	16.01	-75.82	-10.29	67.24	9.43
	Composite	27.91	19.81	-39.67	-11.25	33.55	10.99
	Stainless Steel	58.38	14.96	-88.9	-11.93	78.45	10

Table 2. Peak mean stress for each model.

Material effect

There is a noticeable (p=0.05) difference between material types with composite miniscrews having a higher average MaxPS and MinPS in cortical bone than Ti or SS. Peak MinPS is lowest on the miniscrew at 60° for Ti and SS miniscrews, and similar at 60° and 90° for composite. MinPS is greatest at 45° for all three materials. Peak MinPS is approximately the same in cortical bone for all three miniscrew materials at 90° (range -10.29 to -11.93MPa) but at 45° and 60° MinPS in cortical bone is higher for composite than Ti or SS (-33.26 & -29.83MPa respectively for composite vs. -11.68/-10.05MPa & -16.23/-12.62MPa for Ti/SS). When comparing the MinPS pattern generated for 45°, 60°, and 90° angulations, the composite miniscrew does not mimic the Ti or SS pattern. Rather, MinMS is substantially lower in both the miniscrew and bone at 90° than Ti or SS. Peak vonMS was lowest on the miniscrew at 60° for all three miniscrew materials relative to the other angulations. As with MinPS, the vonMS for the composite miniscrew differs from the Ti and SS pattern generated for 45°, 60°, and 90° angulations and is substantially lower in both miniscrew and bone at 90° than Ti or SS.

5.1.4 Discussion

One of the primary areas of interest in the present study related to the construction of a human model that is both realistic and of sufficient detail to clinically valuable. Comparing these results to other orthodontic studies using FEA is challenging due to several differences between models. Many of the studies available in the literature do not model human anatomy (Gracco et al., 2009; Motoyoshi et al., 2005) and are not 3-dimensional (Brettin et al., 2008), or require additional resolution (Jones et al., 2001). Cattaneo et al (2009) produced a similar high-quality model of two teeth and surrounding bone for evaluation of orthodontic tooth movement and resulting periodontal stresses. Both linear and non-linear PDL mechanical properties were simulated. In a different study, Motoyoshi found peak bone vonMS in their model between 4-33MPa, similar to the levels in the current study (9.43-15.13MPa) However, Motoyoshi used a 2N (~203gm) force applied obliquely at 45°, different in both magnitude and orientation from that in this study.

From the results of the present study, an angulation of 60° is more favorable than either 45° or 90° for all three stress types generated relative to the stress on the miniscrew. Conversely, all three stress types have levels at 60° which are comparable to 45° and 90° in bone, so varied angulation within the range evaluated in the present study may not have a marked effect on the bone. However, miniscrews at 60° do have slightly higher MinPS (compressive) values than either 45° or 90°, which could have an effect on the rate or extent of biological activity and remodeling.

A third area of focus in the present study was the effect of using different miniscrew materials by comparing stress levels generated by popular titanium miniscrews with rarely- or never used stainless steel and composite miniscrews. Although no studies were found that compare Ti and/or SS miniscrews to composite miniscrews, one published study compared some of the mechanical properties of Ti and SS miniscrews (Carano et al., 2005). Carano et al reported that Ti and SS miniscrews could both safely be used as skeletal anchorage, and that Ti and SS miniscrew bending is >0.02mm at 1.471N (150gm) equivalent to the load applied in the present study. There was deformation of >0.01mm noted in the present study. However, the geometry in the study by Carano et al was otherwise not comparable to that in the present study.

There are no studies available which compare Ti and SS stresses or stress pattern generation. Therefore, to compare stresses generated from the use of composite to Ti and SS in the present study, the modulus of the miniscrew in each Ti model was varied to reflect that of SS and composite, with a subsequent test at each angulation. The fact that the general stress pattern for composite is dissimilar to Ti and SS when the miniscrew angulation is changed may be of clinical importance (Pollei 2009). Because composite has a much lower modulus than Ti or SS, it may be that stress shielding does not happen as much in composite miniscrews, and therefore more stress is transferred to the adjacent cortical bone in both compression and tension scenarios. As a result of increased compressive and tensile forces, especially in the 45° model, biological activity related to remodeling may be increased relative to other models with lower stress levels, and therefore have a more significant impact on long-term miniscrew stability and success. Another potential undesirable effect of using composite miniscrews in place of either Ti or SS is the increased deformation and distortion that is inherent due to the decreased modulus of composite relative to titanium or steel. Mechanical or design improvements need to be made to allow for composite miniscrews to be a viable alternative in clinical practice.

5.2 Orthodontic archwire

5.2.1 Introduction

Currently, biomechanical analysis of orthodontic force systems is typically limited to simple 2D force diagrams with only 2 or 3 teeth. Beyond this point, the system often becomes indeterminate. Recent laboratory developments (Badawi et al., 2009) allow investigation of the forces and moments generated with continuous archwires. However, this laboratory technique has 3 significant limitations: interbracket distance is roughly doubled, the PDL is ignored, and only a single resultant force and moment is calculated for each tooth.

The complete dentition CAD model assembled in our lab includes the PDL for each tooth and calculates the resultant stress-strain at any point in the model, improving on the limitations of the laboratory technique. However, the accuracy of this technique depends on the 3 factors mentioned previously: material properties, boundary conditions, and loading conditions. The considerations for material properties and boundary conditions are similar to the other models discussed above, but loading conditions with orthodontic archwires deserves closer attention.

Previously studies in orthodontics have typically used point-forces to load teeth, but fixed appliances rarely generate pure point-forces. In order to properly model a wide range of orthodontic movement, a new technique was developed which stores residual stresses during the insertion (loading) stage of the archwire, followed by a deactivation stage where the dentition is loaded equivalently to intraoral archwire loading. This method provides a new way to investigate orthodontic biomechanics (Canales 2011).

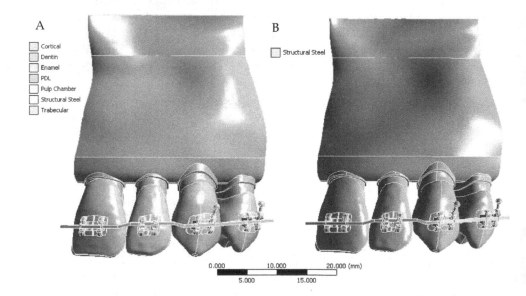

Fig. 9. Four tooth model used for FEA of continuous orthodontic archwires. A. Model with accurate material properties assigned to each body. B. Model with all bodies assumed to be stainless steel.

5.2.2 Methods

To assess the effect of proper PDL modeling, two separate models were generated from our master model. In both models, the upper right central incisor, lateral incisor, canine, and first premolar were segmented from the full model. Brackets (0.022" slot) were ideally placed on the facial surface of each tooth and an archwire was created that had a 0.5 mm intrusive step on the lateral incisor. Passive stainless steel ligatures were placed on each bracket keep the archwire seated. For one model, all bodies were assumed to be stainless steel (mimicking the laboratory setup by Badawi and Major 2009), while suitable material properties were assigned to each body in the second model (Figure 9). Each model was meshed using tetrahedral elements, except for swept hexahedral elements in the archwire, and consisted of 238758 nodes and 147747 elements. The ends of the archwire and the sectioned faces of bone were rigidly fixed. The contacts between the wire and the brackets were assumed frictionless.

5.2.3 Results and discussion

The static equilibrium equations were solved under large displacement assumptions. The final displacement in each model is shown in Figure 10, showing dramatically increased displacement in the PDL model. Notice that in both models, the lateral experienced unpredicted distal displacement due to the interaction of the archwire. This highlights the importance of accurate loading conditions in FEA. In addition to increased overall displacement, the center of rotation of the lateral incisor also moves apically and facially in the stainless steel model (Figure 11). Therefore, any results generated without properly modeling the PDL should be taken with caution – this includes laboratory testing of continuous archwire mechanics (Badawi and Major 2009).

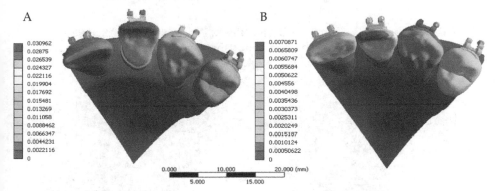

Fig. 10. Displacement viewed from the occlusal in the A) PDL model and B) stainless steel model after placement of a wire with 0.5 mm intrusive step bend. Note the different color scales and that both models have 7.1 times the actually displacement visually displayed.

The stress and strain distributions in the PDL also show variations in both magnitude and distribution (Figures 12 and 13). Unsurprisingly, the PDL shows increased strain when accurately modeled as a less stiff material than stainless steel. In this PDL model, the strain is also concentrated to the PDL, as opposed to more broadly distributed in the stainless steel model. Due to the increased rigidity in the stainless steel model, higher stresses were generated by the same displacement in the archwire.

The results clearly show the importance of the PDL in modeling orthodontic loading. We aim to further improve this model, adding additional teeth, active ligatures, and friction.

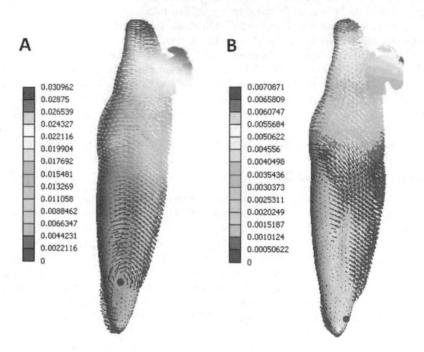

Fig. 11. Displacement viewed from the distal in the A) PDL model and B) stainless steel model after placement of a wire with 0.5 mm intrusive step bend. Note the center of rotation (red dot) in the stainless steel model moves apically and facially.

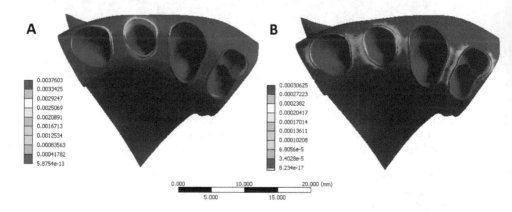

Fig. 12. Equivalent (von-Mises) elastic strain for the A) PDL model and B) stainless steel model after placement of a wire with 0.5 mm intrusive step bend.

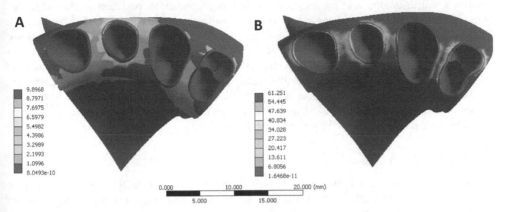

Fig. 13. Equivalent (von-Mises) stress in MPa for the A) PDL model and B) stainless steel model after placement of a wire with 0.5 mm intrusive step bend.

6. Future of dental FEA

Although FEA techniques have greatly improved over the past few decades, further developments remain. More robust solid models, like the one demonstrated in Figure 14, with increased capability to manipulate CAD objects would allow increased research in this area. The ability to fix minor problematic geometry and easily create models with minor variations would greatly reduce the time required to model different biomechanical

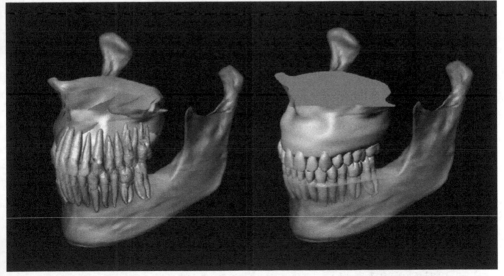

Fig. 14. Full Jaw Orthodontic Dentition (FJORD, UNC Copyright) Model: Isometric view of the solid models of mandible and maxillary arches and dentitions that was reconstructed and combined in SolidWorks software in our lab. The left image renders transparency of the gingiva and bone to reveal internal structures.

situations. Additionally, adding frictional boundaries conditions between teeth and active ligations for orthodontic appliances will continue to increase the accuracy of these models. Three dimensional dynamic simulations for assessing tooth injury, similar to those demonstrated in 2D studies (Huang et al., 2006; Miura and Maeda, 2008), should be re-evaluated. While techniques will continually be optimized to improve numerical approximations, this does not negate the value of finite element techniques in dentistry. These techniques use proven engineering principles to model aspects of dentistry that are unable to be efficiently investigated using clinical techniques, and will continue to provide valuable clinical insights regarding dental biomechanics.

7. Acknowledgement

This study was supported, in part, by AAOF, NIH/NIDCR K08DE018695, NC Biotech Center, and UNC Research Council. We also like to thank Geomagic for providing software and technique supports to our studies.

8. References

Anusavice KI, DeHoff PH, Fairhurst CW. Comparative evaluation of ceramic-metal bond tests using finite element stress analysis. 1980; J Dent Res 59:608-613.

Badawi HM, Toogood RW, Carey JP, Heo G, Major PW. Three-dimensional orthodontic force measurements. Am J Orthod Dentofacial Orthop. 2009;136(4):518-28.

Brettin BT, Grosland NM, Qian F, Southard KA, Stuntz TD, Morgan TA, et al. Bicortical vs monocortical orthodontic skeletal anchorage. Am J Orthod Dentofacial Orthop. 2008 Nov;134(5):625-35.

Bright JA, Rayfield EJ. The response of cranial biomechanical finite element models to variations in mesh density. Anat Rec (Hoboken). 2011 Apr;294(4):610-20.

Brosh T, Machol IH, Vardimon AD. Deformation/recovery cycle of the periodontal ligament in human teeth with single or dual contact points. Archives of Oral Biology. 2002; 47:85-92.

Byrne DP, Lacroix D, Prendergast PJ. Simulation of fracture healing in the tibia: mechanoregulation of cell activity using a lattice modeling approach. J Orthop Res. 2011; 29:1496-1503.

Canales CH. A novel biomechanical model assessing orthodontic, continuous archwire activation in incognito lingual braces. Master Thesis. University Of North Carolina-Chapel Hill. 2011.

Carano A, Lonardo P, Velo S, Incorvati C. Mechanical properties of three different commercially available miniscrews for skeletal anchorage. Prog Orthod. 2005;6(1):82-97.

Cattaneo PM, Dalstra M, Melsen B. Strains in periodontal ligament and alveolar bone associated with orthodontic tooth movement analyzed by finite element. Orthod Craniofac Res. 2009 May;12(2):120-8.

Chi L, Cheng M, Hershey HG, Nguyen T, Ko CC. Biomechanical Re-evaluation of Orthodontic Asymmetric Headgear, In press. Angle Orthodontist. 2011. DOI: 10.2319/052911-357.1

Farah JW and Craig RG. Distribution of stresses in porcelain-fused-to-metal and porcelain jacket crowns. J Dent Res. 1975; 54:225-261.

Farah JW, Craig RG, Sikarskie DL. Photoelastic and finite element stress analysis of a restored axisymmetric first molar. J Biomech. 1973; 6(5):511-20.

Field C, Ichim I, Swain MV, Chan E, Darendeliler MA, Li W, et al. Mechanical responses to orthodontic loading: a 3-dimensional finite element multi-tooth model. Am J Orthod Dentofacial Orthop. 2009;135:174-81.

Geng J-P, Tan KBC, Liu G-R. Application of finite element analysis in implant dentistry:A review of the literature. J Prosthet Dent 2001; 85:585-98.

Gracco A, Cirignaco A, Cozzani M, Boccaccio A, Pappalettere C, Vitale G. Numerical/experimental analysis of the stress field around miniscrews for orthodontic anchorage. Eur J Orthod. 2009 Feb;31(1):12-20.

Hollister SJ, Ko CC, Kohn DH. Bone density around screw thread dental implants predicted using topology optimization, Bioengineering Conference ASME BED 1993; 24:339-342.

Huang H-L, Chang C-H, Hsu J-T, Fallgatter AM, Ko CC. Comparisons of Implant Body Designs and Thread Designs of Dental Implants: A Three-Dimensional Finite Element Analysis. The Int. J Oral & Maxillofac Implants. 2007; 22: 551-562.

Huang HM, Tsai CY, Lee HF et al. Damping effects on the response of maxillary incisor subjected to a traumatic impact force: A nonlinear finite element analysis. J Dent 2006;34:261-8.

Hughes TJR. The finite element method: linear static and dynamic finite element analysis. Prentice-Hall, Inc. 1987.

Huiskes R, Chao EYS. A survey of finite element analysis in orthopedic biomechanics: The first decade. J. Biomechanics. 1983; 6:385-409.

Jones ML, Hickman J, Middleton J, Knox J, Volp C. A validated finite element method study of orthodontic tooth movement in the human subject. J Orthodontics. 2001; 28:29-38.

Kim T, Suh J, Kim N, Lee M. Optimum conditions for parallel translation of maxillary anterior teeth under retraction force determined with the finite element method. Am J Orthod Dentofacial Orthop. 2010 May;137(5):639-47.

Ko CC, Chu CS, Chung KH, Lee MC. Effects of posts on dentin stress distribution in pulpless teeth. J Prosthet Dent 1992; 68:42 1-427.

Ko CC, Kohn DH, and Hollister SJ: Effective anisotropic elastic constants of bimaterial interphases: comparison between experimental and analytical techniques, J. Mater. Science: Materials in Medicine. 1996; 7: 109-117.

Ko CC. Mechanical characteristics of implant/tissue interphases. PhD Thesis. University of Michigan, Ann Arbor. 1994.

Ko CC. Stress analysis of pulpless tooth: effects of casting post on dentin stress distribution. Master Thesis. National Yang-Ming Medical University. 1989.

Korioth TWP, Versluis A. Modeling the mechanical behavior of the jaws and their related structures by finite element analysis. Crit Rev Oral Biol Med. 1997; 8(l):90-104.

Korioth TWP. Finite element modelling of human mandibular biomechanics (PhD thesis). Vancouver, BC, Canada: University of British Columbia. 1992.

Lanczos C. The variational principles of mechanics. 2nd ed. University of Toronto Press. 1962.

Liang W, Rong Q, Lin J, Xu B. Torque control of the maxillary incisors in lingual and labial orthodontics: a 3-dimensional finite element analysis. Am J Orthod Dentofacial Orthop. 2009 Mar;135(3):316-22

Lin, C-L, Chang C-H, Ko CC. Multifactorial analysis of an MOD restored human premolar using auto-mesh finite element approach. *J. Oral Rehabilitation.* 2001; 28(6): 576-85.

Magne P, Versluis A, Douglas WH. Effect of luting composite shrinkage and thermal loads on the stress distribution in porcelain laminate veneers. J Prosthet Dent. 1999 Mar; 81(3):335-44.

Miura J, Maeda Y. Biomechanical model of incisor avulsion: a preliminary report. Dent Traumatol 2008;24:454-57

Motoyoshi M, Yano S, Tsuruoka T, Shimizu N. Biomechanical effect of abutment on stability of orthodontic mini-implant. A finite element analysis. Clin Oral Implants Res. 2005 Aug;16(4):480-5.

O'Brien, WJ. Dental Materials and Their Selection 2nd edition. Quintessence Publishing. 1997.

Peter MCRB, Poort HW, Farah JW, Craig RG. Stress analysis of tooth restored with a post and core. 1983; 62(6):760-763.

Pollei J. Finite element analysis of miniscrew placement in maxillary bone with varied angulation and material type. Master Thesis. University of North Carolina- Chapel Hill. 2009.

Proffit WR, Fields HW, Nixon WL. Occlusal forces in normal and long face adults. J Dent Res 1983; 62:566-571.

Proffit WR, Fields HW. Occlusal forces in normal and long face children. J Dent Res 1983; 62:571-574.

Reinhardt RA, Krejci RF, Pao YC, Stannarrd JG. Dentin stress in post-reconstructed teeth with diminished bone support. J Dent Res. 1983; 62(9):1002-1008.

Rieger MR, Mayberry M, Brose MO. Finite element analysis of six endosseouos implants. J Prosthetic Dentistry. 1990; 63:671-676.

Schmidt H, Alber T, Wehner T, Blakytny R, Wilke HJ. Discretization error when using finite element models: analysis and evaluation of an underestimated problem. J Biomech. 2009; 42(12):1926-34.

Segerlind LJ. Applied finite element analysis. 2nd ed. John Wiley & Sons, Inc. 1984.

Tanne K, Sakuda M, Burstone CJ. Three-dimensional finite element analysis for stress in the periodontal tissue by orthodontic forces. Am J Orthod Dentofacial Orthop. 1987; 92(6):499-505.

Thresher RW and Saito GE. The stress analysis of human teeth. J Biomech. 1973; 6:443-449.

Wang T, Ko CC, Cao Y, DeLong R, Huang CC, Douglas WH: Optical simulation of carious tooth by Monte Carlo method. Proceedings of the Bioengineering Conference, ASME, BED 1999; 42: 593-594.

Yoshida N, Jost-Brinkmann PG, Miethke RR, Konig M, Yamada Y. An experimental evaluation of effects and side effects of asymmetric face-bows in the light of in vivo measurements of initial tooth movements. Am J Orthod Dentofacial Orthop. 1998; 113:558-566.

Yushkevich PA, Piven J, Hazlett HC, Smith RG, Ho S, Gee JC, and Gerig G. User-guided 3D active contour segmentation of anatomical structures: Significantly improved efficiency and reliability. Neuroimage. 2006; 31(3):1116-28.

Zienkiewicz OC, Kelly DW. Finite elements-A unified problem-solving and information transfer method. In: Finite elements in biomechanics. Gallagher RH, Simon. 1982.

Part 2

Cardiovascular and Skeletal Systems

7

Finite Element Modeling and Simulation of Healthy and Degenerated Human Lumbar Spine

Márta Kurutz[1] and László Oroszváry[2]
[1]Budapest University of Technology and Economics,
[2]Knorr-Bremse Hungaria Ltd, Budapest,
Hungary

1. Introduction

Numerical analyses are able to simulate processes in their progress that are impossible to measure experimentally, like aging and spinal degeneration processes. 3D finite element (FE) simulations of age-related and sudden degeneration processes of compression-loaded human lumbar spinal segments and their weightbath hydrotraction treatment are presented here. The goal is to determine the effect of the main mechanical degeneration parameters on the deformation and stress state of the lumbar functional spinal units (FSU) during long- and short-term degeneration processes. Moreover, numerical analysis of the effect of the special underwater traction method, the so-called weightbath hydrotraction therapy (WHT) for treating degenerated lumbar segments is also presented in this chapter.

About 60-85 % of the human population is afflicted by lumbar disc diseases, by low back pain (LBP) problems, most of them the young adults of 40-45 years, due to the degeneration of the lumbar segments. Degeneration means an injurious change in the structure and function of FSU, caused by normal aging or by sudden accidental, often unexpected effects yielding mechanical overloading. Degeneration starts generally in the intervertebral discs: in the central kernel of it, in the nucleus pulposus. Aging degeneration of the disc is manifested in loss of hydration, a drying and stiffening procedure in the texture of nucleus, and a hardening of annulus as well, accompanied by the appearance of buckling, lesions, tears, fiber break in the annulus or disruption in the endplates, collapse of osteoporotic vertebral cancellous bone (Adams et al., 2002). Sudden accidental degeneration of overload yields the instant loss of hydrostatic compression in nucleus, accompanied or due to some other failures mentioned above. Several recent studies concluded that light degeneration of young discs leaded to instability of lumbar spine and the stability restored with further aging (Adams et al., 2002; Rohlmann et al., 2006), however, sudden accidental degeneration may be also dangerous in young age, thus, the present study aims to understand the biomechanical function of these degeneration processes.

The first goal of this study was to obtain numerical conclusions for the mechanical effects of age-related and sudden degeneration processes of the lumbar spine under compression. The question is answered also why the young adults of 40-45 years have the most vulnerable lumbar discs, and why increases the stability of lumbar segments with further aging.

On the other hand, for the degenerated lumbar spine, for the LBP problems, traction might be an effective treatment. Traction treatments have been well-known for a long time, however, as often happens, instead of stress relaxation, the compression increases in the discs due to the muscle activity (Andersson et al., 1983; White and Panjabi, 1990; Ramos and Martin, 1994). These observations verify the importance of the special Hungarian traction method, the so-called weightbath hydrotraction therapy (WHT), introduced by Moll (1956, 1963), indications and contraindications described by Bene (1988). In WHT patients are suspended in water on a cervical collar for a period of 20 minutes, loaded by extra lead weights on ankles, when the patients are practically in sleeping position in the lukewarm water with completely relaxed muscles. WHT consists of instant elastic and viscous creeping phases. In this study the elastic phase will be analyzed. As far as the authors know, finite element analyses of lumbar FSUs in pure axial tension without the effect of muscles cannot be found in the literature. As White and Panjabi (1990) and Bader and Bouten (2000) established, tensile stresses and deformations are analyzed associated with flexion and extension only, without mentioning axial tension as a dominating loading effect.

Thus, in this study, the latter case will be analyzed. By means of this study of WHT, the beneficial clinical impacts of the treatment are supported by numerical mechanical evidences.

2. Short structural anatomy of the lumbar spine and spine segments

The lumbar spine is the section of spinal column between the thorax and the sacrum. It consists of five *vertebrae* named L1 to L5 with their *posterior elements* and *articular facet joints*, of *intervertebral discs, ligaments* and the surrounding *muscles*. The clinical anatomy of the lumbar spine can be studied in the books of Bogduk and Twomey (1987), White and Panjabi (1990), Dvir (2000), Benzel (2001) and Adams et al. (2002).

The lumbar *vertebrae* are quasi cylindrical with a lateral width of 40-50 mm and sagittal depth of 30-35 mm. The height of a lumbar vertebra is about 25-30 mm. The lumbar vertebrae are thicker anteriorly than posteriorly resulting in anteriorly convex curvature of the spine known as the lumbar lordosis. The *vertebral body* consists of an outer shell of high strength *cortical bone* and of the internal *cancellous bone* as a network of vertical and horizontal bone struts called trabeculae. The superior and inferior surface of the vertebral body is covered by the *bony endplates* of thin cortical bone perforated by many small holes as the passage of metabolites from bone to the discs. Towards the upper end of the posterior surface of the vertebral body the pedicles support the posterior elements, the lamina the neural arch, the vertebral foramen, the spinous process, the transverse processes, the superior and inferior articular processes, the synovial joints called articular facet joints.

The *intervertebral discs* separate the adjacent vertebrae. They are quasi cylindrical with a lateral width of 40-45 mm and sagittal depth of 35-40 mm. The height of the lumbar disc is about 10 mm. The disc consists of three components: the nucleus pulposus, the annulus fibrosus and the superior and inferior endplates. The *nucleus pulposus* is a hydrated gel, a semi-fluid mass, an incompressible sphere that exerts pressure in all directions. The *annulus fibrosus* consists of 15-25 concentric laminated layers of collagen lamellae tightly connected to each other in a circumferential form around the periphery of the disc. Each lamella consists of ground substance and collagen fibers. Within each lamella the collagen fibers are arranged in parallel, running at an average direction of 30° to the discs horizontal plane. In adjacent lamellae they run in opposite directions and are therefore oriented at 120° to each other. The *cartilaginous endplates* separate the nucleus and annulus from the vertebral bodies, they cover almost the entire surface of the adjacent vertebral bony endplates. The plates have a mean thickness of 0.6 mm.

Seven types of *ligaments* are distinguished in the lumbar spine. The *anterior longitudinal ligament* covers the anterior surfaces of the vertebral bodies and discs, attached strongly to the vertebral bone and weakly to the discs. The *posterior longitudinal ligament* covers the posterior aspects of the vertebral bodies and discs, attached strongly to the discs and weakly to the bone. The *ligamentum flavum*, the most elastic ligament of the spine connects the lower and upper ends of the internal surfaces of the adjacent laminae. The *intertransverse ligaments* connect the transverse processes by thin sheets of collagen fibers. The *interspinous ligaments* connect the opposing edges of spinous processes by collagen fibers, while the *supraspinous ligaments* connect the peaks of adjacent spinous processes by tendinous fibers. The *capsular ligaments* connect the circumferences of the joining articular facet joints, being perpendicular to the surface of the joints.

The *muscles* of the lumbar spine can be distinguished by their location around the spine. The *postvertebral muscles* can be divided to deep, intermediate and superficial categories. The *prevertebral muscles* are the abdominal muscles. The postvertebral deep muscles consist of short muscles that connect the adjacent spinous and transverse processes and laminae. The intermediate muscles are more diffused, arising from the transverse processes of each vertebra and attaching to the spinous process of the vertebra above. The superficial postvertebral muscles collectively are called the *erector spinae*. There are four *abdominal muscles*, three of them encircle the abdominal region, and the fourth is located anteriorly at the midline.

A *motion segment* or *functional spinal unit* (FSU) is the smallest part of the spine that represents all the main biomechanical features and characteristics of the whole spine. Thus the entire spinal column can be considered as a series of connecting motion segments. The motion segment is a three dimensional structure of six degree of statical/kinematical freedom, consisting of the two adjacent vertebrae with its posterior elements and facet joints, and the intervertebral disc between them, moreover the seven surrounding ligaments, without muscles.

3. Biomechanics of the lumbar spine and spine segments

The spinal column is the *main load bearing structure* of the human musculoskeletal system. It has three fundamental biomechanical functions: *to guarantee the load transfer* along the spinal column without instability; *to allow sufficient physiologic mobility* and flexibility; and *to protect the delicate spinal cord* from damaging forces and motions.

The lumbar spine is the most critical part of the spine in aspect of instability and injury since this part is the maximally loaded part of spine and this part has the maximal mobility at the same time, moreover, this part has the minimal stiffening support from the surrounding organs.

3.1 Loads acting on the lumbar spine

The spinal loads based on biomechanical studies are summarized by Dolan and Adams (2001). The loads acting on the spine can be physiologic and traumatic loads. The *physiologic loads* due to the common, normal activity of the spine have further classes: short-term loads (in flexion, extension), long-term loads (in sitting, standing), repeated or cyclic loads (in gait, walk), dynamic loads (in running, jumping). The *traumatic loads* generally occur suddenly, unexpectedly with great amplitude (impact, whiplash).

Each part of the body is subjected to *gravity load*, proportionally to its mass. The compressive gravity load increases towards the support of the body. This load can be multiplied in acceleration, during a fall, or other effects with acceleration or deceleration.

Muscle loads depend on the muscle activity. The muscles are active tissues, they can contract, and their ability of contraction is governed by the nervous system. The back and abdomen muscles stabilize the spine in upright standing; moreover, they prevent the spine from extreme movements. At the same time, the muscle contraction causes high compressive forces to the lumbar spine. Nachemson (1981) and Sato et al. (1999) classified the muscle forces in different body postures.

The *intra-abdominal pressure* decreases generally the spinal compression due to the abdominal muscle activity. Wide abdominal belts help to reduce the spinal compressive forces.

Ergonomic loads afflict mostly the lumbar spine. By lifting and holding weights the lumbar spine is subjected to high compressive load, depending on the horizontal distance of the load from the lumbar spine. Long-term vibration and cyclical effects may increase the compression in the lumbar spine leading to structural changes and fatigue effects in the tissue of discs and vertebrae.

Traumatic overload of the spine may cause damage in the discs and facet joints. Although muscles can save the spine from excessive injurious loads and movements, this protection works only if the neural system has time enough to activate the muscles.

3.2 Internal forces arising in the lumbar spine

The main internal force acting on the lumbar spine is the *compressive normal force* acting perpendicularly to the middle plane of the discs, causing the compression of the discs. It is accompanied by mainly sagittal and less lateral *shear forces* acting in the middle plane of the discs, causing the slope of the discs to each other. The moment components causing the forwards/backwards bending (flexion/extension) and the lateral bending of the spine are the sagittal and lateral *bending moments*, respectively; and the component that causes the spine to rotate about its long axis is the *torque or torsional moment*. The *tensile force* is also a normal force acting perpendicularly to the middle plane of the discs and causing the elongation of it. Although from physiologic loads there is no pure tensile force acting on the spine, since it acts generally to a part of the discs only as a side effect of other internal forces, however, the aim of traction therapies is even to apply pure tensional force to the lumbar spine.

3.3 Mobility of the lumbar spine

The range of *spinal movements* can be measured both in vivo and in vitro. The spinal movement has six components: three deflections and three rotations. The physiologic movements are the flexion and extension in the sagittal plane, the lateral bending in the frontal plane and the rotation around the long axis of the spine. The spinal motions are generally characterized by three parameters: the *neutral zone* in which the spine shows no resistance, the *elastic zone* in which the spinal resistance works, and the *range of motion*, the sum of the two latter zones.

The mobility of the spine depends on several factors. It depends first of all on the state of the intervertebral discs: the geometry, the stiffness, the fluid content, the degeneration and aging of it. The lumbar region of the spine has greater mobility than the thoracic spine. The range of motion is influenced also by the state of ligaments, the articular facet joints and the posterior bony elements. Viscoelastic properties of discs and ligaments also have an effect on the mobility.

3.4 Biomechanics of the lumbar functional spinal units

The three-dimensional FSU has six force and six motion components that depend highly on the mechanical properties, stiffness or flexibility, load bearing capacity of each structural component of the motion segment, that all depend on the age and degeneration state of them.

Lumbar *vertebral bodies* resist most of the compressive force acting along the long axis of the spine. Most of this load must resisted by the dense network of trabeculae, and less by the cortical shell. The state of the cancellous bone is the main factor of compressive failure tolerance of vertebrae (McGill, 2000). Moreover, the cancellous bone of vertebrae acts as shock absorber of the spine in accidental injurious effects. The load bearing capacity of vertebrae depends on the geometry, mass, bone mineral density (BMD) and the bone architecture of the vertebra, which all are in correlation with aging sex and degeneration. Mosekilde (2000) demonstrated that age is the major determinant of vertebral bone strength, mass, and micro-architecture. The posterior elements of vertebrae have also important role in the load bearing capacity and mobility of segments. Facet joints work as typical contact structures, by limiting the spinal movements. They stabilize the lumbar spine in compression, and prevent excessive bending and translation between adjacent vertebrae, to protect the disc and the spinal cord.

The *intervertebral discs* provide the compressive force transfer between the two adjacent vertebrae, and at the same time they allow the intervertebral mobility and flexibility. The arrangement of the collagen fibers in the annulus fibrosus is optimal for absorbing the stresses generated by the hydrostatic compression state nucleus pulposus in axial loading of the disc, and they play an important role in restricting axial rotation of the spine. Axial compressive stiffness is higher in the outer and posterior regions than in the inner and anterior regions. Tensile stiffness is higher in the anterior and posterior part than in the lateral and inner regions. Thus, the inner annulus near the nucleus seems to be the weakest area of annulus, and the outer posterior part the strongest region. In sustained loading the spine shows viscoelastic features. In quasi-static compression the disc creep is 5-7 times higher than the creep in the bony structures of the segment. Thus, the main factor of segment viscosity is the disc, mainly the disc annulus. The creep of the disc depends on the fluid content of it, mainly on the diurnal variation of it, namely on the fluid loss of daily activity and the overnight bed rest with fluid recovery.

The *ligaments* are *passive tissues* working against tension only. The primary action of the spinal ligaments lying posterior to the centre of sagittal plane rotation is to protect the spine by preventing excessive lumbar flexion. However, during this protection the ligaments may compress the discs by 100% or more. Indeed, the effectiveness of a ligament depends mainly on the moment arm through which it acts. The most elastic ligament, the ligamentum flavum being under pretension throughout all levels of flexion prevents any forms of buckling of spine. The interspinous and supraspinous ligaments may protect against excessive flexion. The capsular ligaments of facet joints restrict joint flexion and distraction of the facet surfaces of axial torsion.

The *muscles* being *active tissues* governed by the *neuromuscular control* are required to provide dynamic stability of the spine in the given activity and posture, and to provide mobility during physiologic activity, moreover to protect the spine during trauma in the post-injury phase. Two mechanical characteristics are necessary to provide these physiologic functions: the muscles must generate forces isometrically and by length change (contraction and elongation), and they must increase the stiffness of the spinal system.

The mechanical behaviour of the whole *functional spinal units* (FSU) depends on the physical properties of its components, mainly on the behaviour of the intervertebral disc, ligaments and articular facet joints. The average *load tolerance* of lumbar segments under quasi-static loading is about 5000 N for compression, 2800 N for tension, 150 N for shear and 20 Nm for axial rotation (Bader and Bouten, 2000). *Flexibility* of the FSU is the ability of the structure to deform under the applied load. Inversely, the *stiffness* is the ability to resist by force to a deformation. The stiffness of the spinal segments increases from cervical to lumbar regions for all loading cases. In lumbar region the stiffness is about 2000-2500 N/mm for compression, 800-1000N/mm for tension, 200-400 N/mm for lateral and 120-200 N/mm for anterior/posterior shear. The rotational stiffness is about 1.4-2.2 Nm/degree for flexion, 2.0-2.8 Nm/degree for extension, 1.8-2.0 Nm/degree for lateral bending and 5 Nm/degree for axial torsion (White and Panjabi, 1990; Bader and Bouten, 2000). The stiffness of the lumbar spine depends on the age and degeneration. In advanced degeneration the stiffness is higher. The stiffness is influenced by the viscous properties of the segments and the load history as well.

4. Degeneration of the lumbar spine and spine segments

The lumbar part is the most vulnerable section of the spine since both the compressive loads and the spinal mobility are maximal in this area. Consequently, the lumbar discs are mostly endangered by degenerations that influence the load bearing capacity of the segments. *Degeneration* means an injurious change in the function and structure of the disc, caused by *aging* or by *environmental effects*, like mechanical overloading (Adams et al., 2000).

Degeneration of FSU starts generally in the intervertebral discs. The first age-related changes of disc occur within the nucleus. Changes to any tissue property of the disc markedly alter the mechanics of load transfer and stability of the whole segment (Ferguson and Steffen, 2003). The mechanical properties of the segment depend also on the state of the bony structures, first of all on the trabecular structure of the internal spongy bone of vertebrae. The osteoporotic changes may also decrease the load bearing capacity of FSU.

Long-term age-related degeneration of the discs is manifested in the loss of hydration, a drying and stiffening procedure in the texture of mainly the nucleus (McNally and Adams 1992; Adams et al., 2002; Cassinelli and Kang, 2000). The functional consequences of aging are that the nucleus becomes dry, fibrous and stiff. The volume of nucleus and the region of hydrostatic pressure of it decrease, consequently, the compressive load-bearing of the disc passes to the annulus. The age-related changes of disc can be accompanied by the appearance of buckling, lesions, tears, fiber break in the annulus or disruption in the endplates, collapse of osteoporotic vertebral cancellous bone. Since the annulus becomes weaker with aging, so the overloading of it can lead to the inward buckling of the internal annulus, or to circumferential or radial tears, fiber break in the annulus, disc prolapse or herniation, or to large radial bulging of the external annulus, reduction of the disc height, or moreover, to endplate damages (Natarajan et al., 2004). The main cause of all these problems is that while the healthy disc has a hydrostatic nucleus, during aging, it becomes fibrous being no longer as a pressurized fluid. Several recent studies concluded that light degeneration of young discs leaded to instability of lumbar spine, while the stability restored with further aging (Adams et al., 2002).

Sudden accidental, often unexpected traumatic degeneration or damage of overload may yield the sudden loss of hydrostatic compression in nucleus, accompanied or due to some other failures mentioned above. In these cases the material of nucleus remains changeless, depending on the actual aging state in which the accidental effect happened. For this reason, sudden traumatic degenerations may be also dangerous in younger age.

Experimental analyses of spinal degenerations are very difficult, sometimes impossible. However, by means of numerical simulation, the effect of the main mechanical factors of degenerations can be analyzed.

5. Finite element modeling of the lumbar spine

Finite element modeling of any structure consists of five main steps: geometrical, material, element type and load modeling, and finally, validating the complete model.

5.1 Geometrical modeling of the lumbar spine segments

Geometrical modeling of the FSU must follow the anatomy of the segment. Beside the topology, additional data such as volume density, surface texture, etc. are needed. Different methods of medical image analyses can be used, like scanners, computer tomography, or magnetic resonance imaging methods.

In geometrical modeling the *vertebral body*, its cortical shell, cancellous core, posterior bony elements and the bony endplates are generally distinguished. For the width of the cylindrical vertebral body, 40-45 mm, for the depth 30-35 mm, and for the height 25-29 mm is generally used. For the thickness of the vertebral cortical wall about 1-1.5 mm, and for the thickness of the cartilaginous endplates 0.5-1 mm, and for the thickness of the cartilage layer of facet joint 0.2 mm, for the area about 1.6 cm^2 are generally used.

In geometrical modeling the *intervertebral disc*, for the height of it about 8-12 mm are generally used, depending on the sex and body height of the subject. In the disc model, the nucleus, the annulus ground substance, the annulus fibers and the cartilaginous endplates are generally distinguished. For the volumetric relation between annulus and nucleus, ratio 3:7 is generally used for the lumbar part L1-S1, and for the area ratio of nucleus 30-50% of the total disc area in cross section is generally used. The sagittal diameter length of the lumbar disc is about 36 mm, the lateral length is about 44 mm. For the orientation of annulus fibers to the mid cross-sectional area of the disc about 30° is used.

As for the cross sectional area of the *ligaments*, for the anterior longitudinal ligaments about 30-70 mm^2, for the posterior longitudinal ligaments about 10-20 mm^2, for the ligamentum flavum 40-100 mm^2, for the capsular ligaments about 30-60 mm^2, for the intertransverse ligaments 2-10 mm^2, for the interspinous ligaments 30-40 mm^2, for the supraspinous ligaments 25-40 mm^2 are generally used.

In modeling the *degenerated segments*, the height of both the vertebrae and the disc is reduced. Volume reduction of the nucleus during aging is also considered.

5.2 Material modeling of the lumbar spine segments

Since FSU is a highly heterogeneous compound structure, the material modeling must be related to the components of it. First the material models of the healthy components are considered.

The detailed data of the material modeling based on the international literature can be studied in Kurutz (2010).

5.2.1 Material models of the healthy lumbar segments

The material models and constants of the components of FSU were generally obtained by experimental mechanical tests of certain specimens obtained from the given component.

The high strength *vertebral cortical shell* is generally considered linear elastic isotropic or transversely isotropic, orthotropic material. In linear elastic isotropic case its Young modulus is considered about 5000-12000 MPa with the Poisson ratio about nu=0.2-0.3. In linear elastic, transversely isotropic case these data are E=12000-22000 MPa and G=3000-5000 MPa with nu=0.2-0.4 in the compressive direction, and E=8000-12000 MPa and G=3000-5000 MPa with nu=0.4-0.5 perpendicularly.

The vertebral *cancellous bone* is modeled generally also as linear elastic isotropic or transversely isotropic, or orthotropic material. In linear elastic isotropic case its Young modulus is considered about 10-500 MPa with the Poisson ratio about nu=0.2-0.3. In linear elastic, transversely isotropic case these data are E=200-250 MPa and G=50-80 MPa with nu=0.3-0.35 in the compressive direction, and E=100-150 MPa and G=50-80 MPa with nu=0.3-0.45 perpendicularly.

The high strength *bony endplate* of vertebrae and the lower strenght *cartiliginous endplate* of disc can hardly be distinguished when specifying material properties. Both bony and cartilaginous endplates are considered generally linear elastic isotropic material, with E=100-12000 MPa and nu=0.3-0.4, and E=20-25 MPa with nu=0.4, respectively.

The *posterior bony elements* are considered linear elastic isotropic material, generally by the same Young's modulus E=2500-3000 MPa and Poisson's coefficient nu=0.2-0.25.

The *articular facet joints* are considered as unilateral friction or frictionless connections with an initial gap of generally 0.5-1 mm.

Disc nucleus pulposus is the most important element in the compressive stiffness of the disc: the hydrostatic compression in it guarantees the stability of the whole disc and segment. The healthy young nucleus is generally modeled as an incompressible fluid-like material. In the case of fluid like linear elastic isotropic solid generally the material moduli E=1-4 MPa with nu=0.49-0.499 are considered. Several authors model the nucleus as incompressible fluid, quasi incompressible fluid, hyperelastic neo-Hookean, or Mooney-Rivlin type material, moreover, poroelastic or viscoelastic or osmoviscoelastic solid with the concerning material data (Kurutz, 2010).

Disc annulus fibrosus is a typical composite-like material with a ground substance of many layers and fiber reinforcements. Material moduli of the ground substance are considered as E=2-10 MPa with nu=0.4-0.45, and of the fibers E=300-500 MPa depending of the radial position, with nu=0.3.

Numerical modeling of *ligaments*, as typical exponentially stiffening soft tissues is not a simple task. Generally, the seven ligaments are incorporated to the FE models as tension only elements. In contrast to its strong nonlinear behaviour (White and Panjabi, 1990), most of the reported FEM studies have adopted linear or bilinear elastic models.

5.2.2 Material models of the degenerated lumbar segments

Aging type degeneration starts generally in the nucleus. A healthy young fluid-like nucleus is in a hydrostatic compression state. During aging, the nucleus loses its incompressibility and becomes even stiffer and stiffer, changing from fluid to solid material. This kind of nucleus degeneration can be modeled by decreasing Poisson's ratio with increasing Young's modulus. This behavior is generally accompanied by the stiffening process of the disc as a whole and by the volume reduction of the nucleus and volume extension of the annulus, furthermore, height reduction of the disc. Moreover, at the same time, annulus tears or internal annulus buckling, or break of the annular fibers, damage and crack or rupture of endplates, osteoporotic defects of vertebral cancellous bone can happen. Consequently, modeling age-related degeneration of FSU is a compound task; it must be done in progress, relating to a lifelong process.

In contrast to the age-related degeneration, in *sudden, often unexpected injurious degeneration* the nucleus may lose its incompressibility without any stiffening and volume change process. In this case the nucleus may quasi burst out and the hydrostatic compression may suddenly stop in it. This kind of nucleus degeneration can be modeled by suddenly decreasing Poisson's ratio with unchanged Young's modulus of nucleus (Kurutz and

Oroszváry, 2010). This behaviour is generally caused or accompanied by the tear or buckling of the internal annulus, break of the annular fibers, fracture of endplates, or collapse of vertebral cancellous bone, depending on the age in which the sudden accidental event happens. Namely, accidental failures can happen in a young disc, as well, or in any age and aging degeneration phases. These effects can be modeled by sudden damage of tissues of the concerning components of the segment. In contrast to the long term aging degeneration, these kinds of damage instability occurs suddenly, generally due to a mechanical overloading (Acaraglou et al., 1995).

The material modeling of segment degeneration can be studied in Kurutz (2010).

5.3 Element type modeling of the lumbar segments

The cancellous core and the posterior bony elements of *vertebrae* can be modeled as 3D solid continuum elements, as isoparametric 8-node hexahedral elements, or as 20- or 27-noded brick elements, moreover, as 10-noded tetrahedral elements. The cortical shell and the endplates can be modeled as thin shell elements, like 4-node shell elements. Quasi-rigid beam elements can connect the posterior vertebra with the medial transverse processes (pedicles) and from the medial transverse processes to the medial spinous process (lamina). Beam elements can also be used to represent the transverse and spinous processes. The bony surface of the facet joints can be represented by shell elements where beam elements link these facets to the lamina, simulating the inferior and superior articular processes. The facet joints can be modeled as 3D 8-noded surface-to-surface contact elements.

The *disc* annulus ground substance is generally modeled as 3D continuum elements. The collagen fibers can be modeled as truss elements or as reinforced bar (rebar) type elements embedded in 3D solid elements. The nucleus pulposus can be modeled as hydrostatic fluid volume elements.

The anterior and posterior longitudinal *ligaments* can be modeled as thin shell elements, or, the ligaments can be modeled as 2-noded axial elements, that is, tension only linear or nonlinear truss or cable or spring elements.

5.4 Load models of the lumbar spine

Loads on lumbar spinal motion segments depend on the aims of the analysis. The segment is generally supported rigidly along the inferior endplate of the lower vertebra, thus, the loads are generally applied on the superior endplate of the upper vertebra.

The loads can be applied as static or dynamic or cyclic loads. Constant static loads or incrementally changing quasi-static loads are generally applied in lumbar spine analyses. In load history analyses the basic loading types are the force or displacement loads, in a load or displacement controlled device.

For different loading, pre-compression is also applied for modeling the upper body weight as 700-1000 N, simulating the intervertebral pressure of standing position and additional compressive forces can be applied to the suitable areas of the endplates to simulate severe motions: 2000 N for lifting a load with straight legs in full flexion; 1000 N for full extension; 1300 N for full lateral bending. For flexion and extension and for axial torsion generally 10-15 Nm is applied.

The loading system of numerical simulation applied by several authors can be studied in Kurutz (2010).

The compressive strength of the lumbar spine varies between 2 kN and 14 kN, depending on the sex, age and bodymass (Adams et al, 2002). For injurious accidental degeneration load for compression about 5 kN can be considered for young, and about 3 kN for old subjects. The lumbar facet joints can resist about 2 kN for shear before fracture occurs. For torsion, damage is initiated when the applied torque rises to about 10-30 Nm. A combination of full backwards bending and 1 kN of compressive load can cause damage in the relating facet joint. In forward bending, injury can occur when the bending moment rises to 50-80 Nm.

5.5 Validation of the finite element models

By using FE models in a numerical simulation, the results should be trustworthy. For this reason, the models to be applied must be checked. Correlation between FE and experimental results can lead to use the FE model predictions with confidence.

In numerical simulations of biomechanics, for FE prediction accuracy assessment, a gold standard can be the experimental validation of numerical results. This enables the analyst to improve the quality and reliability of the FE model and the modeling methodology. If there is very poor agreement between the analytical and experimental data, by using certain numerical techniques for model updating allow the user to create improved models which represent reality much better than the original ones.

Kurutz and Oroszváry (2010) validated the lumbar segment model for both compression and tension and for both healthy and degenerated disc. Distribution of vertical compressive stresses of healthy and degenerated discs in the mid-sagittal horizontal section of the disc was compared with the experimental results of Adams et al. (1996, 2002), obtained by stress profilometry. In axial tension, the calculated disc elongations were compared with the in vivo measured elongations of Kurutz et al. (2003) and Kurutz (2006a) for healthy and degenerated segments in weightbath hydrotraction treatment (Kurutz and Bender, 2010).

6. Finite element simulation of the behaviour of healthy and degenerated lumbar spine and underwater spinal traction therapy

Finite element analyses are able to simulate processes in progress that are impossible to measure experimentally, like spinal degeneration processes. 3D FE simulations of long-term age-related and sudden accidental degeneration processes of human lumbar spinal segments are presented to analyze the compression-related degeneration processes, moreover, to analyze the efficiency of the so-called weightbath hydrotraction treatment.

A 3D geometrical model of a typical lumbar segment L4-5 was created (Fig. 1a). The geometrical data of the segment were obtained by the anatomical measures of a typical lumbar segment (Denoziere, 2004). Cortical and cancellous bones of vertebrae were separately modeled, including posterior bony elements. The thickness of vertebral cortical walls and endplates were 0.35 and 0.5 mm, respectively. For this simple model, we kept the disc height constant by applying 10 mm height. Annulus fibrous consisted of ground substance and elastic fibers (Fig. 1b). Annulus matrix was divided to internal and external ring; with three layers of annulus fibers of 0.1 mm² cross section. The geometry and orientation of facet joints were chosen according to Panjabi et al. (1993).

FE mesh was created in three steps (Fig. 1c). First the geometrical model of FSU was created by using Pro/Engineer code; then the FE mesh was generated by ANSYS Workbench; finally, the several components were integrated to the FE model by ANSYS Classic.

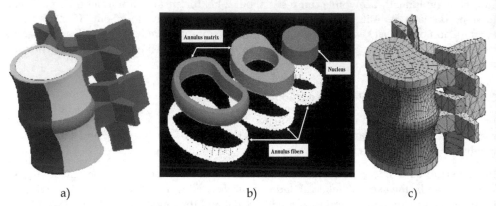

a)   b)   c)

Fig. 1. The geometrical model of a) the segment and b) the intervertebral disc and c) the finite element mesh of the segment

The material moduli of the healthy segment (Table 1) were obtained from the literature (Rohlmann et al., 2006; Goel et al., 1995; Denoziere, 2004; Denoziere and Ku, 2006; Cheung et al., 2003; Antosik and Awrejcewicz, 1999; Noailly et al., 2007; Williams et al., 2007; Shirazi-Adl et al, 1984, 1986; Shirazi-Adl, 1989; Lavaste et al., 1992; Zander et al., 2004). For the bony elements and endplates we considered linear elastic isotropic materials for both tension and compression. Annulus ground substance and nucleus were considered linear elastic for both compression and tension. For the fluid-like healthy nucleus and for the annulus matrix also linear elastic material was considered.

Components of FSU	Young's mod [MPa]	Poisson's ratio
Vertebral cortical bone	12000	0,3
Vertebral cancellous bone	150	0,3
Posterior elements, facet	3500	0,3
Endplate	100	0,4
Annulus ground substance	4	0,45
Annulus fibers	500/400/300*	-
Nucleus	1	0,499
Anterior longitud. ligament	8**	0,35
Posterior longitud. ligament	10**	0,35
Other ligaments	5**	0,35

*external/middle/internal fibers, tension   **tension only

Table 1. Material moduli of the components of healthy segment

Collagen fibers of the annulus were considered as bilinear elastic isotropic tension-only material. To simulate the radial variation of collagen in the fibers, the stiffness of them was increased outwards. All seven ligaments were integrated in the model with bilinear elastic tension-only material.

For our numerical experiments, the basic data of Table 1 were modified.

6.1 Finite element simulation of age-related degeneration processes of lumbar spine

Age-related degeneration starts generally in the nucleus. A healthy young nucleus is in hydrostatic compression state. During aging, the nucleus loses its incompressibility by changing gradually from fluid-like to solid material. This kind of nucleus degeneration was modeled by decreasing Poisson's ratio with increasing Young's modulus. This behavior is generally accompanied by the hardening process of the disc as a whole and by tears, buckling or fiber break of the annulus, and damage of endplates or vertebral cancellous bone. The data of five grades of normal aging degeneration process from healthy to fully degenerated case are seen in Table 2, by gradually decreasing Poisson's ratio with gradually increasing Young's modulus of nucleus, accompanied by aging of other tissues of FSU.

Grades of age-related degeneration* (Young's mod/Poisson's ratio)	grade 1 (healthy)	grade 2	grade 3	grade 4	grade 5 (fully deg.)
nucleus	1/0.499	3/0.45	9/0.4	27/0.35	81/0.3
annulus ground substance	4/0.45	4.5/0.45	5/0.45	5.5/0.45	6/0.45
cancellous bone	150/0.3	125/0.3	100/0.3	75/0.3	50/0.3
endplate	100/0.4	80/0.4	60/0.4	40/0.4	20/0.4

* Bony elements and annulus fibers are seen in Table 1

Table 2. Modeling of age-related degeneration process: material moduli of components of segments from healthy (1) to fully degenerated (5) phases

Simulating aging degeneration processes, 1000 N axial compression load was applied, by considering that the lumbar compression load is about 60% of the total body weight, completed by the muscle forces being nearly the same (Nachemson, 1981; Sato et al., 1999).

The compression load was distributed along the superior and inferior surface of the upper and lower vertebra of FSU, by applying rigid load distributor plates at both surfaces.

Figure 2 illustrates the maximum stress rearrange in the mid-sagittal section of the disc during the degeneration process. While in a healthy disc in Fig. 2a, due to the hydrostatic compression stress state, the maximum compressive stresses occur in the middle of nucleus, in a fully degenerated disc in Fig. 2b, due to the lost hydrostatic compression, the maximum compressive stresses move outwards to the edge of nucleus, towards the annulus ring.

Figure 3a shows the change of mid-sagittal vertical compressive stresses for aging degeneration models of Table 2, in the middle and border of nucleus and in the internal and external nucleus, for 1000 N axial compressive load. The vertical stresses in the center and border of nucleus first decreased and later increased with aging, yielding stress minimum in the nucleus in mildly degenerated state. In the internal annulus monotonous stress decrease was observed, demonstrating the possible internal annulus buckling. In the external annulus the stresses slightly changed.

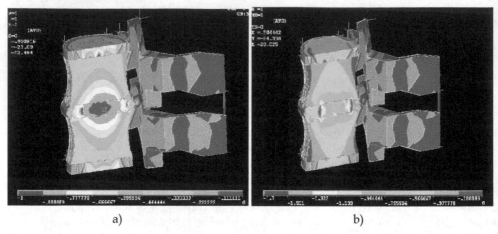

a)   b)

Fig. 2. Mid-sagittal vertical compressive stresses for a) healthy and b) fully degenerated disc

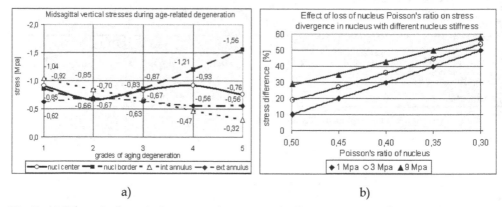

a)   b)

Fig. 3. a) Mid-sagittal vertical compressive stresses in disc components during aging degeneration and b) stress divergence in nucleus center during the loss of hydrostatic compression state in nucleus (Kurutz and Oroszváry, 2010).

In degenerated disc the pressure in the nucleus is not hydrostatic any more, being non-uniform and direction-dependent. In Fig. 3b the stress divergence is seen in the nucleus center between vertical and horizontal stresses, increasing rapidly and quasi linearly with the loss of hydrostatic compression in the nucleus. The initial stress divergence between vertical and horizontal stresses in hydrostatic state at nu=0.499 for the fluid-like nucleus of E=1 MPa was 8-10%, and naturally, for harder nucleus it was higher. By applying more fluid-like material for nucleus (E=0.1 MPa), the initial hydrostatic stress divergence could be decreased to 1-2%.

Fig. 4a illustrates the disc shortening with aging, demonstrating that the maximum disc deformability occurred in mild degeneration. Similar behaviour was observed for the maximum tensile forces in the outermost posterolateral annulus fibers in Fig 4b, that is, the maximum fiber forces belonged also to the mildly degenerated state. Fig. 5a and 5b illustrate the tensile forces in healthy and fully degenerated annulus fibers, respectively.

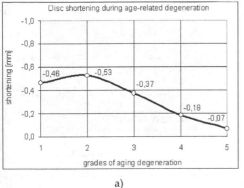

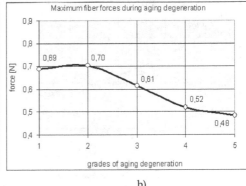

a)   b)

Fig. 4. a) Height loss of disc and b) maximum fiber forces during aging degeneration (Kurutz and Oroszváry, 2010)

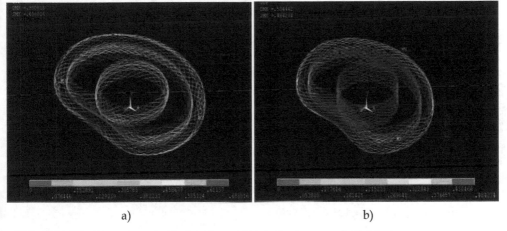

a)   b)

Fig. 5. Annulus fiber forces for a) healthy and b) fully degenerated case

Fig. 6a illustrate the change of posterior, anterior and lateral disc bulging during aging degeneration process, demonstrating that the bulging deformability is maximum in young age or mild degeneration, and it decreases with aging. Fig. 6b shows the change of the mean vertical compressive stiffness of disc components during the aging degeneration process. The minimum of each stiffness function belonged to the mildly degenerated state.

The stiffness of the whole disc depends mainly on the stiffness of the nucleus. In the first period of aging, the dominant effect is the loss of incompressibility of nucleus when the other disc components are considerable soft. This yields that the stresses and the vertical load transfer through the nucleus is minimum (Fig. 3a) and the deformability of disc is maximum in young age (Fig. 4a) leading to the minimum vertical compressive stiffness of discs and the risk of instability of FSUs in mildly degenerated state (Fig. 6b.) Consequently, the lumbar segments are most vulnerable in young age, and the segmental stability increases with further aging and degeneration.

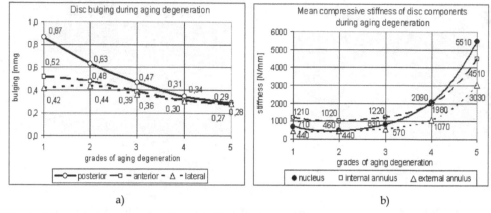

a)   b)

Fig. 6. a) Posterior, anterior and lateral disc bulging and b) mean compressive stiffness of disc components during aging degeneration (Kurutz and Oroszváry, 2010).

6.2 Finite element simulation of sudden degeneration processes of lumbar spine

In contrast to the age-related degeneration process that lasts during a lifelong time, the sudden degeneration has very short, sometimes unexpected, instant processes. For modeling sudden degeneration accompanied by other damaging phenomena, the data of the concerning tissues were modified in Table 1 and Table 2, depending on the actual aging degeneration phase in which the sudden accidental degeneration happened, considering also five phases of accidental degeneration process.

For example, data in Table 3 show the model of sudden degeneration in young age with annulus fiber breaks and tears. In this case the sudden degeneration happened in the first aging degeneration phase of Table 2, when the sudden loss of hydrostatic compression was modeled by rapid immediate decrease of Poisson ratio of nucleus with changeless Young's modulus of it. Annulus tears and fiber breaks were modeled by weakened annulus matrix and fibers, respectively, seen also in Table 3.

Phases of sudden degeneration* (Young's mod/Poisson's ratio)	grade 1	grade 2	grade 3	grade 4	grade 5
nucleus	1/0.499	1/0.45	1/0.40	1/0.35	1/0.3
annulus ground substance	4/0.45	3.5/0.45	3/0.45	2.5/0.45	2/0.45
annulus fibers	500/400/300	375/300/225	250/200/150	125/100/75	5/4/3

* Bony elements are seen in Table 1

Table 3. Modeling of sudden degeneration process with annulus tears and fiber break in young age: modified material moduli of components of segments

Simulating sudden accidental degeneration processes, the 1000 N axial compression load was completed by an unexpected sudden overload of 1000 N, thus, 2000 N axial compression load was applied.

Fig. 7a shows the sudden change of the mid-sagittal vertical compressive stresses in disc components, Fig. 7b illustrates the mid-sagittal central, posterior and anterior disc shortening during the sudden degeneration process in young age with annulus tears and fiber break (Table 3). The mean stress decreased strongly in the nucleus (by 70%), and moderately in the internal annulus (by 25%), while slightly increased in the external annulus (by 8-10%). Disc shortening radically increased during the process, finally by about 200-230%.

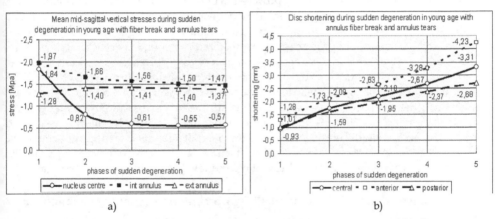

a)   b)

Fig. 7. a) Mid-sagittal vertical compressive stresses in disc components and b) disc shortening during sudden degeneration process

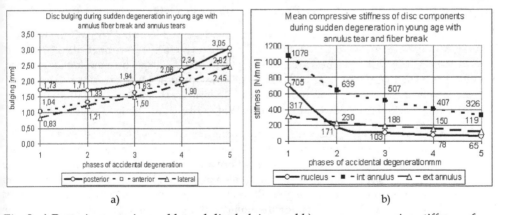

a)   b)

Fig. 8. a) Posterior, anterior and lateral disc bulging and b) mean compressive stiffness of disc components during sudden degeneration process

Fig. 8a shows the sudden mid-sagittal anterior, posterior and lateral disc bulging. In Fig. 8b the mean compressive stiffness of disc components during the sudden degeneration process are illustrated. Disc bulging radically increased in all directions (by 80-200%) due to the fiber breaks with radically decreasing maximum fiber forces from 1.38 N to 0.06 N. Significant stiffness loss was observed for the disc components (nucleus 91%, internal annulus 70%, external annulus 62%). The stiffness loss of the whole disc was 76%.

During sudden degeneration process the vertical compressive load transfer moves from inside to outside, from the nucleus to the external annulus during the progress of degeneration. In the case of internal annulus buckling, the sudden overload of the external annulus may lead to annulus tears and injury. In the case of fiber break and annulus tears, the internal annulus is overloaded.

In contrast to age-related degeneration processes where the disc deformability (shortening, bulging) decreases during aging, in sudden degeneration processes the deformability increases strongly that may lead to injury and pain mainly in younger age. Also in contrast to age-related degeneration processes where the vertical compressive stiffness of discs increases during aging, in sudden degeneration processes it decreases significantly leading to segmental instability and injury. In agreement with the literature we have found by numerical simulation of age-related degeneration that the young and mildly degenerated segments had the smallest stiffness and later the stiffness increased rapidly (Adams et al., 2002; Rohlmann et al., 2006; Schmidt et al., 2006, 2007). Similarly to the age-related degeneration, accidental degeneration may be the most dangerous in young age, due to the sudden stiffness loss starting at the smallest stiffness level, consequently, accidental disc shortening and bulging may cause sudden injury and low back pain in young age again.

6.3 Finite element simulation of weightbath hydrotraction treatment of lumbar spine

Kurutz and Oroszváry (2010) analyzed by FE simulation the stretching effect of a special underwater traction treatment applied for treating degenerative diseases of the lumbar spine, when the patients are suspended cervically in vertical position in the water, supported on a cervical collar alone, loaded by extra lead weights on the ankles.

The biomechanics of WHT has been reported first by Bene and Kurutz (1993). Elongations of lumbar segments during WHT have been measured in vivo by Kurutz et al. (2003). The clinical impacts of WHT have been analyzed by Oláh et al. (2008). The complex description of WHT has been given by Kurutz and Bender (2010) with its application, biomechanics and clinical effects. This numerical study aims to determine the disc elongation, bulging contraction, stress and fiber relaxation effects of WHT during age-related degeneration.

In this underwater cervical suspension, the traction load consists of two parts: (1) the removal of the compressive preload of body weight and muscle forces in water, named *indirect traction load*; and (2) the tensile force of buoyancy with the applied extra lead loads, named *direct traction load*. Based on mechanical calculations, for the standard body weight of 700 N, and the applied extra lead weights 40 N, the indirect and direct traction loads yields 840 N and 50 N, respectively. Thus, for the numerical analysis of WHT we applied 840 N indirect and 50 N direct traction loads.

In the finite element analysis of WHT the above detailed material, geometric and finite element model has been used. Annulus ground substance and nucleus were considered linear elastic in compression, and bilinear elastic in traction. Thus, for the five grades of age-related degeneration process, in the indirect phase of traction, the compressive material moduli were used, seen in Table 2; while for the direct phase of traction, special tensile Young's moduli were applied, obtained by parameter identification (Kurutz and Tornyos, 2004), seen in Table 4. For healthy nucleus we applied fluid-like incompressible material both for tension and compression.

Grades of age-related degeneration for direct traction* (Young's mod/Poisson's ratio)	grade 1 (healthy)	grade 2	grade 3	grade 4	grade 5 (fully deg.)
nucleus	0.4/0.499	1.0/0.45	1.6/0.4	2.2/0.35	2.8/0.3
annulus ground substance	0.4/0.45	1.0/0.45	1.6/0.45	2.2/0.45	2.8/0.45
cancellous bone	150/0.3	125/0.3	100/0.3	75/0.3	50/0.3
endplate	100/0.4	80/0.4	60/0.4	40/0.4	20/0.4

* Bony elements and annulus fibers are seen in Table 1

Table 4. Modeling of age-related degeneration process for direct traction in WHT: material moduli of components of segments from healthy (1) to fully degenerated (5) phases

The unloading effect of WHT in its instant elastic phase is illustrated in Fig. 9 and 10 in terms of aging degeneration. The term 'unloading' is relative: it is related to the compressed state of segments just before the treatment. Due to the bilinear behaviour of the disc during hydrotraction, the results of deformation and stress unloading are distinguished: caused by indirect and direct traction loads.

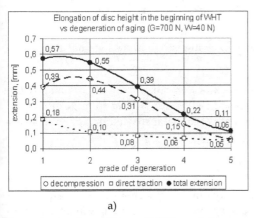
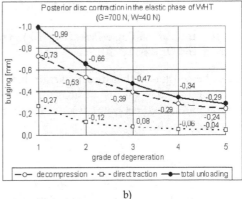

a)   b)

Fig. 9. Unloading effect of WHT on a) disc compression and b) posterior disc bulging (Kurutz and Oroszváry, 2010).

In Fig. 9a the relative elongations of discs are seen compared with their compressed state before the treatment. These initial elastic elongations will be quasi doubled in the creeping phase of WHT (Kurutz, 2006b). The ratio of direct traction extensions versus total extensions is 32%, 19%, 20%, 28% and 47% for degeneration grades 1 to 5, respectively. The ratios of direct/indirect extensions are: 47%, 24%, 25%, 39% and 89%. The minimum ratio belongs to the mild degeneration and increases rapidly with advanced degeneration and aging. The maximum ratio belongs to the fully degenerated cases.

The unloading of posterior bulging, namely, the relative disc contractions can be seen in Fig. 9b. The ratio of direct/total traction contractions decreases monotonously from 27% to 15% for posterior bulging for degeneration grades 1 to 5. The ratio of direct/indirect contractions changes from 37% to 18% in posterior bulging. The minimum ratio belongs to the fully degenerated case.

Fig. 10a and 10b show the vertical stress relaxation in the centre of nucleus and in the annulus external ring. The ratio of direct/total and direct/indirect stress unloading is equally small, 7-8% in healthy and 2-4% in fully degenerated cases. Thus, in stress relaxation the dominant effect is the indirect traction load.

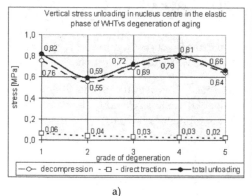
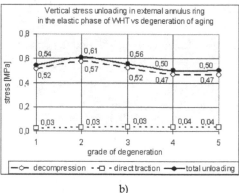

a)   b)

Fig. 10. Stress unloading effect of WHT a) in nucleus centre and b) in the external annulus (Kurutz and Oroszváry, 2010).

It can be concluded that direct traction load with extra lead weights influences mainly the deformations that are responsible for nerve release, while stress relaxation is influenced mainly by the indirect traction load. The traction effect can be increased by applying larger extra lead weights.

7. Conclusion

After a short survey of the structural anatomy and biomechanics of healthy and degenerated lumbar spine, FE modeling and a systematic numerical analysis of the main mechanical features of lumbar spine degenerations was investigated to study the age-related and sudden degeneration processes of it. The fact that mildly degenerated segments have the smallest stiffness both in aging and sudden degeneration processes was numerically proved by answering the question why the LBP problems insult so frequently the young adults.

At the beginning of aging degeneration process, the effect of loss of incompressibility of nucleus, later the hardening of nucleus dominated, yielding the smallest compressive stiffness of disc at mildly degenerated state. In sudden degeneration processes the smallest stiffness happened also in mildly degenerated state. The 2100 N/mm stiffness suddenly decreased by 75-80% to 400-500 N/mm for mild, and the 3600 N/mm stiffness decreased by 60-65% to 1300 N/mm for severely degenerated case. Vertical intradiscal stresses showed significant change during aging degeneration, between 0.6-1.6 MPa. Disc deformability and bulging was maximum in mildly degenerated state and decreased during aging by 30-85%, while in sudden degeneration increased suddenly by 200-300%.

As for conclusion, FE simulations of degeneration processes of lumbar segments and discs may help clinicians to understand the initiation and progression of disc degeneration and will help to improve prevention methods and treating tools for regeneration of disc tissues.

In WHT, discs show a bilinear material behaviour with higher resistance in indirect and smaller in direct traction phase. Consequently, although the direct traction load is only 6% of the indirect one, direct traction deformations are 15-90% of the indirect ones, depending on the grade of degeneration. Moreover, the ratio of direct stress relaxation remains equally about 6-8% only. Consequently, direct traction controlled by extra lead weights influences mostly the deformations being responsible for the nerve release; while the stress relaxation is influenced mainly by the indirect traction load coming from the removal of the compressive body weight and muscle forces in the water. A mildly degenerated disc in WHT shows 0.15 mm direct, 0.45 mm indirect and 0.6 mm total extension; and 0.2 mm direct, 0.6 mm indirect and 0.8 mm total posterior contraction. A severely degenerated disc exhibits 0.05 mm direct, 0.05 mm indirect and 0.1 mm total extension; 0.05 mm direct, 0.25 mm indirect and 0.3 mm total posterior contraction. These deformations are related to the instant elastic phase of WHT that are doubled during the creep period of the treatment.

As for conclusion, WHT unloads the compressed disc: extends disc height, decreases bulging, stresses and fiber forces, increases joint flexibility, relaxes muscles, unloads nerve roots, relieves pain and may prevent graver problems. WHT is an effective non-invasive method to treat lumbar discopathy. By the presented numerical analysis its beneficial clinical impacts can be supported, moreover, the treatment could be planned, the magnitudes of extra loads could be determined by considering the patient's clinical status.

8. Acknowledgment

The present study was supported by the Hungarian National Science Foundation projects OTKA T-022622, T-033020, T-046755 and K-075018. The authors are grateful to Knorr Bremse Hungaria for the help in FE modeling of FSU.

9. References

Acaroglu, E.R., Iatridis, J.C., Setton, L.A., Foster, R.J., Mow, V.C., Weidenbaum, M. (1995). Degeneration and aging affect the tensile behaviour of human lumbar anulus fibrosus, *Spine*, 20(24), 2690-2701.

Adams, M.A., Bogduk, N., Burton, K., Dolan, P. (2002). *The Biomechanics of Back Pain*, Churchill Livingstone, London.

Adams, M.A., Freeman, B.J., Morrison, H.P., Nelson, I.W., Dolan, P. (2000), Mechanical initiation of intervertebral disc degeneration, *Spine*, 25(13), 1625-1636.

Adams, M.A., McNally, D.S., Dolan, P. (1996). Stress distributions inside intervertebral discs. The effects of age and degeneration. *J. Bone Joint Surg. Br.* 78(6), 965-972.

Andersson, G.B., Schultz, A.B., Nachemson, A.L., (1983). Intervertebral disc pressures during traction, *Scand. J. Rehabil. Med.* Suppl. 9, 88-91.

Antosik, T., Awrejcewicz, J., (1999). Numerical and Experimental Analysis of Biomechanics of Three Lumbar Vertebrae, *Journal of Theoretical and Applied Mechanics*, 37(3). 413-434.

Bader, D.L., Bouten, C. (2000). Biomechanics of soft tissues. In: Dvir, Z. (Ed.), *Clinical Biomechanics*, Churchill Livingstone, New York, Edinburgh, London, Philadelphia, 35-64.

Bene, É., (1988). Das Gewichtbad, Zeitschrift für Physikalische Medizin, Balneologie, *Medizinische Klimatologie*. 17, 67-71.

Bene É., Kurutz, M., (1993). Weightbath and its biomechanics, (in Hungarian), *Orvosi Hetilap*, 134. 21. 1123-1129.

Benzel, E.C.: *Biomechanics of Spine Stabilization*, (2001). American Association of Neurological Surgeons, Rolling Meadows, Illinois.

Bogduk, N., Twomey, L.T. (1987). *Clinical Anatomy of the Lumbar Spine*, Churchill Livingstone, New York.

Cassinelli, E., Kang, J.D. (2000). Current understanding of lumbar disc degeneration, *Operative Techniques in Orthopaedics*, 10(4), 254-262.

Cheung, J.T.M., Zhang, M., Chow, D.H.K., (2003). Biomechanical Responses of the Intervertebral Joints to Static and Vibrational Loading: a Finite Element Study, *Clinical Biomechnaics*, 18(9), 790-799.

Denoziere, G., (2004). *Numerical Modeling of Ligamentous Lumbar Motion Segment*, Master thesis, Georgia Institute of Technology, 148 p.

Denoziere, G., Ku, D.N., (2006). Biomechanical Comparison Between Fusion of Two Vertebrae and Implantation of an Artificial Intervertebral Disc, *Journal of Biomechanics*, 39(4), 766-775.

Dolan, P., Adams, M.A. (2001). Recent advances in lumbar spinal mechanics and their significance for modelling, *Clinical Biomechanics, 16(Suppl.)*, S8-S16.

Dvir, Z. (2000). *Clinical Biomechanics*, Churchill Livingstone, New York, Edinburgh, London, Philadelphia.

Ferguson, S.J., Steffen, T. (2003). Biomechanics of the aging spine, *European Spine Journal*, Suppl 2, S97-S103.

Goel, V.K., Monroe, B.T., Gilbertson, L.G., Brinckmann, P. (1995). Interlaminar shear stresses and laminae separation in the disc. Finite element analysis of the L3-L4 motion segment subjected to axial compressive loads. *Spine*, 20(6), 689-698.

Kurutz, M. (2006a). Age-sensitivity of time-related in vivo deformability of human lumbar motion segments and discs in pure centric tension, *Journal of Biomechanics*, 39(1), 147-157.

Kurutz, M., (2006b). In vivo age- and sex-related creep of human lumbar motion segments and discs in pure centric tension, *Journal of Biomechanics*, 39(7), 1180-9.

Kurutz, M. (2010). Finite element modeling of the human lumbar spine, In: Moratal, D. (ed.): *Finite Element Analysis*, SCIYO, Rijeka, 690 p., 209-236.

Kurutz, M, Bender, T. (2010). Weightbath hydrotraction treatment: application, biomechanics and clinical effects, *Journal of Multidisciplinary Healthcare*, 2010(3), 19-27.

Kurutz, M., Oroszváry, L. (2010). Finite element analysis of weightbath hydrotraction treatment of degenerated lumbar spine segments in elastic phase, *Journal of Biomechanics*, 43(3), 433-441.

Kurutz, M., Tornyos, Á., (2004). Numerical simulation and parameter identification of human lumbar spine segments in traction, In: Bojtár I. (ed.): *Proc. of the First Hungarian Conference on Biomechanics*, Budapest, Hungary, June 10-11, 2004, ISBN 963 420 799 5, 254-263.

Kurutz, M., Bene É., Lovas, A., (2003). In vivo deformability of human lumbar spine segments in pure centric tension, measured during traction bath therapy, *Acta of Bioengineering and Biomechanics*, 5(1), 67-92.

Lavaste, F., Skalli, W., Robin, S., Roy-Camille, R., Mazel, C., (1992). Three-dimensional Geometrical and Mechanical Modelling of Lumbar Spine, *Journal of Biomechanics*, 25(10), 1153-1164.

McGill, S.M. (2000). Biomechanics of the thoracolumbar spine, In: Dvir, Z. (Ed.), *Clinical Biomechanics*, Churchill Livingstone, New York, Edinburgh, London, Philadelphia, 103-139.

McNally, D.S., Adams, M.A. (1992). Internal intervertebral disc mechanics as revealed by sress profilometry, *Spine*, 17(1), 66-73.

Moll, K., (1956). Die Behandlung der Discushernien mit den sogenannten "Gewichtsbadern", *Contempl. Rheum.*, 97, 326-329.

Moll, K., (1963). The role of traction therapy in the rehabilitation of discopathy, *Rheum. Balneol. Allerg.*, 3, 174-177.

Mosekilde, L. (2000). Age-related changes in bone mass, structure, and strength effects of loading, *Zeitschrift für Rheumatologie*, 59(Suppl.1), 1-9.

Nachemson, A.L. (1981). Disc pressure measurements, *Spine*, 6(1), 93-97.

Natarajan, R.N., Williams, J.R., Andersson, G.B., (2004). Recent advances in analytical modelling of lumbar disc degeneration. *Spine*, 29(23), 2733-2741.

Noailly, J., Wilke, H.J., Planell, J.A., Lacroix, D., (2007). How Does the Geometry Affect the Internal Biomechanics of a Lumbar Spine Bi-segment Finite Element Model? Consequences on the Validation Process, *Journal of Biomechanics*, 40(11), 2414-2425.

Oláh, M., Molnár, L., Dobai, J., Oláh, C., Fehér, J., Bender, T., (2008). The effects of weightbath traction hydrotherapy as a component of complex physical therapy in disorders of the cervical and lumbar spine: a controlled pilot study with follow-up, *Rheumatology International*, 28(8), 749-756.

Panjabi, M.M., Oxland, T., Takata, K., Goel, V., Duranceau, J., Krag, M., (1993). Articular Facets of the Human Spine, Quantitative Three-dimensional Anatomy, *Spine*, 18(10), 1298-1310.

Ramos, G., Martin, W., (1994). Effects of vertebral axial decompression on intradiscal pressure, *Journal of Neurosurgery.* 81(3), 350-353.

Rohlmann, A., Zander, T., Schmidt, H., Wilke, H.J., Bergmann, G., (2006). Analysis of the influence of disc degeneration on the mechanical behaviour of a lumbar motion segment using the finite element method, *Journal of Biomechanics*, 39(13), 2484-2490.

Sato, K., Kikuchi, S., Yonezawa, T. (1999). In vivo intradiscal pressure measurement in healthy individuals and in patients with ongoing back problems, *Spine*, 24(23), 2468-2474.

Schmidt, H., Heuer, F., Simon, U., Kettler, A., Rohlmann, A., Claes, L., Wilke, H.J., (2006). Application of a New Calibration Method for a Three-dimensional Finite Element Model of a Human Lumbar Annulus Fibrosus, *Clinical Biomechanics*, 21(4), 337-344.

Schmidt, H., Kettler, A., Rohlmann, A., Claes, L., Wilke, H,J.. (2007). The Risk of Disc Prolapses With Complex Loading in Different Degrees of Disc Degeneration - a Finite Element Analysis, *Clinical Biomechanics*, 22(9), 988-998.

Shirazi-Adl, A., (1989). On the Fibre Composite Material Models of the Annulus – Comparison of Predicted Stresses. *Journal of Biomechanics*, 22(4), 357-365.

Shirazi-Adl, S.A., Shrivastava, S.C., Ahmed, A.M., (1984). Stress Analysis of the Lumbar Disc-body unit in Compression. A Three-dimensional Nonlinear Finite Element Study, *Spine*, 9(2), 120-34.

Shirazi-Adl, A., Ahmed, A.M., Shrivastava, S.C., (1986). A Finite Element Study of a Lumbar Motion Segment Subjected to Pure Sagittal Plane Moments, *Journal of Biomechanics*, 19(4), 331-350.

White, A. A., Panjabi, M. M. (1990). *Clinical Biomechanics of the Spine*, Lippincott Williams and Wilkins, Philadelphia, etc.

Williams, J.R., Natarajan, R.N., Andersson, G.B.J., (2007). Inclusion of Regional Poroelastic Material Properties Better Predicts Biomechanical Behaviour of Lumbar Discs Subjected to Dynamic Loading, *Journal of Biomechanics*, 40(9), 1981-1987.

Zander, T., Rohlmann, A., Bergmann, G., (2004). Influence of Ligament Stiffness on the Mechanical Behaviour of a Functional Spinal Unit, *Journal of Biomechanics*, 37(7), 1107-1111.

8

Finite Element Analysis to Study Percutaneous Heart Valves

Silvia Schievano, Claudio Capelli, Daria Cosentino,
Giorgia M. Bosi and Andrew M. Taylor
UCL Institute of Cardiovascular Science, London
UK

1. Introduction

Percutaneous valve implantation is an innovative, successful alternative to open-heart surgery for the treatment of both pulmonary and aortic heart valve dysfunction (Bonhoeffer et al., 2000; Cribier et al., 2002; Leon et al., 2011; Lurz et al., 2008; McElhinney DB et al., 2010; Rodes-Cabau et al., 2010; Smith et al., 2011; Vahanian et al., 2008). However, this minimally-invasive procedure still presents limitations related to device design: stent fracture, availability to a limited group of patients with very specific anatomy and conditions, and positioning and anchoring issues (Delgado et al., 2010; Nordmeyer et al., 2007; Schievano et al., 2007a). Computational simulations (Taylor & Figueroa, 2009), together with advanced cardiovascular imaging techniques can be used to help understand these limitations in order to guide the optimisation process for new device designs, to improve the success of percutaneous valve implantation, and ultimately to broaden the range of patients who could benefit from these procedures.

In this Chapter, we review the applications of finite element (FE) analyses to study percutaneous heart valve devices. The current percutaneous pulmonary valve implantation (PPVI) device is first presented: FE analyses were used to understand stent fractures in the patient specific setting and to study a potential solution to this problem. The shortcomings of the present PPVI stent have created a need for the next generation device. The optimisation study performed on this device using FE analyses, together with the first-in-man clinical application are described. The last part of this Chapter focuses on transcatheter aortic valve implantation (TAVI) and the innovative use of FE analyses to model patient specific implantation procedures. Computational analyses were used as a tool to assess the feasibility and safety of TAVI in patients who are currently considered unsuitable for this procedure.

2. Percutaneous pulmonary valve implantation

In the late 1990's, Professor Philipp Bonhoeffer developed a new technique of heart valve replacement that avoided the need for surgery (Bonhoeffer et al., 2000). This was based on the concept that a heart valve sewn inside a stent could be reduced in size, by crimping it onto a balloon catheter, and then introduced through a peripheral vessel to the desired implantation site in the heart (Fig. 1). Inflation of the balloon deployed the valved stent and anchored it within the old dysfunctional valve. Over the last 11 years, this simple technique has shown a marked learning curve with safety and efficacy improvements that have led to CE mark in 2006, Food and Drug Administration (FDA) approval in 2010, and successful, worldwide clinical use of this procedure in the pulmonary position (>2,000 implants to date) and thousands of implants in the aortic position.

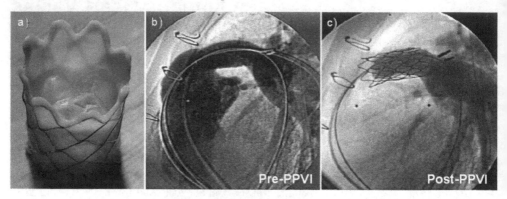

Fig. 1. a) Current PPVI device: Melody™, Medtronic, MN USA. Pulmonary angiograms b) pre- and c) post-PPVI in a patient. In this patient significant pulmonary regurgitation (contrast in the anterior right ventricle in b) has been completely abolished

2.1 Current device

The current PPVI device (Melody™, Medtronic, MN, USA) is constructed with a valve from a bovine jugular vein, sewn into a balloon-expandable, platinum-10%iridium stent (Fig. 1a). The stent is created by 6 wires, formed into a zig-zag shaped pattern, and welded together. Since the platinum welds were prone to fracture, this component of the PPVI was modified during the early clinical experience by introducing a gold braising process to reinforce the crowns. The bovine vein is attached to the stent by sutures around the proximal and distal rings and at each strut intersection. Bovine jugular venous valves are available only up to 22 mm of diameter, thus limiting the use of this minimally-invasive technique to those patients who have a small implantation site. During the procedure in the catheterisation laboratory, the valved stent assembly is hand-crimped to a size of 6 mm onto a custom made delivery system, before being re-expanded inside the patient's implantation site.

2.1.1 Understanding and predicting stent fractures

During the development phases of the current PPVI device, simple bench and animal experiments were performed as part of routine preclinical testing. These *in-vitro* and *in-vivo* experiments predicted valve degeneration with no stent fracturing. However, in our clinical practice, the opposite occurred – good valve function, ~20% stent fractures (Nordmeyer et al., 2007). These discrepancies were most likely due to the fact that in patients, the *in-vivo* loading conditions could not be correctly reproduced with experimental set-ups, where the boundary conditions are simplified and not representative of the real situation. For example, in the bench experiments for fatigue assessment, PPVI stents were placed in distensible cylindrical tubes. However, the final shape of the PPVI stent *in situ* is a long way away from being uniformly cylindrical.

In order to better understand and predict the stent mechanical performance and durability, the implantation site morphology has to be included into *in-vitro* testing. To demonstrate this concept, and to test that computational analysis can indeed predict where stent fractures will occur, we reproduced patient-specific procedures using computational analysis.

Five patients (I to V) who underwent PPVI and experienced fracture between 1 and 6 months after implantation were selected (Cosentino et al., 2011). Biplane, orthogonal, fluoroscopy images (Axiom Artis Flat Detector system, Siemens Medical Systems, Germany) from the actual PPVI procedures of these patients were used to reconstruct the 3D geometry of the stents *in-situ* (Rhinoceros CAD software, McNeel, WA, USA; Schievano et al., 2010b) at the end of balloon inflation, and at early systole and diastole (Fig. 2). From these reconstructions, the displacements of every strut junction of the stent from the end of balloon inflation through systole and diastole were calculated. Circumferential, radial and longitudinal asymmetries were measured for each time step and every patient. The stent expanded into the patients' implantation sites resembled the outline of their arterial walls, resulting in asymmetry in all directions (Fig. 3top). In particular, some of the stent struts were at an early stage of expansion, while others were over-deployed compared to a uniformly deployed stent in a cylinder. The most expanded cells were located in correspondence to the fracture positions as detected from x-rays images in the patients. Furthermore, in all studied patients, the stents were non-circular in cross-section and the terminal rings were more expanded if compared to the central portions.

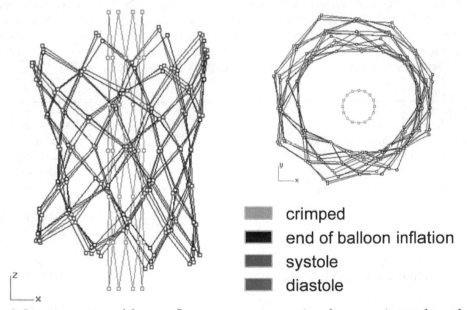

crimped
end of balloon inflation
systole
diastole

Fig. 2. Superimposition of the stent fluoroscopy reconstructions for one patient at the end of balloon inflation, systole, and diastole compared to the initial crimped configuration (lateral and top views)

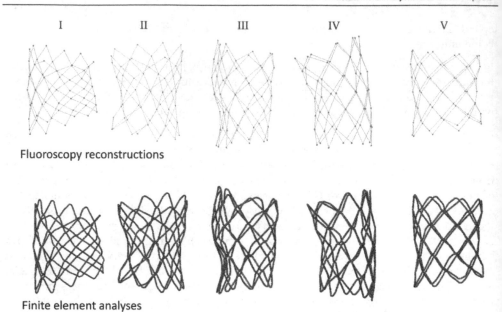

Fluoroscopy reconstructions

Finite element analyses

Fig. 3. Stent fluoroscopy reconstructions for the analysed patients at diastole (top panel) and corresponding FE deployed configurations (bottom panel)

A FE model of the crimped stent in the catheter was created, with a structured hexahedral mesh of 119,360 elements to model the platinum-iridium wires. To reproduce the golden coverings, an additional set of elements was modelled around the junctions and a structured hexahedral mesh was generated using 36,320 elements. Stent geometrical and material properties were provided by the manufacturer (Table 1; Schievano et al., 2007c).

PPVI stent	
Geometry	
Wire diameter	0.33 mm
Crimped configuration internal diameter	4.00 mm
Crimped configuration overall length	34.32 mm
Central zig-zag segment length	5.78 mm
Terminal zig-zag segment length	5.62 mm
Platinum-10% Iridium	
Young modulus	224 GPa
Poisson ratio	0.37
Yield stress	285 MPa
Ultimate strength	875 MPa
Fatigue endurance strength	263 MPa
Gold	
Young modulus	80 GPa
Poisson ratio	0.42

Table 1. Stent geometrical and material properties

Patient-specific stent deployments were replicated in Abaqus/Standard (Simulia, RI, USA) using nodal displacement boundary conditions (Fig. 3bottom). These displacements were those previously calculated from fluoroscopy reconstructions and were applied to the central node of each stent strut junction. Three displacement steps were performed to account for the full loading history of the stent and potential residual stresses: the first step resulted in the stent deployment from its initial crimped status to the end of balloon inflation; the second and the third steps replicated a cardiac cycle from systole to diastole. The elements surrounding the nodes of displacement application were subtracted from the model in the analysis of stress outcomes because they resulted distorted due to the type of displacement condition adopted. This was considered acceptable as fractures at the strut junctions have not been reported for the current stent after golden reinforcements were introduced.

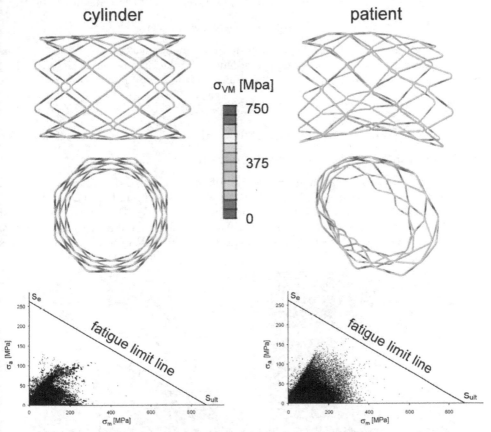

Fig. 4. Comparison of the stress distribution and Goodman diagrams between the stent deployed in a cylindrical configuration, as performed during conventional preclinical testing, and the stent deployed according to individual *in-situ* geometry from 1 of the studied patients. Goodman diagrams are a graphical representation of fatigue analysis (σ_m and σ_a = mean and alternating stresses, while S_{ult} and S_e = material ultimate and fatigue endurance strengths)

Stresses in the stents after deployment in patients' configurations were higher if compared to the same stent uniformly expanded in a cylindrical geometry as done during conventional preclinical testing, and, therefore, the risk of reaching the endurance limit of the material was higher (Fig. 4). The highest stresses occurred close to the strut junctions, which were the most highly bent portions of the device. The peak σ_{VM} was reached during diastole in every patient, and its values ranged between 516.1 and 612.8 MPa.

A fatigue analysis was performed by applying the Goodman method (Beden et al., 2009; Marrey et al., 2006) and the Sines criterion (Sines & Ohgi, 1981). The first uses a graphical approach to relate alternating stress (σ_a) and mean stress (σ_m) during the cardiac cycle with the material strength limits (Fig. 4), where σ_a and σ_m are calculated as follows:

$$\sigma_m = \frac{\sigma_{sys} + \sigma_{dia}}{2} \text{ and } \sigma_a = \frac{|\sigma_{sys} - \sigma_{dia}|}{2} \text{ with } \sigma_{sys} \text{ and } \sigma_{dia} \text{ equal to the maximum principal}$$

stresses at the end of the systolic and diastolic phases respectively.

The second is a multi-axial fatigue criterion that uses the equivalent Von Mises stress as control parameter. To evaluate the fatigue resistance with the Sines criterion, the equivalent Sines stress, for the comparison with the material fatigue endurance strength S_e, was calculated as in the first member of the following disequation: $\sqrt{3}\left(\sqrt{J_2}\right)_a + 3\frac{S_e}{S_{ult}}\sigma_{H,m} \leq S_e$

where $\left(\sqrt{J_2}\right)_a$ and $\sigma_{H,m}$ are the amplitude of the mean square root of the second deviatoric stress invariant and the hydrostation pressure, respectively.

During the cyclic loading condition, the stent in the patients worked at high σ_m and σ_a, if compared to a uniformly deployed stent, thus indicating a high number of areas at risk of fatigue failure as shown by the Goodman distributions (Fig. 4). The portions at highest potential for fracture were non-uniformly distributed amongst the different studied cases, with highest values in the areas close to the strut junctions, where indeed such fractures

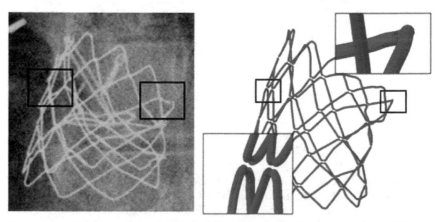

Fig. 5. X-ray image of the stent *in-vivo* in 1 of the analysed patients at 3 months follow-up compared to the Sines stress map in the corresponding patient FE simulation with areas at highest risk for fracture highlighted in red

occurred in the patients (Fig. 5). The Sines criterion predicted fractures in every case, with the most stressed regions in 3 cases close to the strut intersections between the first and the second ring from the proximal end; in one case it was at the lowest terminal crown and in the last patient it was detected between the second and the third crown.

Reproducing patient-specific FE analysis with more realistic loading conditions can provide more accurate information regarding the stent mechanical performance and prediction of fatigue life compared to the conventional bench methodologies/FE analyses/ of free/cylindrically constrained expansion. Indeed, this computational analysis predicted failure and the locations of fractures were similar to the sites where the patients ultimately had stent damage.

2.1.2 Stent-in-stent as a potential solution to stent fractures

FE analysis can also give insight into improved clinical management. This is demonstrated by a FE analysis of stent mechanical performance for various combinations of stent designs. Inserting a separate stent into the pulmonary artery prior to the PPVI device (stent-in-stent technique, Nordmeyer et al., 2008) may reduce valved stent fractures. An optimised combination of different stent mechanical properties can help increase the strength of the device structure while at the same time reduce the stresses. This was computationally tested by expanding 2 PPVI stents 1 inside the other (Schievano et al, 2007c). The mechanical performance of the coupled device was compared with that of a single valved stent, expanded and cyclic loaded at the same conditions. The stresses in the outer stent of the coupled device were similar to those of the single valved stent. However, the stresses in the coupled device inner stent, which holds the valve, were lower when compared to the stresses in the single valved stent (Fig. 6), thus decreasing the risk of fracture for the valved stent.

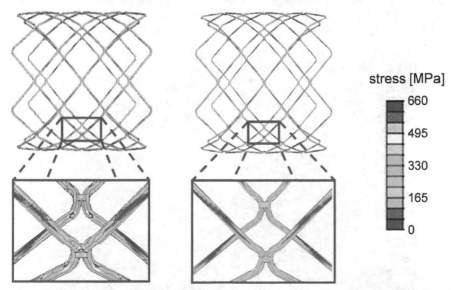

stress [MPa]

660

495

330

165

0

Fig. 6. Stress distributions in the single valved stent and in the inner valved stent when 2 devices are coupled together (stent-in-stent)

The implantation of a previous device before the valved one acts functionally to bolster the vessel and reduce the stresses on the stent. This information, in combination with the computational modelling described above, suggests that pre-stenting may ensure the integrity of the valved device, thus enhancing the success of the percutaneous procedure. Indeed this is the case when we look at patients who undergo pre-stenting, where the risk of developing a stent fracture is reduced (Nordmeyer et al., 2011). However, this solution is still limited to those patients with suitable implantation sites and the current device does not fit all the sizes of patients requiring pulmonary valve treatment. New devices have to be developed to offer this minimally invasive procedure to the entire patient population.

2.2 Next generation device

Despite the success of the PPVI device, <15% of patients requiring pulmonary valve replacement can be treated with the current PPVI device, with the remaining 85% requiring open-heart surgery (Schievano et al., 2007a). The majority of these patients are those with dilated, dynamic pulmonary arteries in whom the current percutaneous device is too small. Importantly, for this morphological problem, animal testing is of very limited use, as there are no models that encompass the wide variations of anatomy seen in these patients with congenital heart disease. Hence development of a new device to deal with the clinical problem and its translation into man is difficult using the conventional pathway of bench followed by animal testing.

Over the last 5 years, a new PPVI device for implantation into the dilated pulmonary artery has been developed in collaboration with Medtronic Cardiovascular. The device is made from self-expandable nitinol zig-zag rings, held together by a polyester graft in an hourglass shape (Fig. 7). The extremities of the stent, with larger diameters, would ensure the anchoring of the device against the pulmonary artery, while the central rings, with smaller diameter, would act as supporting structures for the valve. Nitinol is a shape memory alloy regularly used in bioengineering applications because it combines important qualities such

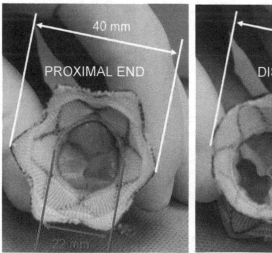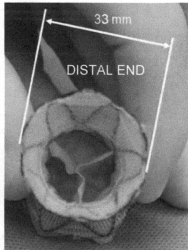

Fig. 7. New PPVI device and dimensions

as biocompatibility, fatigue resistance, and magnetic resonance (MR) compatibility, with the peculiar mechanical behaviour to undergo large completely recoverable deformations (Petrini et al., 2005). The valve is made of porcine pericardium and is sutured in the central portion of the stent-graft. Conventional animal testing and bench testing had been undertaken for this device and demonstrated a good performance in the animal model (Bonhoeffer et al., 2008). This stent-graft design should guarantee a greater adaptability of the device to the wide range of possible implantation site morphologies, with long-term fatigue behaviour that dramatically outperforms conventional metals.

2.2.1 First-in-man application

During the final stages of the preclinical testing of the new device, a patient presented to our Institution requiring pulmonary valve replacement (Schievano et al., 2010a). This patient (42-year-old man with congenital heart disease) had previously undergone 4 open-heart operations and 2 additional thoracic procedures, and remained highly symptomatic with severe pulmonary insufficiency. A further cardiothoracic surgery was considered too high risk. In addition, the pulmonary artery was too dilated for the current PPVI device. Therefore, the patient was considered for implantation of the new device. An integrated strategy to pre-clinical testing, using patient imaging data, computer modelling and biomedical engineering was developed to influence the final device design for implantation.

4D cardiovascular computed tomography (CT) was performed to acquire 3D volumes of the implantation site over 10 frames of the cardiac cycle (CT-SOMATOM Definition, Siemens Medical Systems) according to previously described methodologies (Schievano et al., 2007a), and to measure the 3D deformations in terms of diameter changes at different sections (Fig. 8a; Schievano et al., 2011). Based on the CT implantation site reconstructions (Mimics, Materialise, Belgium), FE models (Fig. 8b) and 3D rapid prototyping (Fig. 8c) models were created (Schievano et al., 2007b; Schievano et al., 2010b). Multiple device shapes and sizes with varying wire stiffness and configurations were tested in the rapid prototyping models and using FE analysis to optimise the anatomical results in the specific patient: the FE analysis identified definitive areas of contact between the computer simulated stent-graft struts of the customized device and implantation site, predicting likely stability and safe anchoring in-vivo. The rapid prototyping models enabled simulation of the clinical implantation procedure and helped the implanters decide the approach for optimal device delivery.

The final device (40 mm diameter in the distal and proximal ring, 22 mm diameter in the central portion holding the valve) underwent acute and chronic animal tests and bench testing. Based on these results, the UK Medicines and Healthcare products Regulatory Agency (MHRA), and the local ethical and industrial committees granted approval for the use of this device on humanitarian grounds. The patient gave informed consent for the procedure and for the anonymous use of his data and images for research.

PPVI was successfully carried out in this patient with the new device in January 2009. During the catheterisation procedure, angiographic studies and balloon sizing of the pulmonary trunk were performed and confirmed that the CT and model dimensions were a true representation of the patient's anatomy. The delivery strategy developed from the pre-procedural trial implantation in the rapid prototyping models was followed and proved safe

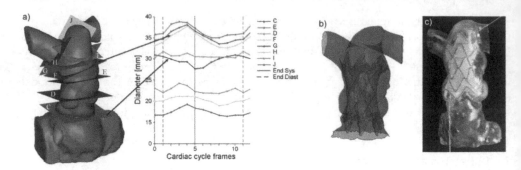

Fig. 8. a) 3D volume reconstruction of the patient's implantation site from 4D CT and diameter change measurements over the cardiac cycle measured in the 8 selected planes (C-L). Examples of b) FE model, and c) plastic rapid prototyping model (guidewire in place) of the patient anatomy with the new PPVI device "implanted" to test different positions, anchoring, and delivery approaches

and successful (Fig. 9a). The patient was symptomatically improved following the procedure. A post-implantation 4D CT confirmed that the device had the position, shape and safe anchoring as predicted by the FE models (Fig. 9b). These post implantation CT images were used to reconstruct the stent geometry *in-situ* and to assess the 3D displacements of the device rings over the cardiac cycle. This information was inputted in a FE analysis as explained above to predict the likelihood of stent fracture. The patient remains well 3 years following the procedure, with no fractures of the stent struts detected to date, as predicted by the FE study.

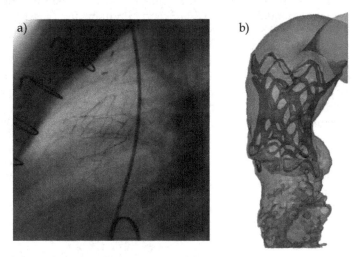

Fig. 9. Lateral view of a) x-ray fluoroscopy, and b) 3D volume CT reconstruction of the new PPVI device, 2 days after implantation in the patient

Whilst such a labour intensive method, which resulted in modifications in the device design prior to this first human case, was essential to enable us to safely transfer this new technology into early clinical practice, this approach would not be necessary, or sustainable, for each patient in routine clinical practice. Ultimately, the aim for this new PPVI technology would be to have a small number of 'off-the-shelf' devices with varying sizes and shapes, which would be suitable for the vast majority of patients. In order to optimise the number of devices necessary to cover the whole range of patients' morphologies, we used FE analysis taking into account patient specific anatomies (Capelli et al, 2010a).

2.2.2 Optimisation of the next generation device

FE modelling is a powerful tool to optimise device design without the need for many prototypes to be created and tested. In pulmonary valve dysfunction, each patient anatomy is completely individual in terms of size, shape and dynamics. FE can help select the most appropriate prosthesis for any individual patient taking into account their specific anatomy, thus enhancing the safety and success of the procedure. Furthermore, FE analysis can help predict how many devices may be needed to cover the whole range of patient morphologies.

Three different potential designs for the new stent-graft were considered, with equivalent central ring diameters (22 mm), but different proximal and distal strut dimensions (Fig. 10; Capelli et al, 2010a): the first stent-graft (SG1) resembled the device that had already been successfully tested in animals (Bonhoeffer et al, 2008). The second stent-graft (SG2) was symmetrical, similar to the device implanted into the patient described in the previous paragraph (Schievano et al, 2010a). The third stent-graft (SG3) was similar to SG2, but with larger proximal and distal diameters. 1D beam elements were chosen to mesh the stent wires (696, 696 and 912 elements for SG1, SG2 and SG3, respectively). Surfaces were created in between the struts to model the polyester fabric, which was meshed using membrane elements that offer strength in the plane of the element, but have no bending stiffness (3850, 3408 and 3909 elements for SG1, SG2 and SG3, respectively). A tight, rigid contact was assumed to simulate the suture between the stent and the graft. The shape memory alloy model implemented in Abaqus code was used to describe the nitinol behaviour of the stent wires. A hyperelastic, isotropic constitutive model based on a reduced polynomial strain

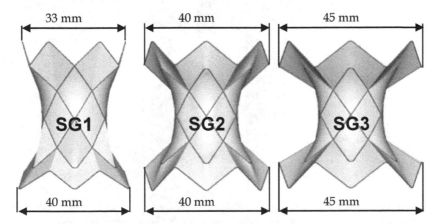

Fig. 10. Models of the 3 stent-grafts SG1, SG2 and SG3 designed and tested using FE analysis

energy density function (C10 = 0.38, C20 = 4.36, C30 = 80.56, C40= -134.72, C50 = 86.24, C60= -19.74) was used for the fabric graft material. The material FE models were validated using experimental tensile tests carried out for the nitinol wires and fabric samples. The valve was neglected in the FE analyses.

Pre-operative 3D MR data (1.5 T Avanto, Siemens Medical Systems) from 62 patients who were morphologically or dimensionally unsuitable for the current Melody™ device were used to reconstruct rigid FE models of the implantation sites – rigid 3D shell elements with number of elements varying between 7,172 and 13,104 according to the complexity of the patient implantation site geometry (Capelli et al, 2010a).

The 3 stent-grafts' FE models were placed inside each patient's outflow tract model, as close as possible to the bifurcation, without obstructing the pulmonary arteries. The implantation of the devices was carried out by crimping the stent-grafts down to 7 mm diameter (catheter dimensions) and then releasing them inside the patients' outflow tract models (Abaqus/Explicit). A general contact algorithm was defined to allow interaction between the devices and the arterial wall. The stent-grafts adapted their shape to the implantation site of each specific patient (Fig. 11). The diameters after deployment were quantified in the proximal, central and distal sections of the stents to judge the safe anchoring of the device inside the artery, which was considered optimal if the proximal and distal diameters measured less than 80% of the original diameters, according to manufacturer's specifications. Furthermore, the central portion of the device should be >18 mm so that the valve sewn inside can be fully deployed.

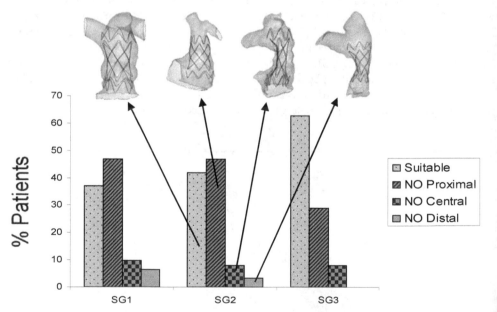

Fig. 11. Percentage of suitable and not suitable patients for SG1, SG2, and SG3 and examples of patients in which device SG2 was virtually implanted. Unsuitable patients are subdivided according to the regions of the stent that did not respect the criteria of safe implantation

According to these criteria, successful implantation was achieved in 37% of the patients with SG1. 42% of the patients would be suitable for the SG2 device and 63% for the SG3 device (Fig. 11). Therefore, 37% of those patients who are currently still treated with surgery would potentially be suitable for PPVI with a new device. Furthermore, if the dimensions of this new device are theoretically increased at the distal end (SG2, as done for the first-in-man case) or both proximal and distal ends (SG3), the number of percutaneous procedures would increase by a further 5% and 36%, respectively. Importantly, these dimensions would be difficult to test in animals because of lack of relevant sizes in these settings. Although animal testing remains important, FE modelling could be integrated into preclinical testing to predict how such devices behave when implanted into the human situation.

With the introduction in clinical practice of the new device in 3 sizes, the total number of patients requiring pulmonary valve replacement who could benefit from a percutaneous approach (current device + 3 sizes of the new device) would be approximately 70%. This would potentially have a big impact on the cost benefit for healthcare, by reducing hospital stay and improving the speed at which patients can get back to normal daily activities.

3. Transcatheter aortic valve implantation

In the past decade, TAVI has been shown to be a feasible and effective option for treatment of patients with symptomatic severe aortic stenosis and high operative risks (Leon et al., 2011; Rodes-Cabau et al., 2010; Smith et al., 2011; Vahanian et al., 2008; Zajarias & Cribier, 2009). Since the first-in-man experience in 2002 (Cribier et al., 2002), several advances in TAVI techniques have led to improved success rates, with acceptable procedure-related complication rates (Delgado et al., 2010). To date, 2 different devices have received CE mark approval – the balloon-expandable Edwards-Sapien® XT Valve (Edwards Lifesciences, CA, USA) and the self-expandable CoreValveReValving System® (Medtronic) – with many other devices emerging into the market (Fig. 12). The encouraging mid- and long-term results of

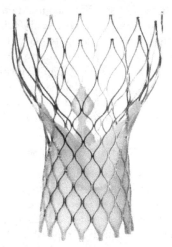

Fig. 12. TAVI Edwards-Sapien® Valve and the self-expandable Medtronic CoreValve ReValving System®
(http://mail.c2i2.org/web10-06/transcatheter_aortic_valve_implantation.asp)

this technique, together with increased patient comfort, shortened intensive care and hospital stay, and the avoidance of cardiopulmonary bypass, make this non-surgical technique extremely appealing (Ussia et al., 2009).

Accurate multidisciplinary pre-procedural evaluation of patients who are considered candidates for TAVI is mandatory to plan the most adequate treatment and to minimise peri- and post-procedural complications (Smith et al., 2011; Vahanian et al., 2008). However, in this emerging field, several issues remain a source of debate. Device sizing and positioning are the main challenges, but vascular complications, electrical conduction abnormalities and post-procedural aortic regurgitation still remain major safety concerns. In addition, according to the current position statement, TAVI is indicated only in patients with severe symptomatic aortic stenosis, and who are considered at high or prohibitive risk for conventional surgery (Ussia et al., 2009, Leon et al., 2011).

Extending TAVI to patients who still undergo conventional surgery poses new challenges, both for clinicians and device manufacturers, but "within 10 years, with further improvement of the devices and procedures, and depending on the long-term results of upcoming controlled trials in a broad population, TAVI may become the treatment of choice in a majority of patients with degenerative aortic stenosis" (Cribier, 2009). Indeed, highly experienced centres have already demonstrated the feasibility of TAVI in failed bioprosthetic heart valves (valve-in-valve procedure) in patients considered at very high-risk or ineligible for surgery (Webb et al., 2010). Previous surgical valve implantations represent a well-defined landmark with rigid boundaries that increase ease of positioning and anchoring, thus making patients with bioprosthetic valves ideal candidates for TAVI (Walther et al., 2011). Conversely, younger patients, with less severe aortic valve stenosis or with valve insufficiency, often present implantation sites with different anatomical and dynamic characteristics that generate procedural and device related hurdles, which means that such patients are currently not suitable for or offered TAVI.

Combining patient-specific imaging data and computational modelling offers a new method to obtain additional, predictive information about responses to cardiovascular device implantation in individual patients (Schoenhagen et al., 2011; Taylor & Figueroa, 2009). FE analyses were performed to explore the feasibility of TAVI using a model of the Edwards-Sapien® device in specific patient morphologies which are currently borderline cases for a percutaneous approach. This method can help in both refining patient selection and characterising device mechanical performance, overall impacting on procedural safety and success in the early introduction of TAVI devices in new patient populations.

3.1 Patients with previously implanted bioprosthetic valves

Four patients with different stenotic bioprosthetic valves previously implanted (patients A – 23 mm Carpentier-Edwards Perimount Magna valve, Edwards Lifesciences; B – 23 mm Soprano™ valve, Sorin Biomedica Cardio, Italy; C – 25 mm Carpentier-Edwards Perimount valve, Edwards Lifesciences; D – 25 mm Epic™ valve, St Jude Medical, MN, USA) and who were referred for surgical replacement of their failing bioprosthetic valve were analysed (Bosi et al., 2010; Migliavacca et al., 2011). Data about the anatomy of the aortic root, coronary arteries, AV leaflets and bioprostheses were acquired using CT imaging (CT-SOMATOM Definition) and these data were used to reconstruct 3D geometries for FE models of the selected patients' implantation sites (Fig. 13). The aortic roots were assumed to be 2 mm thick, with density equal to 1,120 kg/m³ (Conti et al., 2010) for all models and meshed with 3D triangular shell general-purpose elements (Table 2). To describe the mechanical behaviour of the aortic roots, Mooney-Rivlin hyperelastic behaviour was adopted incorporating experimental stress-strain data for the ascending aorta (Okamoto et al., 2002) and taking into account the pre-stretching of the aortic root due to the aortic pressure during the cardiac circle.

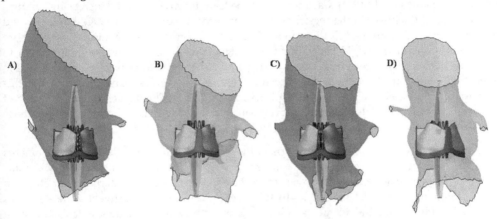

Fig. 13. CT image reconstruction of the selected patients' aortic wall with previously implanted bioprosthetic valves. FE models of the deployment balloon and TAVI stent are placed inside the implantation site models

Part	Elements
Patient-specific model	
A	6,875
B	8,675
C	11,325
D	8,347
Bioprosthesis (stent + leaflets)	
Perimount Magna	44,048 + 29,021
Soprano	51,372 + 28,667
Perimount	44,048 + 29,021
Epic	51,372 + 28,667
TAVI device	
Sapien stent	192,920
Balloon	8,160

Table 2. Number of mesh elements for the different parts involved in the FE simulations

The metal frames of the 4 bioprosthetic valves were reconstructed from the CT images to identify their position inside the patients' outflow tracts. The bioprosthetic valve geometries were re-drawn using CAD software (Rhinoceros) to recreate a complete model of the corresponding commercial device used in the patients, and then placed in the same position as that identified from CT images. Connector elements were used to link the bioprosthetic valves with the aortic vessels to mimic the suture between the ring and the aortic root. Each prosthesis included different frame structures, sewing rings and artificial valve leaflets. The frames (metallic and polymeric) were meshed with 8-node linear hexahedral elements with reduced integration (Table 2), and modelled using an elasto-plastic constitutive behaviour obtained from the manufacturer's data. Bovine and porcine valve leaflets were meshed with 4-node, quadrilateral, shell elements with reduced integration and large-strain formulation (Table 2), while the material properties were simplified using a linear, elastic model (Lee et al., 1984; Zioupos et al., 1994). Calcification of the leaflets was created by increasing Young's modulus and thickness in the region of the commissures (Loree et al., 1994), where a weld constraint was also applied (Schievano et al., 2009): 16 connector elements per commissure were added. An axial, rigid behaviour was assigned to the connectors, allowing them to maintain a fixed distance between the 2 nodes until a threshold force was reached – equal to 0.92 N (Loree et al., 1994) along the direction joining the 2 nodes – and then fail after this threshold value was reached, mimicking calcification failure. The thickening and welding of the failing valve leaflets produced aortic valve geometric orifice areas equal to those measured in the selected patients (1.2 cm^2 for A and B, and 1.4 cm^2 for C and D).

The Edwards-Sapien® stent is characterised by 12 units, each formed by 4 zigzag elements (Fig. 12). A vertical bar divides each unit and a perforated bar is positioned every 4 units. The CAD geometry of this stent was reproduced with a height, internal diameter and thickness of the expanded stent equal to 16.0, 25.4 and 0.3 mm, respectively. The stent was mashed with hexahedral elements, following mesh sensitivity analysis. It is made of stainless steel which was modelled using Von Mises plasticity behaviour from manufacturer data.

A semi-compliant balloon was designed to resemble the commercial balloon used in clinical practice (Z MED II™, NuMed Inc, NY, USA) and placed inside the TAVI stent in order to deploy the device (Capelli et al, 2010b; Gervaso et al., 2008). The balloon was meshed with 0.03 mm thick membrane elements, and a homogeneous, isotropic, linear-elastic Nylon11 was adopted according to manufacturer data. The stent was crimped down onto the balloon, from its original configuration to catheter dimension (8 mm diameter), using a coaxial cylindrical surface.

The stent-balloon system was placed into the aortic root models according to the judgment of 2 clinicians (Fig. 13). Large deformation analysis of stent deployment in the patients' implantation sites, performed with Abaqus/Explicit, was divided in 2 different steps: balloon pressurisation with resulting stent expansion in the aortic root, and balloon deflation with subsequent stent recoil. In all patients, the implantation site model extremities (upper and lower aortic sections and coronary terminal sections) were constrained in all directions (circumferential, radial and longitudinal) in order to mimic the connection with biological structures. Boundary conditions on the balloon were placed to mimic the bond with the catheter.

Contact properties were defined to describe the interactions encountered in these multi-part analyses. Interactions included contact between surfaces belonging to balloon and TAVI stent, balloon and bioprosthetic aortic valve leaflets, TAVI stent and bioprosthetic aortic valve, bioprosthetic aortic valve and aortic wall. Friction between Nylon and stainless steel was included in the model with coefficient equal to 0.25 (De Beule et al., 2008).

Mechanical performance of the stent and the impact of the TAVI device into patients' implantation sites were evaluated by analysing: stent configurations at the end of balloon inflation, stent recoil after balloon deflation, stent and arterial stress distribution and peak values at the end of the expansion and recoil phases, degree of aortic valve stenosis assessed according to the post-TAVI geometrical orifice area, and evaluation of coronary artery obstruction.

At the end of all simulations, the TAVI stent was virtually implanted inside the bioprostheses of the patient-specific aortic root model, with a position that was found in good agreement with available images from a TAVI follow-up case (Fig. 14, Webb et al., 2010).

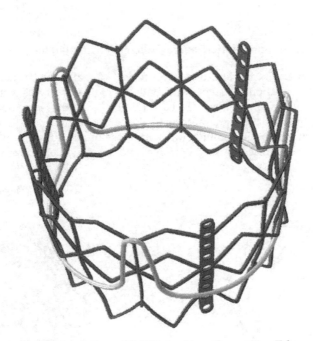

Fig. 14. Result from the FE simulation of TAVI stent in a Carpentier-Edwards bioprosthesis, resembling the CT image reconstruction of a patient's TAVI in the same bioprosthetic valve published by Webb et al., 2010

After deployment, the stent assumed an asymmetrical configuration in the longitudinal direction, more expanded in the distal part. There was no contact with the native aortic wall and/or other cardiac structure. This was also demonstrated by the low stresses (<0.1 MPa) measured in the aortic wall, showing how the previously implanted bioprostheses act as a scaffold for the TAVI stent. After balloon deflation, the bioprosthetic valve leaflets partially recoiled forcing the TAVI stent to a more symmetrical shape, which could be defined almost cylindrical in all cases (Fig. 15). Maximum recoil was measured in the distal sections for all patients (A = 17.0%; B = 13.2%; C = 11.7%; D = 10.8%) with absent or low recoil at the level of the bioprosthetic valves' annulus.

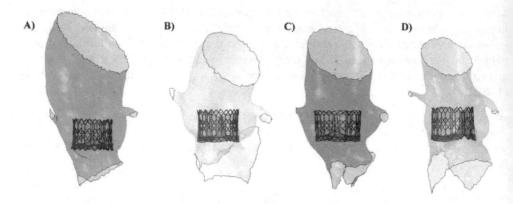

Fig. 15. TAVI stent final configuration (end of balloon deflation) in the patients' models

At the end of both the expansion and recoil phases, for all models, non-uniform stress distribution was recorded on the stent struts with the highest Von Mises stresses occurring at the strut junctions with the vertical bars (Fig. 16). At the same locations material plasticisation was reached that guaranteed final open configurations. If we compare the results from the patient-specific simulations with those of a stent uniformly deployed in a cylinder, which is the idealised implantation site used in conventional preclinical testing of

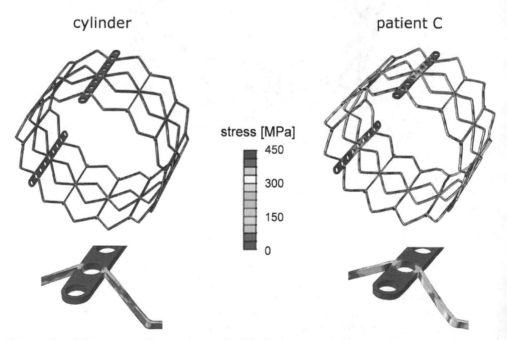

Fig. 16. Von Mises stress distribution in the TAVI stent after balloon deployment and deflation in a cylindrical configuration and in patient C

such devices, the maximum stress reached in the stent deployed in the patients was significantly higher than the stress in the stent deployed in the cylinder (443 vs 270 MPa). In addition, the stent was uniformly deployed in the cylinder during the entire simulation: recoil between the steps of balloon inflation and deflation was low (1.5%) and uniform along the stent length. The non-uniform shape of the deployed configuration in the patients and, therefore, the asymmetric stress distribution might cause long-term stent failure due to the pulsatile loading conditions during the cardiac cycle that are not seen during conventional pre-clinical tests (Schievano et al., 2010b).

After simulated TAVI, the connectors modelling calcification of the leaflet commissures were all broken, thus increasing the aortic valve geometrical orifice area to 3.43, 3.60, 3.72, and 3.73 cm^2 for patient A, B, C and D respectively. Minimum distances between implanted TAVI device and coronary arteries (right and left) after the deflation of the balloon were 10.9, 11.4, 10.5, and 5.5 mm for patient A, B, C and D respectively, with no direct obstruction of the ostia.

In all cases, the virtual implantation of the TAVI device predicted a successful procedural outcome with an orifice area larger than the pre-implantation stenotic area, and no interference with other cardiac structures. Furthermore, patient-specific FE analyses showed no fractures of the stent immediately post-implantation. These methodologies might help engineers to better understand the mechanical behaviour of the stent when interacting with a wide variety of potential anatomies and, therefore, to optimise the device design for different potential clinical applications before actual procedures are performed.

3.2 Patients with aortic incompetence

A patient (E) diagnosed with congenital, moderate aortic valve stenosis, which was treated after birth with aortic balloon valvuloplasty, was selected. At 21 years of age, this patient was referred for surgical repair of her severe aortic valve regurgitation. FE model of TAVI in this patient specific native aortic root and valve leaflets was performed using the same TAVI stent model described above.

CT image data (CT-SOMATOM Definition) from the patient corresponding to mid-systole (i.e. open aortic valve leaflets) were elaborated in the image post-processing software Mimics. A 3D model of the aortic root, coronary arteries and valve leaflets was obtained for the patient (Fig. 17). The native structures were meshed with 3D triangular shell general-purpose elements (18,890 elements). Aortic wall and native valve leaflets were assumed to be respectively 2 and 0.5 mm thick (Conti et al., 2010) with density equal to 1,120 kg/m^3, and Mooney-Rivlin hyperelastic constitutive law was adopted to describe the material behaviour.

In model E, the TAVI stent/balloon was placed in 3 different positions within the aortic root to test the influence of the landing zone into safe anchoring, the interference with other cardiac structures, such as the mitral valve and the atrioventricular node, and potential occlusion of the coronary arteries. First, the central section of the stent was placed aligned to the leaflet commissures (E_M; Fig 17), then it was moved 4.2 mm proximally towards the left ventricle (E_P; Fig. 17), and 4.2 mm distally towards the aortic arch (E_D; Fig. 17).

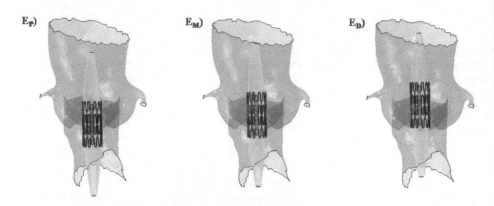

Fig. 17. Balloon and TAVI stent in the 3 analysed positions inside the native aortic valve: in correspondence of the leaflet commissures (E_M); moved 4.2 mm proximally towards the left ventricle (E_P); and, 4.2 mm distally towards the aortic arch (E_D).

The same loading and boundary conditions as those described above for the study of TAVI in bioprosthetic valves were adopted. Also, the same quantities of interests were measured.

In this patient, at the end of stent deployment and balloon deflation, the device assumed an asymmetrical configuration for all tested positions, more open distally than proximally (Fig. 18). In all 3 cases, the interaction between the TAVI device and the native implantation site was well confined within the left ventricular outflow tract and aortic root portion of the patient's morphology, thus reducing the potential risk of heart block and mitral valve leaflet entrapment. The stent implanted in this patient in the most distal position was the closest to the right coronary artery (3.1 mm). The TAVI device caused no direct obstruction of the coronary orifice; however, further fluid-dynamic studies could enhance our understanding of the effects of the TAVI device placement to the coronary flow.

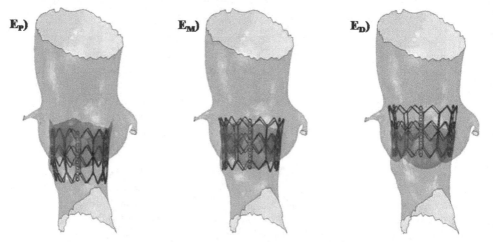

Fig. 18. TAVI stent final configuration (end of balloon deflation) in the 3 simulated positions.

At balloon deflation, the interaction with the native aortic valve and wall caused high recoil in the proximal section of the stent for all 3 tested positions E_M = 14.9%; model E_P = 20.2%; model E_D = 11.4%. This may be considered a potential cause for obstruction of the left ventricular outflow tract and result in potential dislodgment of the TAVI device from its original position. However, the final geometric orifice areas for the analysed patient in the 3 positions E_M, E_P, and E_D – 4.7, 3.7 and 5.3 cm^2 respectively – were same size or larger that the initial orifice area = 3.7 cm^2.

Direct interaction of TAVI stent with native tissue had impacts both on the device and on the implantation site. After recoil, the geometrical configuration of the stent was not uniform, with an asymmetrical stress distribution (max 477 MPa for E_P). The ultimate stress value for the stent material (stainless steel AISI 316L ultimate strength = 515 MPa) was not reached; this means that no stent fractures were seen immediately after deployment. However, pulsatile compressions during cardiac cycle could affect the long-term performance of this stent due to the dynamic nature of the native, non-calcified implantation site. This may be particularly relevant for model E_P where the distal portion of the device was in direct contact with the active muscular portion of the left ventricular outflow tract.

The expansion of the TAVI stent within native tissue also caused high maximum principal stresses (1 MPa) in the arterial wall, in particular in the region of the leaflets for model E_D, at commissure level (Fig. 19). This might induce damage or stimulate remodelling (e.g. stenosis).

A careful balance between interrelated and sometimes contradictory requirements has to be achieved for optimal TAVI positioning and outcomes such as relief of valve dysfunction and safe anchoring with no tissue damage, no coronary obstruction, no interference with other cardiac structures and no device failure in both short- and long-term. Patient specific FE analysis can be used to help in this process of optimisation and represent an additional assessment tool for the selection of patients for TAVI.

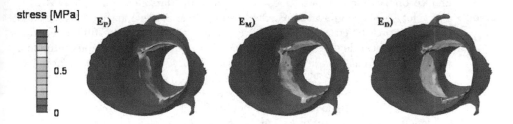

Fig. 19. Maximum principal stress distribution in the native aortic valve leaflets and aortic root after balloon deflation in models E_P, E_M and E_D.

4. Conclusion

Engineering and computational methodologies, together with state of the art imaging technologies, can be used to replicate patient specific patho-physiological conditions in virtual models. Combining patient specific imaging data and computational modelling can improve our understanding of heart structures and the way devices interact with them.

Therefore, advanced FE analyses can become a fundamental tool during preclinical testing of new biomedical devices that could shorten the time for device development, minimise animal experimentation, and ensure greater patient safety during the delicate phase of testing clinical feasibility of novel cardiovascular technologies.

5. Acknowledgments

This work has been supported by grant funding from the British Heart Foundation (BHF) (Dr. Schievano & Mr. Capelli – PhD Studenships), the Royal Academy of Engineering and the Engineering and Physical Sciences Research Council (EPSRC) (Dr. Schievano – Post-doctoral Fellowship), the National Institute of Health Research (NIHR) (Professor Taylor – Senior Research Fellowship), the European Union (Ms. Cosentino – MeDDiCA PhD Studentship 7th Framework Programme, grant agreement 238113) and the Fondation Leducq.

6. References

Beden, S.M., Abdullah, S., Ariffin, A.K., AL-Asady, N.A., & Rahman, M.M. (2009). Fatigue life assessment of different steel-based shell materials under variable amplitude loading. *European Journal of Scientific Research*, Vol.29, No.1, pp. 157-169

Bonhoeffer, P., Boudjemline, Y., Saliba, Z., Merckx, J., Aggoun, Y., Bonnet, D., Acar, P., Le Bidois, J., Sidi, D., & Kachaner, J. (2000). Percutaneous replacement of pulmonary valve in a right-ventricle to pulmonary-artery prosthetic conduit with valve dysfunction. *Lancet*, Vol.356, No.9239, (21 October 2000), pp. 1403-1405

Bonhoeffer, P., Huynh, R., House, M., Douk, N., Kopcak, M., Hill, A., & Rafiee, N. (2008). Transcatheter pulmonic valve replacement in sheep using a grafted self-expanding stent with tissue valve. *Circulation*. Vol.118, pp. S_812

Bosi, G.M., Cerri, E., Capelli, C., Migliavacca, F., Bonhoeffer, P., Taylor, A.M., & Schievano, S. (2010). Patient-specific study of transcatheter aortic valve implantation. Endocoronary Biomechanics Research Symposium, Marseille, France, May 2010

Capelli, C., Taylor, A.M., Migliavacca, F., Bonhoeffer, P., & Schievano, S. (2010a). Patient-specific reconstructed anatomies and computer simulations are fundamental for selecting medical device treatment: application to a new percutaneous pulmonary valve. *Philosophical Transactions of the Royal Society A: Mathematical, Physical and Engineering Sciences*, Vol.368, No.1921, (28 June 2010), pp. 3027-3038

Capelli, C., Nordmeyer, J., Schievano, S., Lurz, P., Khambadkone, S., Lattanzio, S., Taylor, A.M., Petrini, L., Migliavacca, F., & Bonhoeffer, P. (2010b). How do angioplasty balloons work: a computational study on balloon expansion forces. *EuroIntervention*, Vol.6, No.5, (November 2010), pp. 638-642

Cribier, A., Eltchaninoff, H., Bash, A., Borenstein, N., Tron, C., Bauer, F., Derumeaux, G., Anselme, F., Laborde, F., & Leon, M.B. (2002). Percutaneous transcatheter implantation of an aortic valve prosthesis for calcific aortic stenosis - first human case description. *Circulation*, Vol.106, No.24, (10 December 2002), pp. 3006-3008

Cribier, A. (2009). Transcatheter Aortic Valve Implantation: What are the perspectives? at Innovations in Cardiovascular Interventions (ICI) Meeting 2009. http://www.paragon-conventions.net/ICI_PDF/Alain%20Cribier.pdf

Conti, C.A., Votta, E., Della Corte, A., Del Viscovo, L., Bancone, C., Cotrufo, M., & Redaelli, A. (2010). Dynamic finite element analysis of the aortic root from MRI-derived parameters. *Medical Engineering & Physics*, Vol.32, No.2, (March 2010), pp. 212-221

Cosentino, D., Capelli, C., Pennati, G., Díaz-Zuccarini, V., Bonhoeffer, P., Taylor, A.M., Schievano, S. (2011). Stent Fracture Prediction in Percutaneous Pulmonary Valve Implantation: A Patient-Specific Finite Element Analysis. *International Conference on Advancements of Medicine and Health Care through Technology, IFMBE Proceedings*, Vol.36, Part 4, (01 January 2011), pp. 288-293

De Beule, M., Mortier, P., Carlier, S.G., Verhegghe, B., Van Impe, R., & Verdonck, P. (2008). Realistic finite element-based stent design: The impact of balloon folding. *Journal of Biomechanics*, Vol.41, No.2, pp. 383-389

Delgado, V., Ewe, S.H., Ng, A.C.T., van der Kley, F., Marsan, N.A., Schuijf, J.D., Schalij, M.J., & Bax, J.J. (2010) Multimodality imaging in transcatheter aortic valve implantation: Key steps to assess procedural feasibility. *Eurointervention*, Vol.6, No.5 (November 2010), pp. 643-652

Gervaso, F., Capelli, C., Petrini, L., Lattanzio, S., Di Virgilio, L., & Migliavacca, F. (2008). On the effects of different strategies in modelling balloon-expandable stenting by means of finite element method. *Journal of Biomechanics*, Vol.41, No.6, pp. 1206-1212

Lee, J.M., Boughner, D. R., & Courtman, D.W. (1984). The glutaraldehyde-stabilized porcine aortic-valve xenograft .2. Effect of fixation with or without pressure on the tensile viscoelastic properties of the leaflet material. *Journal of Biomedical Materials Research*, Vol.18, No.1, (January 1984), pp. 79-98

Leon, M.B., Piazza, N., Nikolsky, E., Blackstone, E.H., Cutlip, D.E., Kappetein, A.P., Krucoff, M.W., Mack, M., Mehran, R., Miller, C., Morel, M.A., Petersen, J., Popma, J.J., Takkenberg, J.J.M., Vahanian, A., van Es, G.A., Vranckx, P., Webb, J.G., Windecker, S., & Serruys, P.W. (2011). Standardized endpoint definitions for transcatheter aortic valve implantation clinical trials a consensus report from the valve academic research consortium. *Journal of the American College of Cardiology*, Vol.57, No.3, (18 January 2011), pp. 253-269

Loree, H.M., Grodzinsky, A.J., Park, S.Y., Gibson, L.J., Lee, R.T. (1994). Static circumferential tangential modulus of human atherosclerotic tissue. *Journal of Biomechanics*, Vol.27, No.2, (February 1994), pp. 195-204

Lurz, P., Coats, L., Khambadkone, S., Boudjemline, Y., Schievano, S., Muthurangu, V., Lee, T.Y., Parenzan, G., Derrick, G., Cullen, S., Walker, F., Tsang, V., Deanfield, J., Taylor, A.M., & Bonhoeffer, P. (2008). Percutaneous pulmonary valve implantation - Impact of evolving technology and learning curve on clinical outcome. *Circulation*, Vol.117, No.15, (15 April 2008), pp. 1964-1972

Marrey, R.V., Burgermeister, R., Grishaber, R.B., & Ritchie, R.O. (2006). Fatigue and life prediction for cobalt-chromium stents: a fracture mechanics analysis. *Biomaterials*, Vol.27, No.9, (March 2006), pp. 1988-2000

McElhinney, D.B., Hellenbrand, W.E., Zahn, E.M., Jones, T.K., Cheatham, J.P., Lock, J.E., & Vincent, J.A. (2010). Short- and medium-term outcomes after transcatheter pulmonary valve placement in the expanded multicenter US melody valve trial. *Circulation*, Vol.122, No.5, (3 August 2010), pp. 507-516

Migliavacca, F., Baker, C., Biglino, G., Bosi, G., Capelli, C., Cerri, E., Corsini, C., Cosentino, D., Hsia, T.Y., Pennati, G., & Schievano, S. (2011). Numerical simulations to study aortic arch pathologies: application to hypoplastic left heart syndrome, aortic coarctation and aortic valve diseases. In "New endovascular technologies: from bench test to clinical practice". Eds. Chakfé N, Durand B and Meichelboeck W. Europrot, Strasbourg, France, 9-22, 2011

Nordmeyer, J., Khambadkone, S., Coats, L., Schievano, S., Lurz, P., Parenzan, G., Taylor, A.M., Lock, J.E., & Bonhoeffer, P. (2007). Risk stratification, systematic classification and anticipatory management strategies for stent fracture after percutaneous pulmonary valve implantation. *Circulation*, Vol.115, No.11 (20 March 2007), pp. 1392-1397

Nordmeyer, J., Coats, L., Lurz, P., Lee, T.Y., Derrick, G., Rees, P., Cullen, S., Taylor, A.M., Khambadkone, S., & Bonhoeffer, P. (2008). Percutaneous pulmonary valve-in-valve implantation: a successful treatment concept for early device failure. *European Heart Journal*, Vol. 29, No.6 (March 2008), pp. 810-815

Nordmeyer, J., Lurz, P., Khambadkone, S., Schievano, S., Jones, A., McElhinney, D.B., Taylor, A.M., & Bonhoeffer, P. (2011). Pre-stenting with a bare metal stent before percutaneous pulmonary valve implantation: Acute and one-year outcomes. *Heart*, Vol.97, No.2, (January 2011), pp. 118-123

Okamoto, R., Wagenseil, J.E., Delong, W.R., Peterson, S.J., Kouchoukos, N.T., & Sundt, T.M. (2002). Mechanical Properties of Dilated Human Ascending Aorta. *Annals of Biomedical Engineering*, Vol.30 No.5, (May 2002), pp. 624–635

Petrini, L., Migliavacca, F., Massarotti, P., Schievano, S., Dubini, G., & Auricchio, F. (2005). Computational studies of shape memory alloy behavior in biomedical applications. *Journal of Biomechanical Engineering*, Vol.127, No.4 (August 2005), pp. 716-725

Rodes-Cabau, J., Webb, J.G., Cheung, A., Ye, J., Dumont, E., Feindel, C.M., Osten, M., Natarajan, M.K., Velianou, J.L., Martucci, G., DeVarennes, B., Chisholm, R., Peterson, M.D., Lichtenstein, S.V., Nietlispach, F., Doyle, D., DeLarochelliere, R., Teoh, K., Chu, V., Dancea, A., Lachapelle, K., Cheema, A., Latter, D., & Horlick, E. (2010). Transcatheter aortic valve implantation for the treatment of severe symptomatic aortic stenosis in patients at very high or prohibitive surgical risk acute and late outcomes of the multicenter canadian experience. *Journal of the American College of Cardiology*, Vol.55, No.11, (16 March 2010), pp. 1080-1090

Schievano, S.; Coats, L.; Migliavacca, F.; Norman, W.; Frigiola, A.; Deanfield, J.; Bonhoeffer, P., & Taylor, A.M. (2007a). Variations in right ventricular outflow tract morphology following repair of congenital heart disease: implications for percutaneous pulmonary valve implantation. *Journal of cardiovascular magnetic resonance*, Vol.9, No.4, (February 2007), pp. 687-695

Schievano, S.; Migliavacca, F.; Coats, L.; Khambadkone, S.; Carminati, M.; Wilson, N.; Deanfield, J.E.; Bonhoeffer, P., & Taylor, A.M. (2007b). Percutaneous pulmonary valve implantation based on rapid prototyping of right ventricular outflow tract and pulmonary trunk from MR data. *Radiology*, Vol.242, No.2, (February 2007), pp. 490-497

Schievano, S., Petrini, L., Migliavacca, F., Coats, L., Nordmeyer, J., Lurz, P., Khambadkone, S., Taylor, A.M., Dubini, G., & Bonhoeffer, P. (2007c). Finite element analysis of stent deployment: understanding stent fracture in percutaneous pulmonary valve implantation. *Journal Interventional Cardiology*, Vol.20, No.6, (December 2007), pp. 546-554

Schievano, S., Kunzelman, K., Nicosia, M.A., Cochran, R.P., Einstein, D.R., Khambadkone, S., & Bonhoeffer, P. (2009). Percutaneous mitral valve dilatation: single balloon versus double balloon. A finite element study. *Journal of Heart Valve Disease*, Vol.18, No.1, (January 2009), pp. 28-34

Schievano, S.; Taylor, A.M.; Capelli, C.; Coats, L.; Walker, F.; Lurz, P.; Nordmeyer, J.; Wright, S.; Khambadkone, S.; Tsang, V.; Carminati, M., & Bonhoeffer, P. (2010a). First-in-man implantation of a novel percutaneous valve: a new approach to medical device development. *Eurointervention*, Vol.5, No.6, (January 2010), pp. 745-750

Schievano, S., Taylor, A.M., Capelli, C., Lurz, P., Nordmeyer, J., Migliavacca, F., & Bonhoeffer, P. (2010b). Patient specific finite element analysis results in more accurate prediction of stent fractures: Application to percutaneous pulmonary valve implantation. *Journal of Biomechanics*, Vol.43, No.4, (3 March 2010), pp. 687-93

Schievano, S., Capelli, C., Young, C., Lurz, P., Nordmeyer, J., Owens, C., Bonhoeffer, P., & Taylor, A.M. (2011). Four-dimensional computed tomography: a method of assessing right ventricular outflow tract and pulmonary artery deformations throughout the cardiac cycle. *European Radiology*, Vol.21,No.1, (January 2011), pp. 36-45

Schoenhagen, P., Hill, A., Kelley, T., Popovic, Z., & Halliburton, S.S. (2011). In vivo imaging and computational analysis of the aortic root. Application in clinical research and design of transcatheter aortic valve systems. *Journal of Cardiovascular Translational Research*, Vol.4, No. 4, (August 2011), pp. 459-469

Sines, G., & Ohgi, G. (1981). Fatigue criteria under combined stresses or strains. *Journal of Engineering Materials and Technology*, Vol.103, No.2, (April 1981), pp. 82-90

Smith, C.R., Leon, M.B., Mack, M.J., Miller, D.C., Moses, J.W., Svensson, L.G., Tuzcu, E.M., Webb, J.G., Fontana, G.P., Makkar, R.R., Williams, M., Dewey, T., Kapadia, S., Babaliaros, V., Thourani, V.H., Corso, P., Pichard, A.D., Bavaria, J.E., Herrmann, H.C., Akin, J.J., Anderson, W.N., Wang, D., Pocock, S.J., & Investigators PT (2011) Transcatheter versus surgical aortic-valve replacement in high-risk patients. *New England Journal of Medidcine*, Vol.364, No.23, (9 June 2011), pp. 2187-2198

Taylor, C.A., & Figueroa, C.A. (2009). Patient-specific modeling of cardiovascular mechanics. *Annual Review of Biomedical Engineering*, Vol.11, pp. 109-134

Ussia, G.P., Mulè, M., Barbanti, M., Cammalleri, V., Scarabelli, M., Immè, S., Capodanno, D., Ciriminna, S., & Tamburino, C. (2009). Quality of life assessment after percutaneous aortic valve implantation. *European Heart Journal*, Vol.30, No.14, (July 2009), pp. 1790-1796

Vahanian, A., Alfieri, O., Al-Attar, N., Antunes, M., Bax, J., Cormier, B., Cribier, A., De Jaegere, P., Fournial, G., Kappetein, A.P., Kovac, J., Ludgate, S., Maisano, F., Moat, N., Mohr, F., Nataf, P., Piérard, L., Pomar, J.L., Schofer, J., Tornos, P., Tuzcu, M., van Hout, B., Von Segesser, L.K., Walther, T. & European Association of Cardio-Thoracic Surgery; European Society of Cardiology; European Association of Percutaneous Cardiovascular Interventions. (2008). Transcatheter valve implantation for patients with aortic stenosis: a position statement from the European Association of Cardio-Thoracic Surgery (EACTS) and the European Society of Cardiology (ESC), in collaboration with the European Association of

Percutaneous Cardiovascular Interventions (EAPCI). *European Heart Journal,* Vol.29, No.11, (June 2008), pp. 1463-1470

Walther, T., Dehdashtian, M.M., Khanna, R., Young, E., Goldbrunner, P.J., & Lee, W. (2011). Trans-catheter valve-in-valve implantation: in vitro hydrodynamic performance of the SAPIEN+cloth trans-catheter heart valve in the Carpentier-Edwards Perimount valves. *European Journal of Cardio-thoracic Surgery,* Vol.40, No.5, (November 2011), pp. 1120-1126

Webb, J.G., Wood, D.A., Ye, J., Gurvitch, R., Masson, J.B., Rodés-Cabau, J., Osten, M., Horlick, E., Wendler, O., Dumont, E., Carere, R.G., Wijesinghe, N., Nietlispach, F., Johnson, M., Thompson, C.R., Moss, R., Leipsic, J., Munt, B., Lichtenstein, S.V., & Cheung, A. (2010). Transcatheter valve-in-valve implantation for failed bioprosthetic heart valves. *Circulation,* Vol.121, No.16, (27 April 2010), pp. 1848-1857

Zajarias, A., & Cribier, A.G. (2009). Outcomes and safety of percutaneous aortic valve replacement. *Journal of the American College of Cardiology,* Vol.53, No.20, (19 May 2009), pp. 1829-1836

Zioupos, P., Barbenel, J.C., & Fisher, J. (1994). Anisotropic elasticity and strength of glutaraldehyde fixed bovine pericardium for use in pericardial bioprosthetic valves. *Journal of Biomedical Materials Research,* Vol.28, No. 1, (January 1994), pp. 49-57

9

Simulation by Finite Elements of Bone Remodelling After Implantation of Femoral Stems

Luis Gracia et al.[*]
Engineering and Architecture Faculty, University of Zaragoza,
Spain

1. Introduction

Degenerative osteoarthritis and rheumatoid diseases lead to a severe destruction of the hip joint and to an important functional disability of the patient. Several attempts by many surgeons were documented, in the history of Orthopedic Surgery, to restore an adequate function of the pathologic joint. All these attempts failed due to the use of inadequate materials or due to technical problems.

In the past 20th century, during the sixties, a successful replacement of a pathologic hip joint was finally achieved. It was the first arthroplasty of the hip providing a good functional outcome. This new technique was described by Charnley in 1961 (Charnley, 1961). Two materials were then introduced in the orthopedic surgery, the polyethylene and the polymethyl methacrylate. This later is known as bone cement, and allowed a good fixation of the prosthetic implants into the femoral canal and the pelvic acetabulum. This technique represented one of the most important advances in the Orthopedic Surgery during the 20th century.

Based on the original model developed by Charnley, total cemented hip implants have been improved with new materials and prosthetic designs. The most important advances have been described in the cements and cementation techniques (Mulroy & Harris, 1990; Noble et al., 1998; Reading et al., 2000), and in the sterilization and manufacture of the prosthetic polyethylene (Medel et al., 2004; Urries et al., 2004; D'Antonio et al, 2005; Oral et al., 2006; Faris et al., 2006; Gordon et al., 2006; Wolf et al., 2006). Nevertheless, the original stem design of Charnley remains unaltered and fully operational.

First generation cemented prosthesis, inserted by manual techniques (first generation cement fixation), were associated with high rates of aseptic loosening and mechanical failures (Olsson et al., 1981; Stauffer, 1982; Harris et al., 1982; Halley & Wroblewski, 1986;

[*]Elena Ibarz[1], José Cegoñino[1], Antonio Lobo-Escolar[2,3], Sergio Gabarre[1], Sergio Puértolas[1],
Enrique López[1], Jesús Mateo[2,3], Antonio Herrera[2,3]
[1]*Engineering and Architecture Faculty, University of Zaragoza, Spain*
[2]*Medicine School, University of Zaragoza, Spain*
[3]*Miguel Servet University Hospital, Zaragoza, Spain*

Mohler et al., 1995). Cementless implants were developed as an alternative for young and active patients. In the cementless hip replacement, there exists a direct contact between the prosthesis and the bone, and a primary rigid fixation of the implant is required for a proper outcome. This is obtained with a press-fit fixation technique, where for a perfect adjustment the implant is slightly larger than the surrounding bone.

In the postoperative first months, a secondary fixation is achieved when the surrounding bone ingrowths into the implant (bone ingrowth fixation) (Herrera et al., 2001). The designs and materials of cementless femoral stems have evolved from the original, in order to obtain more physiologic load transmissions and a better fixation. In the earlier femoral stem models, the goal was to achieve a great fixation into the femoral diaphysis. Examples of these models were AML and Lord femoral implants, with large porous coating surfaces along their diaphyseal areas. Over the years, the osteointegration rates of these large porous coating stems were found to be only around 35% (Hennessy et al., 2009). A strong devitalization of the proximal femoral metaphysis, and the proximal diaphysis in the case of Lord stems, was also demonstrated with the use of these implants (Grant & Nordsletten, 2004).

These prostheses were found to be stable in the long term, but their distal diaphyseal fixation produced a removal of normal stress on the proximal bone, being the main cause of the proximal devitalization. This situation is known as *stress shielding*. New models were then designed taking into account not only mechanical concepts but also bone biology. In order to preserve proximal bone stock, modern femoral stems have a lesser diameter, and are coated only proximally with hydroxyapatite. Despite these improvements, the stress shielding is still found in the long term in all total hip replacements.

Bone is living tissue undergoes a constant process of replacement of its structure, characterized by bone resorption and new bone formation, without changing their morphology. This process is called bone remodeling. On the other hand, bone adapts its structure, according to Wolff's Law, to the forces and biomechanical loads that receives (Buckwalter et al., 1995). In a normal hip joint, loads from the body are transmitted to the femoral head, then to the medial cortical bone of femoral neck towards the lesser trochanter, where they are distributed by the diaphyseal bone (Radin, 1980).

Body weight is transmitted to the femoral head in a normal hip joint. This load goes through the cortical bone of the medial femoral neck down to the lesser trochanter, where it is distributed to the diaphyseal bone. The implantation of a cemented or cementless femoral stem involves a clear alteration of the physiological load transmission. Loads are now passed through the prosthetic stem in a centripetal way, from the central marrow cavity to cortical bone (Herrera & Panisello 2006). This alteration of the normal biomechanics of the hip results in a phenomenon called *adaptive bone remodeling* (Huiskes et al, 1989), which means that physiological remodeling takes place in a new biomechanical environment.

The implantation of a cemented or cementless femoral stem produced a clear alteration of the physiological transmission of loads, as these are now passed through the prosthetic stem, in a centripetal way, from the central marrow cavity to the cortical bone (Marklof et al., 1980). These changes of the normal biomechanics of the hip bone leads to a phenomenon called adaptive remodeling (Huiskes et al., 1989), since bone has to adapt to the new biomechanical situation. Remodeling is a multifactorial process dependent on both mechanical and biological factors. Mechanical factors are related to the new distribution of loads caused by implantation of the prosthesis in the femur, the physical characteristics of the implant (size, implant design and alloy), and the type of anchoring in the femur: metaphyseal, diaphyseal, hybrid, etc. (Summer & Galante, 1992; Sychter & Engh, 1996; Rubash et al., 1998; McAuley et al., 2000; Gibson et al., 2001; Glassman et al., 2001). Biologics are related to age and weight of the individual, initial bone mass, quality of primary fixation and loads applied to the implant. Of these biological factors, the most important is initial bone mass (Sychter & Engh, 1996).

Orthopedic surgeons have taken many years to learn the biomechanics and biology of bone tissue. We began to focus on these sciences when long-term revisions of cemented and cementless femoral stems proved extensive atrophy of the femoral cortical bone, a pathological phenomenon caused by *stress-shielding*. Different models of cementless stems have tried to achieve perfect load transfer to the femur, mimicking the physiological transmission from the femoral calcar to the femoral shaft. The main objective was to avoid stress-shielding, since in absence of physiological transmission of loads, and lack of mechanical stimulus in this area, causes proximal bone atrophy.

Cemented stem fixation is achieved by the introduction of cement into bone, forming a bone-cement interface. Inside the cement mantle a new interface is made up between cement and stem. It might seem that the cement mantle enables better load distribution in the femur; however the design, material and surface of prostheses, play an important role in transmission and distribution of charges, influencing bone remodeling (Ramaniraka et al., 2000; Li et al., 2007)

Long-term follow-up of different models of cementless stems have shown that this is not achieved, and to a greater or lesser extent the phenomenon of stress-shielding is present in all the models, and therefore the proximal bone atrophy. It is interesting to know, in cemented stems, not only the stress-shielding and subsequent proximal bone atrophy, but also the long-term behavior of cement-bone and stem-cement interfaces. This requires long-term studies monitoring the different models of stems.

Research in different fields concerning Orthopaedic Surgery and Traumatology requires a methodology that allows, at the same time, a more economic approach and the possibility of reproducing in an easy way different situations. Such a method could be used as a guide for research on biomechanics of the locomotor system, both in healthy and pathologic conditions, along with the study of performance of different prostheses and implants. To that effect, the use of simulation models, introduced in the field of Bioengineering in recent years, can undoubtedly mean an essential tool to assess the best clinical option, provided that it will be accurate enough in the analysis of specific physiological conditions concerning certain pathology.

Finite element (FE) simulation has proved to be specially suitable in the study of the behaviour of any physiological unit, despite its complexity. Nowadays, it has become a powerful tool in the field of Orthopaedic Surgery and Traumatology, helping the surgeons to have a better understanding of the biomechanics, both in healthy and pathological conditions. FE simulation let us know the biomechanical changes that occur after prosthesis or osteosynthesis implantation, and biological responses of bone to biomechanical changes. It also has an additional advantage in predicting the changes in the stress distribution around the implanted zones, allowing preventing future pathologies derived from an unsuitable positioning of the prostheses.

In this sense, finite element simulation has made easier to understand how the load is transmitted after the implantation of a femoral stem, and to predict how the stem impacts on the biomechanics in the long-term. Finite Element method can find out the long-term behavior and the impact on biomechanics of any prosthetic models. Up to now, long clinical trials, with a follow-up of at least 10 years, were needed to achieve this knowledge. Design of new femoral stem models is another important application of the Finite Element simulation. New models can be pre-tested by simulation in order to improve the design and minimize the stress-shielding phenomenon.

The Finite Element Method (FEM) was originally developed for solving structural analysis problems relating to Mechanics, Civil and Aeronautical Engineering. The paternity of this method is attributed to Turner, who published his first, historic, job in 1956 (Turner et al., 1956). In 1967, Zienkiewicz OC published the book "The finite element method in structural and continuum mechanics" (Zienkiewicz, 1967) which laid down mathematical basis of the method. Other fundamental contributions to the development of Finite Element Method (FEM) took place on dates nearest (Imbert, 1979; Bathe, 1982; Zienkiewicz & Morgan, 1983; Hughes, 1987).

2. Methodology for the finite element analysis of biomechanical systems

One of the most significant aspects of biomechanical systems is its geometric complexity, which greatly complicates the generation of accurate simulation models. Classic models just suffered from this lack of geometrical precision, present even in recent models, which challenged, in most studies, the validity of the results and their extrapolation to clinical settings.

Fig. 1. Real model of an implanted femur, 3D laser scanner and femoral stem

Currently, there are methodologies developed over recent years that avoid such problems, allowing the generation of models with the desired precision in a reasonable time and cost is not excessive. Thus, the use of 3D laser scanners (Fig. 1) together with three-dimensional images obtained by CT allow making geometric models that combine high accuracy in the external form with an excellent definition of internal interfaces. The method requires not only appropriate software tools, capable of processing images, but also its compatibility with the programs used later to generate the finite element model. For example, in Fig. 1 (left) is shown the real model of an implanted femur and in Fig. 2 the result obtained by a three-dimensional laser scanner Roland Picza (after processing by Dr. Picza 3 and 3D Editor programs).

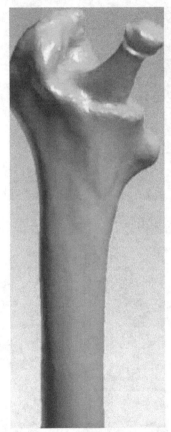

Fig. 2. 3D scanning of the implanted femur shown in Fig. 1

In these models, the characterization of the internal structure is made by 3D CT, from images like that shown in Figs. 3 and 4. An alternative to the above procedure is the use of 3D geometrical reconstruction programs, for example, MIMICS (Mimics, 2010). In any case, the final result is a precise geometrical model which serves as a basis for the generation of a finite elements mesh.

In view of the difficulties experienced in living subjects, FE simulation models have been developed to carry out researches on biomechanical systems with high reproducibility and versatility. These models allow repeating the study as many times as desired, being a non-aggressive investigation of modified starting conditions

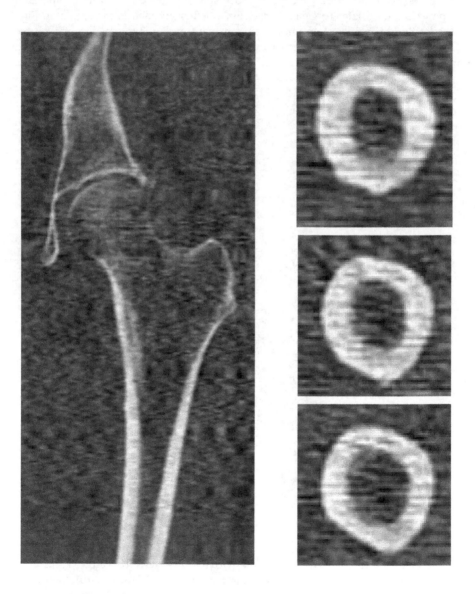

Fig. 3. CT images of a healthy femur

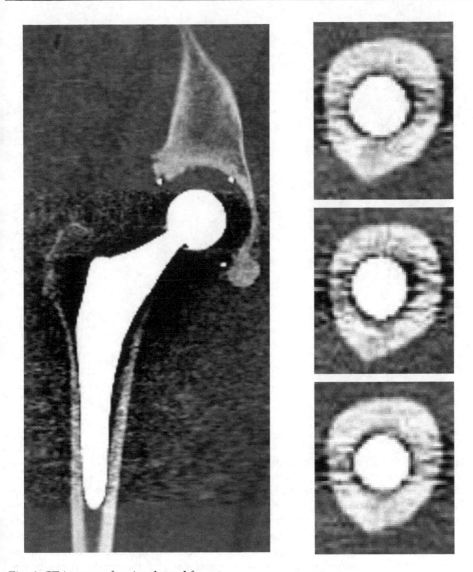

Fig. 4. CT images of an implanted femur

However, work continues on the achievement of increasingly realistic models that allow putting the generated results and predictions into a clinical setting. To that purpose it is mainly necessary the use of meshes suitable for the particular problem, as regards both the type of elements and its size. It is always recommended to perform a sensitivity analysis of the mesh to determine the optimal features or, alternatively, the minimum necessary to achieve the required accuracy. In Fig. 5 is shown a FE mesh of healthy and implanted femurs, using tetrahedron type elements. It can be seen that the element size allows depicting, with little error, the geometry of the implanted femur, compared with Fig. 2.

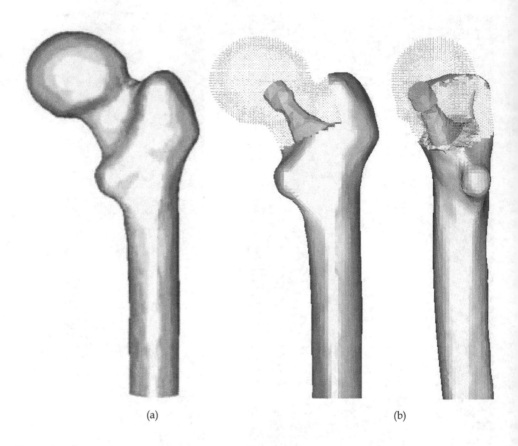

(a)   (b)

Fig. 5. Meshing of proximal femur: a) Healthy femur; b) Implanted femur

A key issue in FE models is the interaction between the different constitutive elements of the biomechanical system, especially when it comes to conditions which are essential in the behaviour to be analyzed. The biomechanical behaviour of a cementless stem depends basically on the conditions of contact between the stem and bone, so that the correct simulation of the latter determines the validity of the model. In Fig 6 can be seen the stem-femur contact interface, defined by the respective surfaces and the frictional conditions needed to produce the press-fit which is achieved at surgery.

Finally, in FE simulation models is essential the appropriate characterization of the mechanical behaviour of the different materials, usually very complex. So, the bone exhibits an anisotropic behaviour with different responses in tension and compression (Fig. 7). Moreover, it varies depending on the bone type (cortical or cancellous) and even along different zones in the same specimen. This kind of behaviour is reproducible in a reliable way in the simulation, but it leads to an excessive computational cost in global models. For this reason, in most cases, and specially in long bones, a linear elastic behaviour in the operation range concerning strains and stresses is considered.

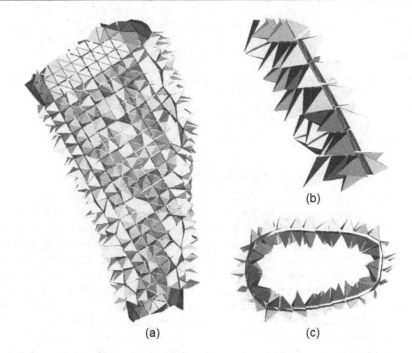

Fig. 6. Contact interface femur-stem

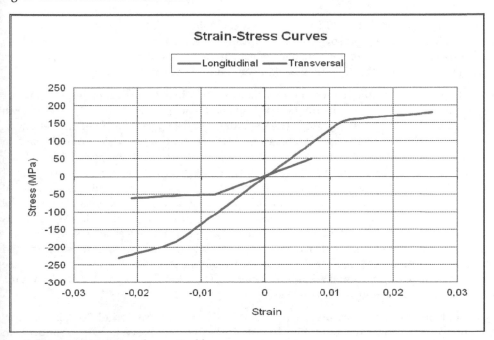

Fig. 7. Strain-stress curves for cortical bone

In soft tissues, the behaviour is even more complex, usually as a hyperelastic material. This is the case of ligaments, cartilages and muscles, also including a reologic effect with deferred strains when the load conditions are maintained (viscoelastic behaviour).

This inherent complexity to the different biological tissues, reproducible in reduced or local models, is very difficult to be considered in global models as the used to analyse prostheses and implants, because the great amount of non-linearities do the convergence practically unfeasible. On the other hand, it leads to a prohibitive computational cost, only possible to undertake by supercomputers.

3. Application to the behaviour of hip prostheses. Cementless stems

In the last three decades, different designs of cementless stem have sought to obtain a physiological load transfer to the proximal femur so as to avoid stress-shielding and bone loss due to proximal atrophy. Nevertheless, most of the reports about this class of stems have proved that they do not completely prevent proximal bone atrophy (Tanzer et al., 1992; Bugbee et al., 1997; Hellmann et al., 1999; Engh et al., 2003; Sinha et al., 2004; Braun et al., 2003; Herrera et al., 2004; Canales et al., 2006). More recently, several stems incorporated a hydroxyapatite coating (HA) in their metaphyseal zone, seeking to obtain a better proximal osteointegration (Tonino et al., 1999) as a way to achieve a better load transfer and avoid stress shielding (Nourissat et al., 1995). However, the reports on this improvement showed a moderate increase in the amount of the implant surface with bone on-growth, which ranged between 35-50% in porous coated implants and between 45-60% in HA coated implants.

Considering that a loss of 30-40% in bone mass is required for it to be observed in a serial X-Ray (Engh et al., 2000), prospective studies using densitometry (DXA) are considered the ideal method for quantifying the changes of bone mass produced by different stems over the years (Kroger et al., 1996; Rosenthall et al., 1999; Schmidt et al., 2004).

Long term studies of different cementless stems show a high incidence of stress-shielding, caused by the change in the distribution of loads on the femur (Engh et al., 2003; Glassman et al., 2006; Wick & Lester, 2004). The monitoring of an anatomic femoral stem with metaphyseal load-bearing and HA coating (ABG-I), that was carried out through a prospective, controlled study that included 67 patients (Group I) in the period 1994-99, has confirmed that even though the clinical results are very favourable, a high percentage of cases with stress-shielding are detected (Herrera et al., 2004). This results in a proximal atrophy which has been quantified with DXA (Panisello et al., 2009a). For that reason the stem has been redesigned (ABG-II) in an attempt to improve the proximal transfer of loads and reduce the phenomenon of stress-shielding. The main differences between both stems concern geometrical design and material. The overall length has been reduced by 8% and the proximal and distal diameters by 10%. The prosthesis shoulder has been modified. The material has changed from Wrought Titanium (Ti 6Al-4V) alloy to TMZF (Titanium, Molybdenum, Zirconium and Ferrous) alloy.

A similar design study was done with the ABG-II stem in the period 2000-05, with 69 patients of comparable demographic characteristics than the previous one (Group II). In both studies the surgical technique, post-operative rehabilitation program, densitometry studies and statistical analysis were identical (Panisello et al., 2009b). The study confirmed less proximal atrophy, therefore one could ask if the new design has effectively improved the load transfer conditions in the proximal femur, producing less stress-shielding. The simulation with Finite Elements (FE) allows us to verify the correlation between the mechanical stimulus and the changes detected in the bone density. In order to do this, the evolution of the mechanical stimulus over a period of 5 years has been analysed, correlating the findings with the quantified Bone Mineral Density (BMD) evolution in the studies using DXA.

Several objectives can be covered by the FE simulation: firstly, to determine the long-term changes of BMD in the femur after the implantation of ABG-I and ABG-II stems throughout the first five postoperative years, analysing the correlation between evolution of BMD and stress level, focussed on the average stresses (tension and compression) in cortical and cancellous bone for each one of the Gruen zones (Gruen et al., 1979) (Fig 8); secondly, to analyse the appropriate transfer of loads through contact between the bone and prosthesis. And finally, to analyse the long term differences between the implantation of an ABG-I and ABG-II prostheses to test if the changes in the design and alloy of the prosthesis produce a better transfer of loads in the proximal zone.

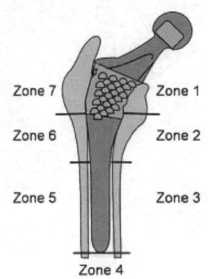

Fig. 8. Gruen zones

The development of the model of a healthy femur is crucial to make accurate the whole process of simulation, and to obtain reliable results. The development of the FE models was made following the same methodology as explained before. A cadaverous femur was used with two hip prostheses type ABG-I ad ABG-II, manufactured by Stryker (Fig. 9). This cadaverous femur had originally belonged to a healthy 60 year old man and was only used in order to define the geometry of the model, without any relation with BMD measures. Firstly, each of the parts necessary to set the final model were scanned using a three dimensional scanner Roland Picza brand. As a result, a cloud of points which approximates the scanned geometry was obtained. These surfaces must be processed through the programs Dr.Picza-3 and 3D-Editor. This will eliminate the noise and performs smooth surfaces, resulting in a geometry that reliably approximates to the actual geometry.

From the scanned femur a geometric model of the outer geometry of the femur was obtained with no distinction between cortical bone, cancellous bone and bone marrow. To determine the geometry of the cancellous bone and medullar cavity 30 transverse direction (5 mm gap) tomographic cross-sections and eight longitudinal direction cross-sections were taken using CT scan (General Electric Brightspeed Elite). A three-dimensional mesh of healthy femur, based on linear tetrahedral elements (Fig. 5, healthy model), was made in I-deas software (I-deas, 2009). So as to develop the pattern with prosthesis, an ABG-I prosthesis was scanned to obtain its geometry. Afterwards we proceeded with the operation on a cadaver femur with a prosthesis being implanted in the same way as a real hip replacement operation would be carried out.

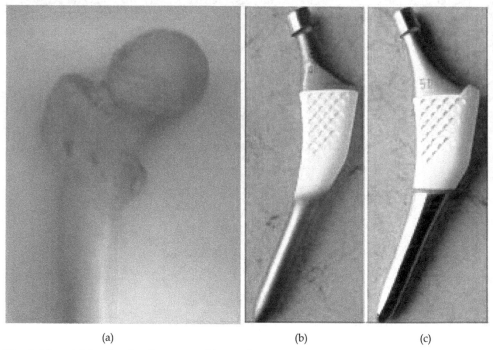

(a)   (b)   (c)

Fig. 9. Material for the development of finite element model: a) Healthy femur; b) ABG-I stem; c) ABG-II stem

Once the three meshes had been generated in I-deas (healthy femur, prosthesis ABG-I and operated femur), the prosthesis was positioned in the femur always taking the mesh of the operated femur as the base. From the previous process of modelling on the cadaveric femur, only the cortical bone was used, from which the cancellous bone was modelled again, in such a way that it fit perfectly to the contact with the prosthesis. Work with the ABG-II prosthesis was undertaken in the same way.

The program *Abaqus 6.7* (Abaqus, 2009) was used for the calculation, with the *Abaqus Viewer* being used for the results post processing. It was necessary to undertake a contact simulation between the prosthesis and the cancellous bone for which a friction coefficient of 0.5 was considered simulating the press-fit setting according with (Shirazi et al., 1993). In light of the former results a sensitivity analysis was carried out in order to determine the appropriate interface conditions, considering several friction coefficient values from 0.2 to 0.5 in steps of 0.05, obtaining significant differences in the analyzed range, but with similar results from 0.4 to 0.5. An analysis with bonded interface was also realized. It was observed that the value of 0.5 for the friction coefficient corresponds practically to a bonded interface concerning the stem mobility, but with the advantage that allows moving apart the stem from the bone in higher tension zones, providing a more realistic stress distribution inside the bone.

The final model for the healthy femur contains a total of 68086 elements (38392 for cortical bone, 27703 for cancellous bone and 1990 for bone marrow) (Fig. 10). The final model with ABG-I stem comprises a total of 60401 elements (33504 for cortical bone, 22088 for cancellous bone and 4809 for ABG-I stem) (Fig. 5b). The final model with the ABG-II stem is made up of 63784 elements (33504 for cortical bone, 22730 for cancellous bone and 7550 for ABG-II stem) (Fig. 11).

Fig. 10. FE model of healthy femur. Coronal section

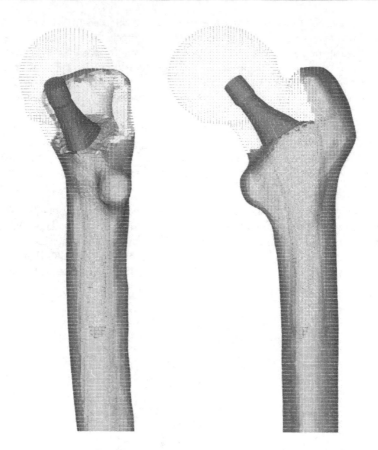

Fig. 11. FE model after removal of the femoral head and positioning of the cementless stems ABG-II.

Three boundary conditions were defined: clamped in the medial diaphyseal part of the femur, force on the prosthetic head due to the reaction of the hip caused by the weight of the person (400% body weight) and force on the greater trochanter (200% body weight) generated by the abductor muscles (Herrera et al., 2007).

The values of the mechanical properties used in the prostheses as well as the biological materials are shown in Table 1. These values have been obtained from the bibliography specializing in the subject (Ashman and Rho, 1988; Ashman et al., 1984; Evans, 1973; Ionescu et al., 2003; Jacobs, 1994; Mat Web, 2009; Meunier et al., 1989; Turner et al., 1999; Van der Val et al., 2006) and they have been simplified considering an isotropic behaviour. In the design of the ABG-II second generation prosthesis a different titanium alloy was used to that of the ABG-I. The prosthetic ABG-I stem is made with a Wrought Titanium (Ti 6Al-4V) alloy, the elasticity modulus of which is 110 GPa. On the other hand the TMZF alloy which is used on the ABG-II stem has a Young's modulus of 74-85 GPa, according to the manufacturer information, using a mean value of 79.5 GPa in the different analyses.

	ELASTIC MODULUS (MPa)	POISSON RATIO	MAXIMUM COMPRESSION (MPa)	MAXIMUM TENSION (MPa)
CORTICAL BONE	20000 [15, 20, 21]	0.30 [16, 18]	150 [16, 18]	90 [16, 18]
CANCELLOUS BONE	959 [14]	0.12 [17]	23 [16, 18]	
BONE MARROW	1 [16, 18]	0.30 [16, 18]		
ABG-I STEM	110000 [22]	0.33 [19]		
ABG-II STEM	74000 - 85000 [22]	0.33 [19]		

Table 1. Mechanical properties of materials

Taking various studies as reference (Kerner et al., 1999; Turner et al., 2005), a linear relationship between the bone mass values, which come from the medical study collected in Panisello (Panisello, 2009a), and the apparent density was established in addition to a cubic relationship between the latter and the elastic modulus, using a maximal Young's modulus of 20 GPa, thereby obtaining the cortical bone modulus of elasticity values for each one of the 7 Gruen zones. To carry out the analysis of the results, the cortical bone of each model is divided into seven zones which coincide with the Gruen zones. The elastic modulus obtained from the values in Panisello (Panisello, 2009a) was used as an input in the cortical bone. These values are being successively adjusted for each one of the models (femur with ABG-I stem and femur with ABG-II stem) in different moments of time: post-operative, 6 months, 1, 3 and 5 years. In addition, the initial data corresponding to the pre-operative moment are used as an input in the healthy model. The mechanical properties of the cortical bone have been calculated from the bone mass data from groups I and II respectively.

For the complete comparative analysis of both stems, all of the possible combinations of bone mass (group I, ABG-I, 67 patients in the period 1994-99 and group II, ABG-II, 69 patients in the period 2000-05) prosthetic geometry (ABG-I and ABG-II) and stem material (Wrought Titanium or TMZF) were simulated. This way it was possible to compare the mechanical performance of both prostheses in what refers to the transmission of loads and the interaction in the bone-prosthesis contact zone. It also makes possible to distinguish the most influential parameter (geometry or material) for the design of future prosthetic stems.

The average von Mises stress is used, given that despite not distinguishing between tension and compression values, it is sufficiently indicative of the tendency of the mechanical stimulus and it is standard in FE software.

The main features of each of the boundary conditions are:

1. Clamped in the middle of the femoral shaft

The middle zone has been clamped instead of distal zone because middle zone is considered enough away from proximal bone (Fig. 12). This model can be compared with other that have been clamped at a distal point, since the loads applied practically coincided with the femoral axis direction thus reducing the differences in final values.

2.   Hip muscles Loads

Forces generated by the abductor muscles are applied on the greater trochanter, in agreement with most authors' opinion (Weinans et al., 1994; Kerner et al., 1999). Generally, muscle strength generated in the hip joint is 2 times the body weight, and this produces a reaction in the femoral head that accounts for 2.75 times the body weight. However, when the heel impacts to the ground, and in double support stage of the gait, the load increases up to 4 times the body weight. The latter case, being the worst one, has been considered to impose the boundary conditions. It has also been considered a body weight of 79.3 kg for cementless stems, and 73 kg for cemented stems. Those were the average values obtained from the clinical sample to be contrasted with the simulation results. The load due to the abductor muscles, accounting for 2 times the corporal weight, is applied to the proximal area of the greater trochanter, at an angle of 21 degrees, as shown in Fig. 12.

3.   Reaction strength on the femoral head due to the body weight.

As already mentioned, we have studied the case of a person to 79.3 kg in cementless stems, and 73 kg in cemented stems, in the worst case of double support or heel impact stages of the gait. The resultant force on the femoral head would be worth 4 times the body weight (Fig. 12).

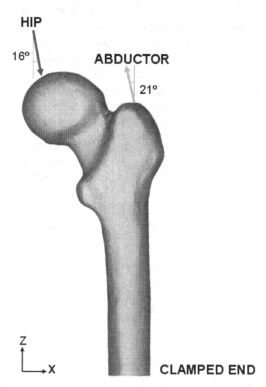

Fig. 12. Boundary conditions applied in the healthy femur model. Similar conditions are used in the implanted models

Both models with prostheses ABG-I and II have been simulated in five different moments of time coincident with the DXA measurements: postoperative, 6 months, 1, 3 and 5 years, in addition to the healthy femur as the initial reference. In both groups of bone mass an increase of stress in the area of the cancellous bone is produced, which coincides with the end of the HA coating, as a consequence of the bottleneck effect which is produced in the transmission of loads, and corresponds to Gruen zones 2 and 6, where no osteopenia can be seen in contrast to zones 1 and 7.

BMD evolution in the operated and healthy hip is shown in Fig. 13 for both prostheses. For ABG-I, the preoperative measurements performed in both hips showed slightly higher BMD rates in the healthy hip, although these were not statistically significant. Postoperative values were taken as a reference for the operated hip. A decrease in BMD was detected in all zones except zone 4, six months after surgery. Between 6 and 12 postoperative months there was a slight additional loss of BMD in zones 1 and 7, but some bone recovery in the middle and distal areas around the implant. No significant changes in BMD were observed in zones 1 to 6 from the end of the first year to the end of the fifth year.

For ABG-II, the preoperative measurements performed in both hips showed again slightly higher bone density rates in the healthy hip, although these were not statistically significant. No changes or a minimal decrease in bone density was detected in zones 1 to 6, six months after surgery, attributed to rest period, partial weight bearing and the later effects of surgery. The bone loss was statistically significant only in zone 7. A slight additional loss of bone density was observed in zone 7, as well as some bone recovery in the middle and distal areas around the implant. Minor changes in bone density were observed in zones 1 to 6 from the end of the first year to the end of the fifth year. The bone mass remains stable in this period, with a little bone recovery in zones 2 and 6. Nevertheless, there was some decrease in zone 7 in the period between the first and fifth year, when a loss of 23.88% can be reached. The bone density in the contra-lateral healthy hip (bone mass group II) showed some slight differences during the follow-up, with decreases between 1.4 and 2.7%, more evident in the proximal part of the femur, richer in cancellous bone. The values obtained for zones 3 to 5 were similar to those of the operated femurs; in zones 2 and 6 they were slightly superior; only zones 1 and 7 showed significant differences.

Fig. 14 shows the results of the average von Mises stress (MPa), corresponding to the combinations of geometry (ABG-I, ABG-II) and prosthesis material for group I of bone mass at five years, and Fig. 15 shows the equivalent results for group II of bone mass. It can be confirmed that the global behaviour of the prostheses is the same in both models; however, in the case of the ABG-II stem higher stress values are reached on both the cancellous and cortical bones, fundamentally in the proximal zones.

A tensional increase is noticeable in the whole area close to the lesser trochanter with the use of the ABG-II stem, as well as the tensional increase that the insertion of the ABG-II prosthesis involves with respect to the ABG-I. These differences can be observed in a more clear way in Figs. 14 and 15. In both figures it is clearly noticeable that the result corresponding to the two material for every stem are practically the same (superimposed lines); however, the results corresponding to both geometrical designs (ABG I, ABG II) are different, with a higher tensional level for the ABG II stem.

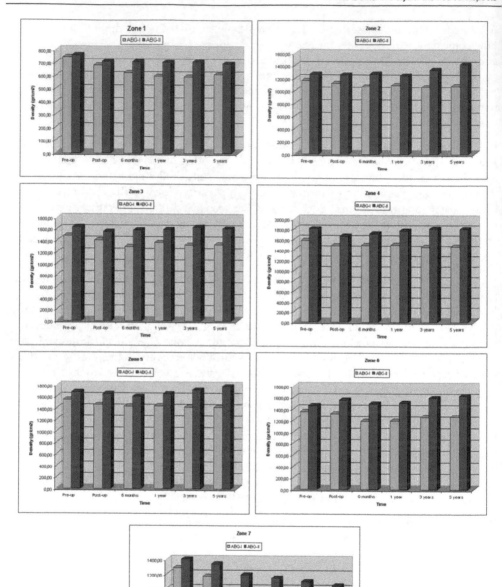

Fig. 13. Evolution of bone mass density for ABG I (bone mass group I) and ABG II (bone mass group II), corresponding to five years, in the Gruen zones.

It could be checked that in every case the stress corresponding to the ABG-II stem is greater than the one resulting from the insertion of the ABG-I stem (Figs. 14 and 15). In the figures it can be seen that in every zone and for any time the stress achieve higher values in ABG-II than in ABG-I stem. This way it is possible to confirm that with the second generation of stem (ABG-II) the stress increases in practically every zone with this increase being most evident in zone 7.

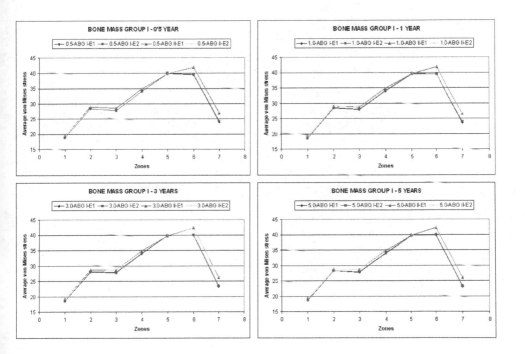

Fig. 14. Comparison of the average von Mises stress as a function of design and material for bone mass group I

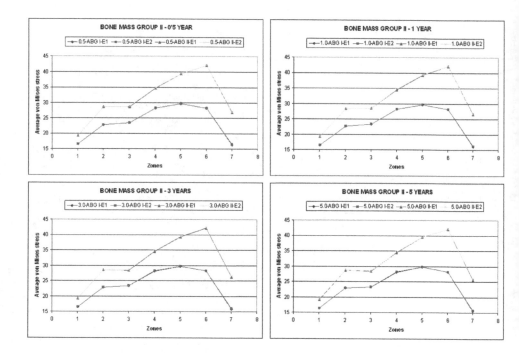

Fig. 15. Comparison of the average von Mises stress as a function of design and material for bone mass group II

Figure 16 shows the evolution of the bone mass (%) and the average von Mises stress (%) for each one of the 7 Gruen zones in both models, considering the corresponding group and material for each stem.

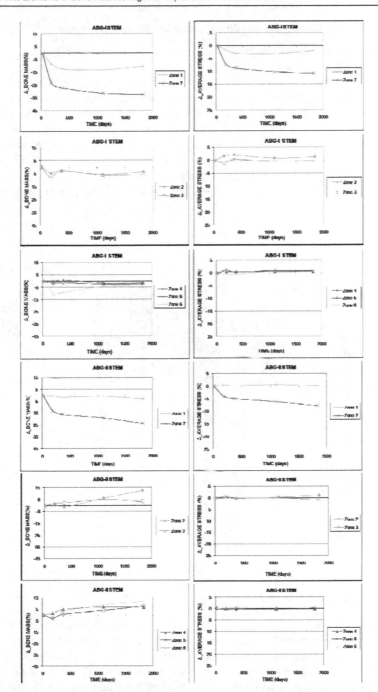

Fig. 16. Bone mass (%) variation versus time and variation in average von Mises (%) stress versus time for the femur with prosthesis ABG-I and ABG-II in the Gruen zones

4. Application to the behaviour of hip prostheses. Cemented stems

In the case of cemented stems the process of modelling was similar, varying the surgical cut in the femoral neck of the healthy femur. Each stem was positioned into the femur, always taking the superimposed mesh of the operated femur as a base (Fig. 17). In the previous process of modelling, on the cadaveric femur, only the cortical bone was used. The cancellous bone was modelled again taking into account the cement mantle surrounding the prosthesis and the model of stem (ABG or Versys), so as to obtain a perfect union between cement and cancellous bone. The cement mantle was given a similar thickness, in mm, which corresponds to that usually achieved in patients operated on, different for each of the stem models studied and each of the areas of the prosthesis, so that the model simulation be as accurate as possible. Cemented anatomical stem ABG, and the Versys straight, polished stem, were used in the study.

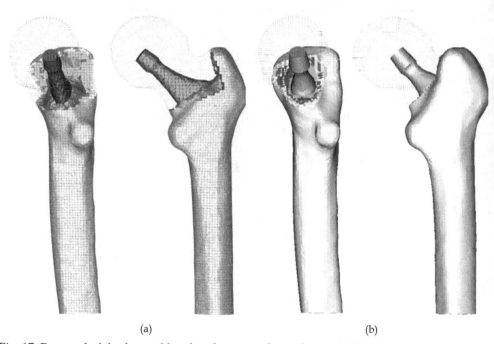

(a)   (b)

Fig. 17. Removal of the femoral head and cemented prosthesis positioning:
(a) ABG-cemented and (b) Versys

In the models of cemented prostheses it is not necessary to define contact conditions between the cancellous bone and the stem. In this type of prosthesis the junction between these two elements is achieved by cement, which in the EF model should simulate conditions of perfect union between cancellous bone-cement and cement-stem. It has also been necessary to model the diaphyseal plug that is placed in actual operations to prevent the spread of the cement down to femoral medullary canal. Fig. 18 shows the coronal sections of the final models.

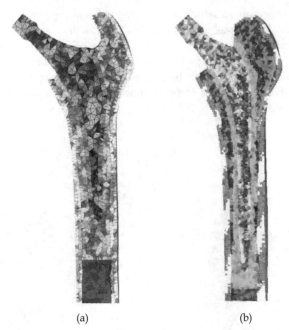

(a) (b)

Fig. 18. Longitudinal section of the FE models with cemented femoral prostheses: (a) ABG-cemented and (b) Versys.

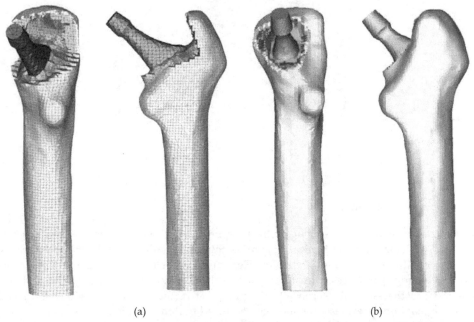

(a) (b)

Fig. 19. FE model with cemented femoral prostheses: (a) ABG-cemented and (b) Versys.

Both models were meshed with tetrahedral solid elements linear type, with a total of 74192 elements in the model for ABG-cemented prosthesis (33504 items cortical bone, cancellous bone 17859, 6111 for the ABG stem-cement, cement 13788 and 2930 for diaphyseal plug), and 274651 in the model for prosthetic Versys (119151 items of cortical bone, cancellous bone 84836, 22665 for the Versys stem, 44661 for the cement and 3338 in diaphyseal plug). In Fig. 19 are shown both models for cemented stems. Boundary conditions were imposed in the same way as in the cementless stems (Fig. 20).

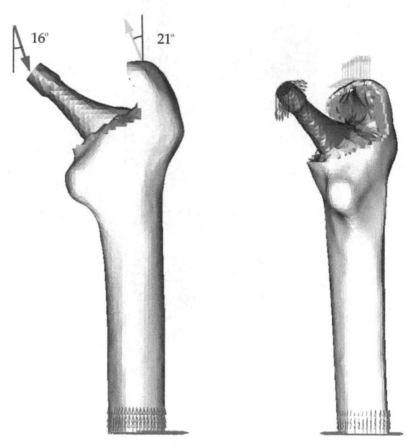

Fig. 20. Boundary conditions

Calculation was performed using again the program Abaqus 6.7. Both prostheses have been simulated with the same mechanical properties, thus, the result shows the influence of stem geometry on the biomechanical behavior. Figs. 21 and 22 show the variation (%) of bone mass and average von Mises stress (%) in each of the Gruen zones for each of the models of cemented prostheses, with reference to the preoperative time. It can be seen that for both stems, the maximum decrease in bone mass occurred in Zone 7. This decrease in bone mass is greater in the Versys model than in the ABG stem

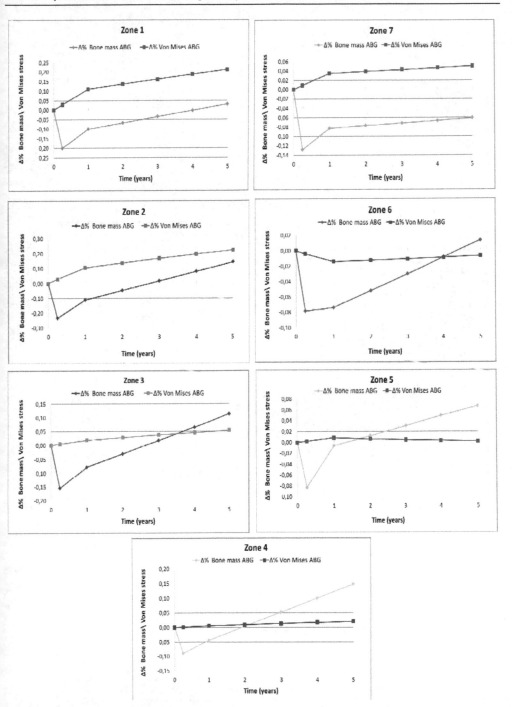

Fig. 21. BMD and average von Mises stress evolution for ABG cemented stem

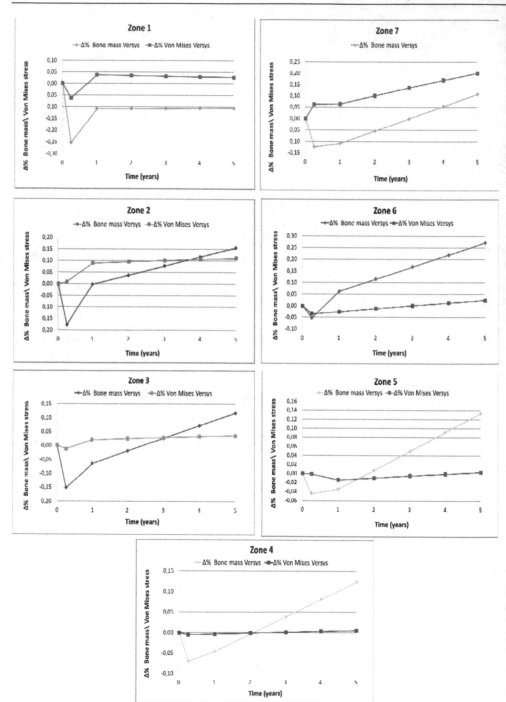

Fig. 22. BMD and average von Mises stress evolution for Versys cemented stem

5. Conclusion

Prior to the development of our FE models several long-term studies of bone remodeling after the implantation of two different cementless stems, ABG I and ABG II, were performed (Panisello et al., 2006; Panisello et al., 2009a; Panisello et al., 2009b). These studies were performed using DXA, a technique that allows an accurate assessment of bone density losses in the different Gruen zones. We take as a reference to explore this evolution the postoperative value obtained in control measurements and those obtained from contralateral healthy hip. New measurements were made at 6 months, one year and 5 years after surgery. The ABG II stem is an evolution of the ABG I, which has been modified both in its alloy and design. The second generation prosthesis ABG-II is manufactured with a different titanium alloy from that used in the ABG-I. The prosthetic ABG-I stem is made with a Wrought Titanium alloy (Ti 6Al-4V) of which elasticity modulus is 110 GPa. Meanwhile, the TMZF alloy, which is used on the ABG-II stem, has a Young's modulus of 74-85 GPa, according to the manufacturer information, using a mean value of 79.5 GPa in the different analyses. On the other hand, the ABG II stem has a new design with less proximal and distal diameter, less length and the shoulder of stem has been redesign to improve osteointegration in the metaphyseal area.

In our DXA studies, directed to know the loss of bone mass in the different zones of Gruen caused by the stress-shielding, we found that ABG II model produces less proximal bone atrophy in post-operative measurements, for similar follow-up periods. In the model ABG II, we keep finding in studies with DXA a proximal bone atrophy, mainly in zones 1 and 7 of Gruen, but with an improvement of 8.7% in the values obtained in ABG I series. We can infer that improvements in the design of the stem, with a narrower diameter in the metaphyseal area, improve the load transfer to the femur and therefore minimizes the stress-shielding phenomenon, resulting in a lower proximal bone atrophy, because this area receive higher mechanical stimuli. These studies for determination of bone mass in Gruen zones, and the comparative study of their postoperative evolution during 5 years have allowed us to draw a number of conclusions: a) bone remodelling, after implantation of a femoral stem, is finished one year after surgery; b) variations in bone mass, after the first year, are not significant.

The importance of these studies is that objective data from a study with a series of patients allow us to confirm the existence of stress-shielding phenomenon, and quantify exactly the proximal bone atrophy that occurs. At the same time they have allowed us to confirm that the improvements in the ABG stem design, mean in practice better load transfer and less stres-shielding phenomenon when using the ABG II stem.

DXA studies have been basic to validate our FE models, because we have handled real values of patients' bone density, which allowed us to measure mechanical properties of real bone in different stages. Through computer simulation with our model, we have confirmed the decrease of mechanical stimulus in femoral metaphyseal areas, having a higher stimulus in ABG II type stem, which corresponds exactly with the data obtained in studies with DXA achieved in patients operated with both models stems.

In the case of cemented stems, densitometric studies were performed with two different types of stem: one straight (Versys, manufactured in a cobalt-chromium alloy) and other anatomical (ABG, manufactured in forged Vitallium patented by Stryker Howmedica). It

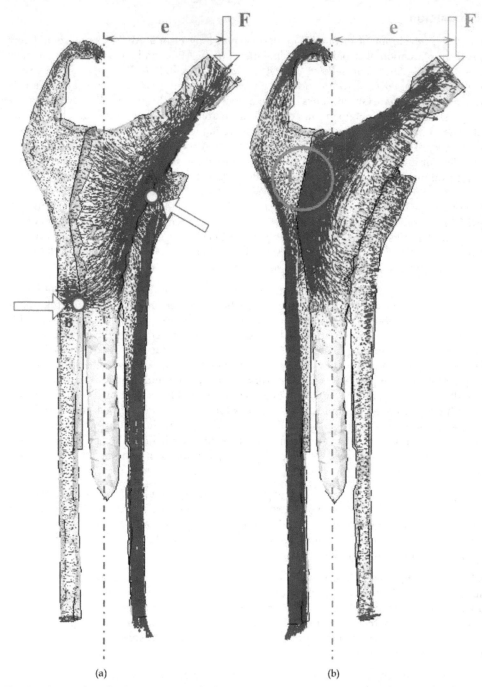

Fig. 23. Stress distributions in coronal plane: a) minimum principal (compression); b) maximum principal (tension)

was carried on the same methodology used in the cemented stems series. Densitometric studies previously made with cementless stems allow us to affirm that bone remodeling is done in the first postoperative year, a view shared by most of the authors. So, we accept that bone mineral density values obtained one year after surgery can be considered as definitive. As in cementless models, densitometric values have been used for comparison with those obtained in the FE simulation models. Our studies confirmed that the greatest loss of bone density affects the zone 7 of Gruen (Joven, 2007), which means that stress-shielding and atrophy of metaphyseal bone also occurs in cemented prostheses. This phenomenon is less severe than in non-cemented stems, therefore we can conclude that the load transfer is better with cemented stems than with cementless stems. The findings of proximal bone atrophy, mainly in the area 7, agree with those published by other authors (Arabmotlagh et al., 2006; Dan et al., 2006). We have also found differences in the rates of decrease in bone density in the area 7 of Gruen, wich were slightly lower in the anatomical ABG stem than in the Versys straight stem. This also indicates that the prosthesis design has influence in the remodeling process, and that mechanical stimuli are different and related to the design.

From a mechanical point of view, in the case of cementless stems, the improvements in the design are limited for the intrinsic behaviour of the mechanical system stem-bone. So if the compression and tension distributions are depicted (Fig. 23), one can see that, at initial steps without osteointegration, compression stresses can be transmitted in the contact interface but not tension stresses. Therefore the bending moment due to the load eccentricity only can be equilibrated by means of a couple of forces acting at points A and B, respectively, as shown in Fig. 23b. This behaviour explain the proximal bone atrophy in zone 7 of Gruen, and depending on the actual position of point A the same effect in zone 1 of Gruen. A similar effect occurs in cemented stems when the debonding appears at the bone-cement interface in the tensioned zones.

6. References

ABAQUS (2009). Web site, http://www.simulia.com/

Arabmotlagh, M.; Sabljic, R. & Rittmeister, M. (2006). Changes of the biochemical markers of bone turnover and periprosthetic bone remodelling after cemented arthroplasty. *J Arthroplasty*, 21, 1, 129-34. ISSN: 0883-5403

Ashman, R. B., Cowin, S. C.; Van Buskirk, W. C. et al. (1984). A continuous wave technique for the measurement of the elastic properties of cortical bone. *J Biomech*, 17, 349–61. ISSN: 0021-9290

Ashman, R. B. & Rho, J. Y. (1988). Elastic modulus of trabecular bone material. *J Biomech*, 21, 177-81. ISSN: 0021-9290

Bathe, K. J. (1982). *Finite element procedures in engineering analysis*, Prentice-Hall, New Jersey. ISBN-10: 0133173054

Braun, A.; Papp, J. & Reiter, A. (2003). The periprosthetic bone remodelling process signs of vital bone reaction. *Int Orthop*, 27, S1, 7-10. ISSN: 0341-2695

Buckwalter, J. A.; Glimcher, M. J.; Cooper, R. R. et al. (1995). Bone biology, *J Bone Joint Surg Am*, 77, 1276-1289. ISSN: 0021-9355

Bugbee, W.; Culpepper, W.; Engh, A. et al. (1997). Long-term clinical consequences of stress-shielding after total hip arthroplasty without cement. *J Bone Joint Surg Am*, 79, 1007-12. ISSN: 0021-9355

Canales, V.; Panisello, J. J.; Herrera, A. et al. (2006). Ten year follow-up of an anatomical hydroxyapatite-coated total hip prosthesis. *Int Orthop*, 30, 84-90. ISSN: 0341-2695

Charnley, J. (1961). Arthroplasty of the hip: a new operation. *Lancet*, 1129-32. ISSN: 0140-6736

D'Antonio, J. A.; Manley, M. T.; Capello, W. N. et al. (2005). Five year experience with Crossfire highly cross-linked polyethylene. *Clin Orthop Relat R*, 441, 143-50. ISSN: 0009-921X

Dan, D.; Germann, D.; Burki, H. et al. (2006). Bone loss alter total hip arthroplasty. *Rheumatol Int*, 26, 9, 792-8. ISSN: 0172-8172

Engh, C. A. Jr.; Mc Auley, J.P.; Sychter, C.J. et al. (2000). The accuracy and reproducibility of radiographic assessment of stress-shielding. *J Bone Joint Surg Am*, 82, 1414-20. ISSN: 0021-9355

Engh, C. A. Jr; Young, A. M.; Engh, C. A. Sr. et al. (2003). Clinical consequences of stress shielding after porous-coated total hip arthroplasty. *Clin Orthop Relat R*, 417, 157-63. ISSN: 0009-921X

Evans, F. G. (1973). *The Mechanical Properties of Bone*. American Lecture Series, n. 881, Springfield, IL. ISBN: 0398027757

Faris, P. M.; Ritter, M. A.; Pierce A. L. et al. (2006). Polyethylene sterilization and production affects wear in total hip arthroplasties. *Clin Orthop Relat R*, 453, 305-8. ISSN: 0009-921X

Gibbons, C. E. R.; Davies, A. J.; Amis, A. A. et al. (2001). Periprosthetic bone mineral density changes with femoral components of different design philosophy. *In Orthop*, 25, 89-92. ISSN: 0341-2695

Glassman, A. H.; Crowninshield, R. D.; Schenck, R. et al. (2001). A low stiffness composite biologically fixed prostheses. *Clin Orthop Relat R,*, 393, 128-136. ISSN: 0009-921X

Glassman, A. H.; Bobyn, J. D. & Tanzer, M. (2006). New femoral designs: do they influence stress-shielding? *Clin Orthop Relat R*, 453, 64-74. ISSN: 0009-921X

Gordon, A. C.; D'Lima, D. D. & Colwell, C. W. (2006). Highly cross-linked polyethylene in total hip arthroplasty. *J Am Acad Orthop Surg*, 14, 9, 511-23. ISSN: 1067-151X

Grant, P. & Nordsletten, L. (2004). Influence of porous coating level on proximal femoral remodelling. *J Bone Joint Surg Am*, 86-A, 12, 2636-41. ISSN: 0021-9355

Gruen, T. A.; McNeice, G. M. & Amstutz H. C. (1979). Modes of failure of cemented stem-type femoral components: a radiographic analysis of loosening. *Clin Orthop Relat R*, 141, 17-27. ISSN: 0009-921X

Halley, D. K. & Wroblewski, B. M. (1986). Long-term results of low-friction arthroplasty in patients 30 years of age or younger. *Clin Orthop Relat R*, 211, 43-50. ISSN: 0009-921X

Harris, W. H.; McCarthy, J. & O'Neill, D. A. (1982). Femoral component loosening using contemporary techniques of femoral cement fixation. *J Bone Joint Surg Am*, 64, 7, 1063-67. ISSN: 0021-9355

Hellman, E. J.; Capello, W. N. & Feinberg, J. R. (1999). Omnifit cementless total hip arthroplasty: A 10-years average follow up. *Clin Orthop Relat R*, 364, 164-174. ISSN: 0009-921X

Hennessy, D. W.; Callaghan, J. J. & Liu, S. S. (2009). Second-generation extensively porous-coated THA stems at minimum 10-year followup. *Clin Orthop Relat R*, 467, 9, 2290-6. ISSN: 0009-921X

Herrera, A.; Domingo, J. J. & Panisello, J. J. (2001). Controversias en la artroplastia total de cadera. Elección del implante. In: *Actualizaciones en Cirugía Ortopedica y Traumatología II*, 141-62, Edit. Masson, Barcelona. ISBN: 8445810944

Herrera, A.; Canales, V.; Anderson, J. et al. (2004). Seven to ten years follow up of an anatomic hip protesis. *Clin Orthop Relat R*, 423, 129-37. ISSN: 0009-921X

Herrera, A. & Panisello, J. J. (2006). Fisiología del hueso y remodelado óseo. In: *Biomecánica y resistencia ósea*, 27-42, Edit MMC, Madrid. ISBN: 8468977116

Herrera, A.; Panisello, J. J., Ibarz, E. et al (2007). Long-term study of bone remodelling after femoral stem: A comparison between dexa and finite element simulation. *J Biomech*, 40, 16: 3615-25. ISSN: 0021-9290

Herrera, A.; Panisello, J. J.; Ibarz, E. et al. (2009). Comparison between DEXA and Finite Element studies in the long term bone remodelling o an anatomical femoral stem. *J Biomech Eng*, 31, 4, 1004-13. ISSN: 0148-0731

Hughes, T. J. R. (1987). *The finite element method*, Prentice-Hall, New Jersey. ISBN-10 013317025X

Huiskes, R.; Weinans, H. & Dalstra, M. (1989). Adaptive bone remodeling and biomechanical design considerations for noncemented total hip arthroplasty. *Orthopedics*, 12, 1255-1267. ISSN: 0147-7447

I-DEAS (2009). Web site, http://www.ugs.com/

Imbert, F. J. (1979). *Analyse des structures par élément finis*, Cepadues Edit, Toulouse. ISBN: 2854280512

Ionescu, I.; Conway, T.; Schonning, A. et al. (2003). Solid modeling and static finite element analysis of the human tibia. *Summer Bioengineering Conference*, 889-90, Florida.

Jacobs, C. R. (1994). *Numerical simulation of bone adaptation to mechanical loading*. Dissertation for the degree of Doctor of Philosophy, Stanford University.

Joven, E. (2007). *Densitometry study of bone remodeling in cemented hip arthroplasty with stem straight and anatomical*. Doctoral Degree Disertation, University of Zaragoza.

Kerner, J.; Huiskes, R. ; van Lenthe, G. H. et al. (1999). Correlation between pre-operative periprosthetic BMD and post-operative bone loss in THA can be explained by strain-adaptative remodelling. *J Biomech*, 32, 695-703. ISSN: 0021-9290

Kroger, H.; Miettinen, H.; Arnala, I. et al. (1996). Evaluation of periprosthetic bone using dual energy X-ray absorptiometry: precision of the method and effect of operation on bone mineral density. *J Bone Miner Res*, 11, 1526-30. ISSN: 0884-0431

Li, M. G.; Rohrl, S. M.; Wood, D. J. et al. (2007). Periprosthetic Changes in Bone Mineral Density in 5 Stem Designs 5 Years After Cemented Total Hip Arthroplasty. No Relation to Stem Migration. *J Arthroplasty*, 22, 5, 698-91. ISSN: 0883-5403

Marklof, K. L.; Amstutz, H. C. & Hirschowitz, D. L. (1980). The effect of calcar contact on femoral component micromovement.A mechanical study. *J Bone Joint Surg Am*, 62, 1315-1323. ISSN: 0021-9355

Mat Web (Material Property Data) (2009). Web site, http://www.matweb.com/

McAuley, J., Sychterz, Ch. & Ench, C. A. (2000). Influence of porous coating level on proximal femoral remodeling. *Clin Orthop Relat R*, 371, 146-153. ISSN: 0009-921X

Medel, F. J.; Gómez-Barrena, E.; García-Álvarez, F. et al. (2004). Fractography evolution in accelerated aging of UHMWPE after gamma irradiation in air. *Biomaterials*, 25, 1, 9-21. ISSN: 0142-9612

Meunier, A.; Riot, O. ; Christel, P. et al. (1989). Inhomogeneities in anisotropic elastic constants of cortical bone. *Ultrasonics Symposium*, 1015-1018.

MIMICS. (2010). Web site, http://www.materialise.com/

Mohler, C. G.; Callaghan, J. J.; Collis, D. K. et al. (1995). Early loosening of the femoral component at the cement-prosthesis interface after total hip replacement. *J Bone Joint Surg Am*, 77, 9, 1315-22. ISSN: 0021-9355

Mulroy, R. D.; Harris, W. H. (1990). The effect of improved cementing techniques on component loosening in total hip replacement. An 11 year radiographic review. *J Bone Joint Surg Br*, 72, 5, 757-60. ISSN: 0301-620X

Noble, P. C.; Collier, M. B.; Maltry, J. A. et al. (1998). Pressurization and centralization enhance the quality and reproductibility of cementless. *Clin Orthop Relat R*, 355, 77-89. ISSN: 0009-921X

Nourissat, C.; Adrey, J.; Berteaux, D. et al. (1995). The ABG Standar hip prosthesis: Five year results. In: *Hydroxyapatite coated hip and knee arthroplasty*, 227-38, Epinette JA, Geesink RGT, Eds. Edit Expansion Cientifique Francaise, Paris. ISBN: 2704614709

Olsson, S. S.; Jenberger, A. & Tryggo, D. (1981). Clinical and radiological long-term results after Charnley-Muller total hip replacement. A 5 to 10 year follow-up study with special reference to aseptic loosening. *Acta Orthop Scand*, 52, 5, 531-42. ISSN: 0001-6470

Oral, E.; Christensen, S. D.; Malhi, A. S. et al. (2005). Wear resistance and mechanical properties of highly cross-linked, ultrahigh-molecular weight polyethylene doped with vitamin E. *J Arthroplasty*, 21, 4, 580-91. ISSN: 0883-5403

Panisello, J. J.; Herrero, L.; Herrera, A. et al. (2006). Bone remodelling after total hip arthroplasty using an uncemented anatomic femoral stem: a three-year prospective study using bone densitometry. *J Orthop Surg*, 14, 122-25. ISSN: 1022-5536

Panisello, J. J.; Herrero, L.; Canales, V. et al. (2009a). Long-term remodelling in proximal femur around a hydroxyapatite-coated anatomic stem. Ten years densitometric follow-up. *J Arthroplasty*, 24, 1, 56-64. ISSN: 0883-5403

Panisello, J. J.; Canales, V.; Herrero, L. et al. (2009b). Changes in periprosthetic bone remodelling alter redesigning an anatomic cementless stem. *Inter Orthop*, 33, 2, 373-80. ISSN: 0341-2695

Radin, E. L. (1980). Bimechanics of the Human hip. *Clin Orthop Relat R*, 152, 28-34. ISSN: 0009-921X

Ramaniraka, N. A.; Rakotomanana, L. R. & Leyvraz, P. F. (2000). The fixation of the cemented femoral component .Effects of stem stiffness, cement thickness and roughness of the cement-bone surface . *J Bone Joint Surg Br*, 82, 297-303. ISSN: 0301-620X

Reading, A. D.; McCaskie, A. W.; Barnes, M. R. et al. (2000). A comparison of 2 modern femoral cementing techniques: analysis by cement-bone interface pressure measurement, computerized image analysis and static mechanical testing. *J Arthroplasty*, 15, 4, 479-87. ISSN: 0883-5403

Rosenthall, L.; Bobyn, J.D. & Tanzer, M. (1999). Bone densitometry: influence of prosthetic design and hydroxyapatite coating on regional adaptive bone remodelling. *Int Orthop*, 23, 325-29. ISSN: 0341-2695

Rubash, H. E.; Sinha, R. K.; Shanbhag, A. S. et al. (1998). Pathogenesis of bone loss after total hip arthroplasty. *Orthop Clin N Am*, 29(2), 173-186. ISSN: 0030-5898.

Schmidt, R.; Nowak, T.; Mueller, L. et al. (2004). Osteodensitometry after total hip replacement with uncemented taper-design stem. *Int Orthop*, 28, 74-7. ISSN: 0341-2695

Shirazi-Adl, A., Dammak, M. & Paiement, G. (1993). Experimental determination of friction characteristics at the trabecular bone/porous-coated metal interface in cementless implants. *J Bio Mat Res A*, 27, 167-75. ISSN: 1549-3296

Sinha, R. K.; Dungy, D. S. & Yeon, H. B. (2004). Primary total hip arthroplasty with a proximally porous-coated femoral stem. *J Bone Joint Surg Am* 86, 6, 1254-61. ISSN: 0021-9355

Stauffer, R. N. (1982). Ten-year follow-up study of total hip replacement. *J Bone Joint Surg Am*, 64, 7, 983-90. ISSN: 0021-9355

Sumner, D.R. & Galante, J. O. (1992). Determinants of stress shielding: design versus materials versus interface. *Clin Orthop Relat R*, 274, 202-12. ISSN: 0009-921X

Sychter, C. J. & Engh, C. A. (1992). The influence of clinical factor on periprosthetic bone remodeling. *Clin Orthop Relat R*, 322, 285-92. ISSN: 0009-921X

Tanzer, M.; Maloney, W. J.; Jasty, M. et al. (1992). The progression of femoral cortical osteolysis in association with total hip arthroplasty without cement. *J Bone Joint Surg Am*, 74, 404-10. ISSN: 0021-9355

Tonino, A. J.; Therin, M. & Doyle, C. (1999). Hydroxyapatite coated femoral stems: Histology and histomorphometry around five components retrieved at post-mortem. *J Bone Joint Surg Br*, 81, 148-54. ISSN: 0301-620X

Turner, M. J.; Clough, R. W.; Martin, M. C. & Topp, L. J. (1956). Stiffness and deflection analysis of complex structures. *J Aeronautical Sciences*, 23, 9, 805-823. ISSN: 0095-9812

Turner, C. H.; Rho, J. ; Takano, Y. et al. (1999). The elastic properties of trabecular and cortical bone tissues are similar: results from two microscopic measurement techniques. *J Biomech*, 32, 437-41. ISSN: 0021-9290

Turner, A. W. L.; Gillies, R.M.; Sekel, R. et al. (2005). Computational bone remodelling simulations and comparisons with DEXA results. *J Orthop Res*, 23, 705-12. ISSN: 0736-0266

Urriés, I; Medel, F. J.; Ríos F. et al. (2004). Comparative cyclic stress-strain and fatigue resistance behaviour of electron-beam and gamma irradiated UHMWPE. *J Biomed Mater Res B*, 70, 1, 152-60. ISSN: 1552-4973

van der Val, B. C. H.; Rahmy, A. ; Grimm, B. et al. (2006). Preoperative bone quality as a factor in dual-energy X-ray absorptiometry analysis comparing bone remodelling between two implant types. *Int Orthop*, On line. ISSN: 0341-2695

Weinans, H.; Huiskes, R. & Grootenboer, H. J. (1994). Effects of fit and bonding characteristics of femoral stems on adaptive bone remodelling. *J Biomech Eng*, 116, 4, 393-400. ISSN: 0148-0731

Wick, M. & Lester D.K. (2004). Radiological changes in second and third generation Zweymuller stems. *J Bone Joint Sur Br*, 86, 8, 1108-14. ISSN: 0301-620X

Wolf, C.; Maninger, J.; Lederer, K. et al. (2006). Stabilisation of crosslinked ultra-high molecular weight polyethylene (UHMWPE)-acetabular components with alpha-tocopherol. *J Mater Sci Mater Med*, 17, 12, 1323-31. ISSN: 0957-4530

Zienkiewicz, O. C. (1967). *The finite element method in structural and continuum mechanics*, Prentice-Hall, New Jersey. ASIN: B000HF38VG

Zienkiewicz, O. C. & Morgan, K. (1983). *Finite element and approximation*, John Wiley & Sons, New York. ISBN 10: 0471982407

10

Tissue Modeling and Analyzing for Cranium Brain with Finite Element Method

Xianfang Yue[*], Li Wang, Ruonan Wang,
Yunbo Wang and Feng Zhou
Mechanical Engineering School,
University of Science and Technology Beijing, Beijing,
China

1. Introduction

Numerous methods for measuring intracranial pressure (ICP) have been described, and many of them are suitable for different clinical disorders [1-5]. One method for ICP monitoring is through the ventricular system [6,7], which requires stereotaxic techniques and may not be practical for surgical experiments in the brain regions. Ventricular monitoring of ICP is also associated with intracerebral hemorrhage and infection [6]. Another method to monitor ICP is through the subarachnoid space at the cisterna magna, in which catheter placement may be difficult and dangerous due to the anatomy [8,9]. Some studies monitored ICP via epidural [10,12], which has limitations in measuring acute changes in ICP and is inaccurate in some cases when compared with ventricular monitoring [8]. These methods have many disadvantages of invasion, low-accuracy, cross-infection, etc [13,14]. Although many efforts have recently been made to improve the minitraumatic or non-invasive methods [15-19], noninvasive means of measuring ICP do not exist unfortunately in clinic [20,21]. With the significance of raised ICP in the studies of intracranial pathophysiology, espacially in neurosurgical disorders, it would be valuable to have a simple and reliable method to monitor intracranial pressure (ICP) in clinic. Therefore, the minitraumatic or non-invasive, accurate and simple method to measure ICP is an important question of research in neurosurgery. In this study, we propose a new, minitraumatic, simple, and reliable measurable system that can be used to monitor ICP from the exterior surface of skull bone.

The 'Monro [22] – Kellie [23] doctrine' states that an adult cranial compartment is incompressible, and the volume inside the cranium is a fixed volume thus creates a state of volume equilibrium, such that any increase of the volumes of one component (i.e. blood, CSF, or brain tissue) must be compensated by a decrease in the volume of another. If this cannot be achieved then pressure will rise and once the compliance of the intracranial space is exhausted then small changes in volume can lead to potentially lethal increases in ICP. The compensatory mechanism for intracranial space occupation obviously has limits. When the amount of CSF and venous blood that can be extruded from the skull has been

---

[*] Corresponding Author

exhausted, the ICP becomes unstable and waves of pressure develop. As the process of space occupation continues, the ICP can rise to very high levels and the brain can become displaced from its normal position. Dr. Sutherland firstly perceived a subtle palpable movement within the bones of cranium. Dr. Upledger [24] discovered that the inherent rhythmic motion of cranial bones was caused by the fluctuation of CSF. Accordingly, the cranium can move and be deformed as the ICP fluctuates. By pasting strain foils on the exterior surface of skull bone, the skull strains can be measured with the strain gauge. ICP variation can be obtained through the corresponding processing based on the strains. So the ICP can be monitored by measuring the strains of skull bone [25].

2. Mechanical analysis of cranial cavity deformation

2.1 Mechanical analysis of deformation of the skull as a whole

There are two aspects of effects on forces on the objects. One can make objects produce the acceleration, another is make objects deform. In discussing the external force effect, the objects are assumed to be a rigid body not compressed. But in fact, all objects will deform under loading, but different with the degrees. Here, we will mechanically analyze the overall deformation of cranial cavity under the external force.

(1) Two basic assumptions

To simplify the analysis of deformation of human skull, we assume:

1.   Uniformly-continuous materials

The human skull is presumed to be everywhere uniform, and the sclerotin is no gap in bone of cranial cavity.

2.   The isotropy

The human skull is supposed to have the same mechanical properties in all directions. The thickness and curvature of human skull vary here and there. The external and internal boards are all compact bones, in which external board is thicker than internal board but the radian of external board is smaller than that of the internal board. The diploe is the cancellous bone between the external board and the internal board, which consists of the marrow and diploe vein. The parietal bone is the transversely isotropic material [26], namely, it has the mechanical property of rotational symmetry in any axially vertical planes of skull [27]. The tensile and compressive abilities of compact bone are strong. The important mechanical characteristics of cancellous bone are viscoelasticity [28], which is generally considered as the construction of semi-closed honeycomb composed of bone trabecula reticulation. The main composition of cerebral duramater, a thick and tough bilayer membrane, is collagenous fiber11, which is viscoelastic material [29]. And the thickness of duramater is obviously variable with the changing ICP [30]. The mechanical performance of skull is isotropic along tangential direction on the skull surface [31], in which the performance of compact bone in the external board is basically the same as that in the internal board [32], thus both cancellous bone and duramater can be regarded as isotropic materials [33]. And the elastic modulus of fresh duramater is variable with delay time [34]. And there are a number of sutures in the cranial cavity. But while a partial skull is studied on the local deformation, we can regard each partial skull as quasi-homogeneous and quasi-isotropic. In addition, the sutures of cranial cavity are also the continuous integration with ages.

(2) Two concepts

1.   stress (σ)

Stress is the internal force per unit area. The calculation unit is kg/cm^2 or kg/mm^2.

2.   Transverse deformation coefficient

When the objects are under tension, not only the length is drawn out from l to l_1, but also the width is reduced from b to b_1. This shows that there are the horizontal compressive stresses in the objects. Similarly, while the object is compressed, not only its length shortens but its width increases. It indicates that the horizontal tensile stress distributes in the objects.

The horizontal absolute deformation is noted as $\Delta b = b - b_1$, and the transverse strain is $\varepsilon_0 = \Delta b/b$.

In mechanics of materials, the transverse strain ε_0 is proportional to the longitudinal strain ε of the same material within the scope of the Hooke theorems' application. The ratio of its absolute value $\mu = |\varepsilon_0|/|\varepsilon|$ is a constant. μ is known as the coefficient of lateral deformation, or Poisson ratio. The Poisson ratio of any objects can be detected by the experiment.

While vertically compressed, the objects simultaneously have a horizontal tension. Therefore, when the head attacked in opposite directions force, the entire human skull will take place the longitudinal compression and transverse tension with the same direction of force. Then the longitudinal compressive stress and the horizontal tensile stress will be generated in the scelrotin. Thus, the stress of arbitrary section along radial direction in shell is just equal to the tangential pulling force along direction perpendicular to the normal vertical when ICP is raised.

2.1.1 The finite-element analysis of strains by ignoring the viscoelasticity of cranial cavity

The geometric shape of human skull is irregular and variable with the position, age, gender and individual. So the cranial cavity system is very complex. Moreover, the cranial cavity is a kind of viscoelactic solid with large elastic modulus, and the brain tissue is also a viscoelastic fluid with great bulk modulus. It is now almost impossible to accurately analyze the brain system. Only by some simplification and assumptions, the complex issues can be made. Considering the special structure of cranial cavity composed of skull, duramater, encephalic substance, etc, here we simplify the model of cranial cavity as a regular geometry spheroid of about 200mm external diameter, which is an hollow equal-thickness thin-wall shell.

The craniospinal cavity may be considered as a balloon. For the purpose of our analysis, we adopted the model of hollow sphere. We presented the development and validation of a 3D finite-element model intended to better understand the deformation mechanisms of human skull corresponding to the ICP change. The skull is a layered sphere constructed in a specially designed form with a Tabula externa, Tabula interna, and a porous Diploe sandwiched in between. Based on the established knowledge of cranial cavity importantly composed of skull and dura mater (Fig2.1), a thin-walled structure was simulated by the composite shell elements of the finite-element software [35]. The thickness of skull is 6mm, that of duramater is 0.4mm, that of external compact bone is 2.0mm, that of cancellous bone is 2.8mm, and that of internal compact bone is 1.2mm.

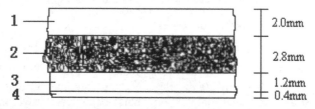

Fig. 2.1. Sketch of layered sphere. The thin-walled structure of cranial cavity is mainly composed of Tabula externa, Diploe, Tabula interna and dura mater.

Above all, we should prove the theoretical feasibility of the strain-electrometric method to monitor ICP. We simplify the theoretical calculation by ignoring the viscoelasticity of cancellous bone and dura mater. And then we make the analysis of the actual deformation of cranial cavity by considering the viscoelasticity of human skull-dura mater system with the finite-element software. At the same time, we can determine how the viscoelasticity of human skull and dura mater influences the strains of human skull respectively by ignoring and considering the viscoelasticity of human skull and dura mater.

2.1.2 The stress and strain analysis of discretized elements of cranial cavity

In order to obtain the numerical solution of the skull strain, the continuous solution region of cranial cavity divided into a finite number of elements, and a group element collection glued on the adjacent node points. Then the large number of cohesive collection can be simulated the overall cranial cavity to carry out the strain analysis in the solving region. Based on the block approximation ideas, a simple interpolation function can approximately express the distribution law of displacement in each element. The node data of the selected field function, the relationship between the nodal force and displacement is established, and the algebraic equations of regarding the nodal displacements as unknowns can be formed, thus the nodal displacement components can be solved. Then the field function in the element collection can be determined by using the interpolation function. If the elements meet the convergence requirements, with the element numbers increase in the solving region with the shrinking element size, and the approximate solution will converge to exact solutions [36].

The solving steps for the strains of cranial cavity with the ICP changes are shown in Fig2.2. The specific numerical solution process is:

1.   The discretized cranial cavity

The three-dimensional hollow sphere of cranial cavity is divided into a finite number of elements. By setting the nodes in the element body, an element collection can replace the structure of cranial cavity after the parameters of adjacent elements has a certain continuity.

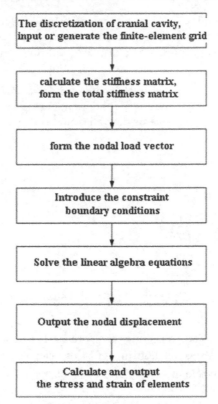

Fig. 2.2. Block diagram of **numerical solution** steps of cranial cavity with the finite-element method

2.   The selection of displacement mode

To make the nodal displacement express the displacement, strain and stress of element body, the displacement distribution in the elements are assumed to be the polynomial interpolation function of coordinates. The items of polynomial number are equal to the freedom degrees number of elements, that is, the number of independent displacement of element node. The orders of polynomial contain the constant term and linear terms.

According to the selected displacement mode, the nodal displacement is derived to express the displacement relationship of any point in the elements. Its matrix form is:

$$\{f\} = [N]\{\delta\}^e \tag{2.1}$$

Where: $\{f\}$ - The displacement array of any point within the element; $[N]$ - The shape function matrix, its elements is a function of location coordinates; $\{\delta\}^e$ - The nodal displacement array of element.

The block approximation is adopted to solve the displacement of cranial cavity in the entire solving region, and an approximate displacement function is selected in an element, where need only consider the continuity of displacement between elements, not the boundary conditions of displacement. Considering the special material properties of the middle cancellous and duramater, the approximate displacement function can adopt the piecewise function.

3.   Analyze the mechanical properties of elements, and derive the element stiffness matrix

a.   Using the following strain equations, the relationship of element strain (2.2) is expressed by the nodal displacements derived from the displacement equation (2.1):

$$\text{Strain equations}\begin{cases} \varepsilon_x = \dfrac{\partial u}{\partial x}, \gamma_{xy} = \dfrac{\partial u}{\partial y} + \dfrac{\partial v}{\partial x} \\[2mm] \varepsilon_y = \dfrac{\partial u}{\partial y}, \gamma_{yz} = \dfrac{\partial u}{\partial z} + \dfrac{\partial w}{\partial y} \\[2mm] \varepsilon_z = \dfrac{\partial u}{\partial z}, \gamma_{zx} = \dfrac{\partial u}{\partial x} + \dfrac{\partial v}{\partial z} \end{cases}$$

$$\{\varepsilon\} = [B]\{\delta\}^e \tag{2.2}$$

Where:[B] — The strain matrix of elements; $\{\varepsilon\}$ — The strain array at any points within the elements.

b.   The constitutive equation reflecting the physical characteristics of material is $\{\sigma\} = [D]\{\varepsilon\}$, so the stress relationship of stress can be expressed with the nodal displacements derived from the strain formula (2.2):

$$\{\sigma\} = [D][B]\{\delta\}^e \tag{2.3}$$

Where: $\{\sigma\}$ - The stress array of any points in the elements; $[D]$ - The elastic matrix related to the element material.

c.   Using the variational principle, the relationship between force and displacement of the element nodes is established:

$$\{F\}^e = [k]^e \{\delta\}^e \tag{2.4}$$

Where: $[k]^e$ - Element stiffness matrix, $[k]^e = \iiint [B]^T [D][B] dx dy dz$; $\{F\}^e$ - Equivalent nodal force array, $\{F\}^e = \{F_V\}^e + \{F_S\}^e + \{F_C\}^e$; $\{F_V\}^e$ - Equivalent nodal force on the nodes transplanted from the element volume force P_V, $\{F_V\}^e = \iiint_V [N]^T \{P_V\} dV$; $\{F_S\}^e$ - Equivalent nodal force on the nodes transplanted from the element surface force, $\{F_A\}^e = \iint_A [N]^T \{P_S\} dA$; $\{F_C\}^e$ - Concentration force of nodes.

4.   Collecting all relationship between force and displacement, and establish the relationship between force and displacement of cranial cavity

According to the displacement equal principle of the public nodes in all adjacent elements, the relationship between force and displacement of overall cranial cavity collected from the element stiffness matrix:

$$\{F\} = [K]\{\delta\} \tag{2.5}$$

Where: $\{F\}$ - Load array; $[K]$ - The overall stiffness matrix; $\{\delta\}$ - The nodal displacement array of the entire cranial cavity.

5.   Solve the nodal displacement

After the formula (2.1) ~ (2.5) eliminating the stiffness displacement of geometric boundary conditions, the nodal displacement can be solved from the gathered relationship groups between force and displacement.

6.   By classifying the nodal displacement solved from the formula (2.2) and (2.3), the strain and stress in each element can be calculated.

In this paper, the studied cranial cavity is a hollow three-dimensional sphere, its external radius $R = 100$ mm , the curvature of hollow shell is 0.01 rad/mm , the thickness of shell wall is 6mm, so the element body of hollow spherical can be treated as the regular hexahedron. The following is the stress and strain analyses in the three-dimensional elements in the cranial space. The 8-node hexahedral element (Fig2.3) is used to be the master element. The origin is set up as the local coordinate system (ξ, η, ζ) in the element. Trough the transformation between rectangular coordinates and local coordinates, the space 8-node isoparametric centroid element can be obtained. The relationship of coordinate transformation is:

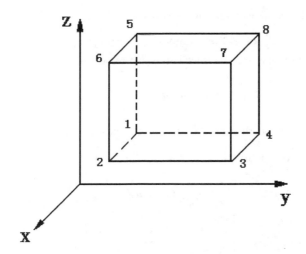

Fig. 2.3. The space 8-node isoparametric centroid element

$$\begin{cases} x = \sum_{i=1}^{8} N_i(\xi,\eta,\zeta)x_i \\ y = \sum_{i=1}^{8} N_i(\xi,\eta,\zeta)y_i \\ z = \sum_{i=1}^{8} N_i(\xi,\eta,\zeta)z_i \end{cases} \tag{2.6}$$

Then the displacement function of element is:

$$\begin{cases} u = \sum_{i=1}^{8} N_i(\xi,\eta,\zeta)u_i \\ v = \sum_{i=1}^{8} N_i(\xi,\eta,\zeta)v_i \\ w = \sum_{i=1}^{8} N_i(\xi,\eta,\zeta)w_i \end{cases} \tag{2.7}$$

Where: x_i, y_i, z_i and u_i, v_i, w_i are respectively the coordinate values and actual displacement of nodes.

The element displacement function with matrix is expressed as:

$$\{\delta\} = \begin{Bmatrix} u \\ v \\ w \end{Bmatrix} = \sum_{i=1}^{8} \begin{bmatrix} N_i & 0 & 0 \\ 0 & N_i & 0 \\ 0 & 0 & N_i \end{bmatrix} \begin{Bmatrix} u_i \\ v_i \\ w_i \end{Bmatrix} = \sum_{i=1}^{8} [N_i]\{\delta_i\} = [N]\{\delta\}^e \tag{2.8}$$

Where: $\{\delta_i\}$ - Nodal displacement array, $\{\delta_i\} = [u_i \quad v_i \quad w_i]^T$ $(i=1,2,\cdots\cdots,8)$; $\{\delta\}^e$ - The nodal displacement array of entire element, $\{\delta\}^e = [\{\delta_1\} \quad \{\delta_2\} \quad \cdots\cdots \quad \{\delta_8\}]^T$; N_i - The uniform shape function of 8 nodes, which can be expressed as:

$$N_i = \frac{1}{8}(1+\xi_i\xi)(1+\eta_i\eta)(1+\zeta_i\zeta) \quad (i=1,2,\cdots\cdots,8) \tag{2.9}$$

Where: ξ_i, η_i, ζ_i is the coordinates of node i in the local coordinate system (ξ,η,ζ).

The derivative of composite function to local coordinates is:

$$\begin{cases} \dfrac{\partial N_i}{\partial \xi} = \dfrac{1}{8}\xi_1(1+\eta_1\eta)(1+\zeta_1\zeta) \\ \dfrac{\partial N_i}{\partial \eta} = \dfrac{1}{8}\eta_1(1+\xi_1\xi)(1+\zeta_1\zeta) \\ \dfrac{\partial N_i}{\partial \zeta} = \dfrac{1}{8}\zeta_1(1+\xi_1\xi)(1+\eta_1\eta) \end{cases} \tag{2.10}$$

The strain relationship of space elements is:

$$\{\varepsilon\} = [B]\{\delta\}^e = \sum_{i=1}^{8}[B_i]\{\delta_i\}$$

(2.11)

The strain matrix $[B]$ of space element:

$$[B_i] = \begin{bmatrix} \dfrac{\partial N_i}{\partial x} & 0 & 0 \\[2mm] 0 & \dfrac{\partial N_i}{\partial y} & 0 \\[2mm] 0 & 0 & \dfrac{\partial N_i}{\partial z} \\[2mm] \dfrac{\partial N_i}{\partial y} & \dfrac{\partial N_i}{\partial x} & 0 \\[2mm] 0 & \dfrac{\partial N_i}{\partial z} & \dfrac{\partial N_i}{\partial y} \\[2mm] \dfrac{\partial N_i}{\partial z} & 0 & \dfrac{\partial N_i}{\partial x} \end{bmatrix}$$

(2.12)

The shape function was derivative to be:

$$\left\{ \begin{array}{c} \dfrac{\partial N_i}{\partial x} \\[2mm] \dfrac{\partial N_i}{\partial y} \\[2mm] \dfrac{\partial N_i}{\partial z} \end{array} \right\} = [J]^{-1} \left\{ \begin{array}{c} \dfrac{\partial N_i}{\partial \xi} \\[2mm] \dfrac{\partial N_i}{\partial \eta} \\[2mm] \dfrac{\partial N_i}{\partial \zeta} \end{array} \right\}$$

(2.13)

The matrix $[J]$ is the three-dimensional Yake ratio matrix of coordinate transformation:

$$[J] = \left\{ \begin{array}{ccc} \dfrac{\partial x}{\partial \xi} & \dfrac{\partial y}{\partial \xi} & \dfrac{\partial z}{\partial \xi} \\[2mm] \dfrac{\partial x}{\partial \eta} & \dfrac{\partial y}{\partial \eta} & \dfrac{\partial z}{\partial \eta} \\[2mm] \dfrac{\partial x}{\partial \zeta} & \dfrac{\partial y}{\partial \zeta} & \dfrac{\partial z}{\partial \zeta} \end{array} \right\} = \begin{bmatrix} \sum\limits_{i=1}^{8}\dfrac{\partial N_i}{\partial \xi}x_i & \sum\limits_{i=1}^{8}\dfrac{\partial N_i}{\partial \xi}y_i & \sum\limits_{i=1}^{8}\dfrac{\partial N_i}{\partial \xi}z_i \\[2mm] \sum\limits_{i=1}^{8}\dfrac{\partial N_i}{\partial \eta}x_i & \sum\limits_{i=1}^{8}\dfrac{\partial N_i}{\partial \eta}y_i & \sum\limits_{i=1}^{8}\dfrac{\partial N_i}{\partial \eta}z_i \\[2mm] \sum\limits_{i=1}^{8}\dfrac{\partial N_i}{\partial \zeta}x_i & \sum\limits_{i=1}^{8}\dfrac{\partial N_i}{\partial \zeta}y_i & \sum\limits_{i=1}^{8}\dfrac{\partial N_i}{\partial \zeta}z_i \end{bmatrix}$$

(2.14)

The stress-strain relationship of space elements is:

$$\{\sigma\} = [D]\{\varepsilon\} = [D][B]\{\delta\}^e$$

(2.15)

The elasticity matrix $[D]$ is:

$$[D] = \frac{E(1-\mu)}{(1+\mu)(1-2\mu)} \begin{bmatrix} 1 & 0 & 0 & 0 & 0 & 0 \\ \dfrac{\mu}{1-\mu} & 1 & 0 & 0 & 0 & 0 \\ \dfrac{\mu}{1-\mu} & \dfrac{\mu}{1-\mu} & 1 & 0 & 0 & 0 \\ 0 & 0 & 0 & \dfrac{1-2\mu}{2(1-\mu)} & \dfrac{\mu}{1-\mu} & \dfrac{\mu}{1-\mu} \\ 0 & 0 & 0 & 0 & \dfrac{1-2\mu}{2(1-\mu)} & \dfrac{\mu}{1-\mu} \\ 0 & 0 & 0 & 0 & 0 & \dfrac{1-2\mu}{2(1-\mu)} \end{bmatrix} \quad (2.16)$$

The element stiffness matrix from the principle of virtual work is:

$$[k]^e = \iiint_V [B]^T [D][B] dx dy dz = \begin{bmatrix} k_{11} & k_{12} & \cdots & k_{18} \\ k_{21} & k_{22} & \cdots & k_{28} \\ & \cdots & \cdots & \\ k_{81} & k_{82} & \cdots & k_{88} \end{bmatrix} \quad (2.17)$$

Where:

$$\left[k_{ij} \right] = \iiint_V [B]^T [D][B] dx dy dz = \int_{-1}^{1}\int_{-1}^{1}\int_{-1}^{1} [B]^T [D][B] J d\xi d\eta d\zeta \quad (2.18)$$

The equivalent nodal forces acting on the space element nodes are:

$$\{F\}^e = [k]^e \{\delta\}^e \quad (2.19)$$

Because the internal pressure in the cranial cavity is surface force, the equivalent load for the pressure acting on the element nodes is:

$$\{F_S\}^e = \iint_S [N]^T \{P_S\} dS \quad (2.20)$$

The relationship between force and displacement in the entire cranial cavity is:

$$\{F\} = [K]\{\delta\} \quad (2.21)$$

Then after obtaining the nodal displacement, the stress and strain in each element can be calculated by combining the formula (2.11) and (2.15).

2.2 The stress and strain analysis for complex structure of cranial cavity deformation

Cranial cavity is the hollow sphere formed by the skull and the duramater. From the Fig2.2, there are obvious interfaces among the various parts of outer compact bone, middle cancellous bone, inner compact bone and duramater, which is consistent with the characteristics of composite materials [37]. Four layered composite structure of cranial cavity is almost lamelleted distribution. Therefore, the lamelleted structure is adopted to establish and analyze the finite-element model of cranial cavity, and the laminated shell element is used to describe the thin cranial cavity made up of skull and duramater. Here the cranial cavity deformation of laminated structure is analyzed as follows:

(1) The stress and strain analysis for the single layer of cranial deformation

Each layers of cranial cavity are all thin flat film. The skulls are transversely isotropic material. The thickness of Tabula externa, diploe, Tabula interna, duramater is all very small. So compared with the components in the surface, the stress components are very small along the normal direction, and can be neglected. So the deformation analysis to single-layer cranial cavity can be simplified to be the stress problems of two-dimensional generalized plane.

The stress-strain relationship of each single-layer structure in the cranial cavity:

$$\{\sigma\} = [Q]\{\varepsilon\} \tag{2.22}$$

Namely:

$$\begin{Bmatrix} \sigma_1 \\ \sigma_2 \\ \tau_{12} \end{Bmatrix} = \begin{bmatrix} Q_{11} & Q_{12} & 0 \\ Q_{21} & Q_{22} & 0 \\ 0 & 0 & Q_{66} \end{bmatrix} \begin{Bmatrix} \varepsilon_1 \\ \varepsilon_2 \\ \mu_{12} \end{Bmatrix} \tag{2.23}$$

Where, 1,2 - The main direction of elasticity in the plane; $[Q]$- Stiffness matrix,

$Q_{11} = mE_1$, $Q_{12} = m\mu_{12}E_2$, $Q_{22} = mE_2$, $Q_{66} = G_{12}$; $m = \left(1 - \dfrac{\mu_{12}{}^2 E_2}{E_1}\right)^{-1}$; E_1, E_2 - The elastic

modulus of four independent surfaces in each layer structure; G_{12} - Shear modulus; μ_{12} - Poisson's ratio of transverse strain along the 2 direction that the stress acts on the 1 direction.

(2) The stress and strain analysis for the laminated deformation of cranial cavity

The cranial cavity is as a whole formed by the four-layer structures. So the material, thickness and elastic main direction are all different. The overall performance of cranial cavity is anisotropic, macroscopic non-uniformity along the thickness direction and non-continuity of mechanical properties. Thus, the assumptions need to be made before analyzing the overall deformation of cranial cavity [38]:

1.   The same deformation in each layer

Each single layer is strong glued. There are the same deformation, and no relative displacement;

2.   No change of direct normal

The straight line vertical to the middle surface in each layer before the deformation remains still the same after the deformation, and the length of this line remains unchanged whether before or after deformation;

3. $\sigma_z = 0$

The positive stress along the direction of thickness is small compared to other stress, and can be ignored;

4.   The plane stress state in each single layer

Each single-layer structure is similar to be assumed in plane stress state.

From the four-layer laminated structure composed of Tabula externa, diploe, Tabula interna, duramater, the force of each single-layer structure is indicated in Fig2.4. The middle surface in the laminated structure of cranial cavity is the xy coordinate plane. z axis is perpendicular to the middle surface in the plate. Along the z axis, each layer in turn will be compiled as layer 1, 2, 3, and 4. The corresponding thickness is respectively t_1, t_2, t_3, t_4. As a overall laminated structure, the thickness of cranial cavity is h, shown in Fig2.5.

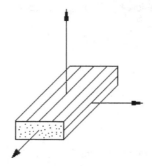

Fig. 2.4. The orientation relationship in each single-layer structure of cranial cavity

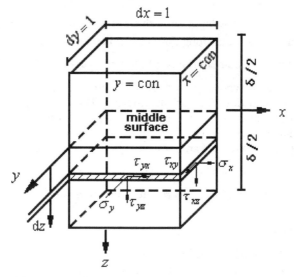

Fig. 2.5. The sketch of four-layered laminated structure of cranial cavity

Then

$$h = \sum_{k=1}^{4} t_k \tag{2.24}$$

The z coordinates is respectively z_{k-1} and z_k, then $z_0 = -h/2$ and $z_4 = h/2$.

The displacement components at any point within the laminated structure of cranial cavity:

$$\left.\begin{array}{l} u = u(x,y,z) \\ v = v(x,y,z) \\ w = w(x,y,x) \end{array}\right\} \tag{2.25}$$

The strain is:

$$\left\{\begin{array}{l} \varepsilon_x = \dfrac{\partial u}{\partial x} = \dfrac{\partial u_0}{\partial x} - z\dfrac{\partial^2 w}{\partial x^2} \\[2mm] \varepsilon_y = \dfrac{\partial v}{\partial y} = \dfrac{\partial v_0}{\partial y} - z\dfrac{\partial^2 w}{\partial y^2} \\[2mm] \gamma_{xy} = \dfrac{\partial u}{\partial y} + \dfrac{\partial v}{\partial x} = \dfrac{\partial u_0}{\partial y} + \dfrac{\partial v_0}{\partial x} - 2\dfrac{\partial^2 w}{\partial x \partial y} \end{array}\right. \tag{2.26}$$

Where: $u = u(x,y,z)$, $v = v(x,y,z)$, $w = w(x,y,z)$ - The displacement components at any point within the cranial cavity; $u_0(x,y)$, $v_0(x,y)$ - The displacement components in the middle surface; $w(x,y)$ - Deflection function, the deflection function of each layer is the same.

Fomular (2.26) can be expressed to be in matrix form:

$$\{\varepsilon\} = \{\varepsilon_0\} + z\{k\} \tag{2.27}$$

Where: ε_0 - Strain array in the plane, $\{\varepsilon_0\} = \left[\dfrac{\partial u_0}{\partial x}, \dfrac{\partial v_0}{\partial y}, \left(\dfrac{\partial u_0}{\partial y} + \dfrac{\partial v_0}{\partial x}\right)\right]^T$; k - Strain array of

bending in the surface, $\{k\} = \left[-\dfrac{\partial^2 w}{\partial x^2}, -\dfrac{\partial^2 w}{\partial y^2}, -2\dfrac{\partial^2 w}{\partial x \partial y}\right]^T$.

The mean internal force and torque acting on the laminated structure of cranial cavity in the unit width is:

$$\left\{\begin{array}{l} N_i = \int_{-h/2}^{h/2} \sigma_i dz \\[2mm] M_i = \int_{-h/2}^{h/2} \sigma_i z dz \end{array}\right. \quad (i = x, y, xy) \tag{2.28}$$

Taking into account the discontinuous stress caused by the discontinuity along the direction of laminated structure in the cranial cavity, the formula (2.28) can be rewritten in matrix form:

$$\begin{cases} \{N\} = \sum_{k=1}^{n} \int_{z_{k-1}}^{z_k} \{\sigma\} dz \\ \{M\} = \sum_{k=1}^{n} \int_{z_{k-1}}^{z_k} \{\sigma\} z dz \end{cases} \qquad (2.29)$$

After substituting the formula (2.22) and (2.27) into equation (2.29), the average internal force and internal moment of the laminated structure in the cranial cavity is:

$$\begin{Bmatrix} N \\ M \end{Bmatrix} = \begin{bmatrix} \sum \int [Q] dz & \sum \int [Q] z dz \\ \sum \int [Q] z dz & \sum \int [Q] z^2 dz \end{bmatrix} \begin{Bmatrix} \varepsilon_0 \\ k \end{Bmatrix} = \begin{bmatrix} A & B \\ B & D \end{bmatrix} \begin{Bmatrix} \varepsilon_0 \\ k \end{Bmatrix} \qquad (2.30)$$

Where: $[A]$ - The stiffness matrix in the plane, $A_{ij} = \sum_{k=1}^{n} Q_{ij}^{(k)}(z_k - z_{k-1})$; $[B]$ - Coupling stiffness matrix, $B_{ij} = \frac{1}{2} \sum_{k=1}^{n} Q_{ij}^{(k)}(z_k^2 - z_{k-1}^2)$; $[D]$ - Bending stiffness matrix, $D_{ij} = \frac{1}{3} \sum_{k=1}^{n} Q_{ij}^{(k)}(z_k^3 - z_{k-1}^3)$, $(i, j = 1, 2, 6)$.

Then the flexibility matrix of laminated structure in the cranial cavity is:

$$\begin{bmatrix} a & b \\ c & d \end{bmatrix} = \begin{bmatrix} A & B \\ B & D \end{bmatrix}^{-1} \qquad (2.31)$$

Where: $[a] = [A]^{-1} + [A]^{-1}[B]\left([D] - [B][A]^{-1}[B]\right)^{-1}[B][A]^{-1}$;

$[b] = -[A]^{-1}[B]\left([D] - [B][A]^{-1}[B]\right)^{-1}$; $[c] = -\left([D] - [B][A]^{-1}[B]\right)^{-1}[B][A]^{-1} = [b]^T$;

$[d] = \left([D] - [B][A]^{-1}[B]\right)^{-1}$

Thus, the stress-strain relationship of laminated structure in the cranial cavity is:

$$\begin{Bmatrix} \varepsilon_0 \\ k \end{Bmatrix} = \begin{bmatrix} a & b \\ b^T & d \end{bmatrix} \begin{Bmatrix} N \\ M \end{Bmatrix} \qquad (2.32)$$

With the changing ICP, to determine how the viscoelasticity of human skull and duramater influences the strains of human skull respectively by ignoring and considering the viscoelasticity of human skull and duramater, we make the analysis of the actual deformation of cranial cavity by considering the viscoelasticity of human skull-duramater system with the finite-element software MSC_PATRAN/NASTRAN and ANSYS. Considering the complexity to calculate the viscoelasticity of human skull and duramater, we can simplify the calculation while on-line analysis only considering the elasticity but ignoring the viscoelasticity of human skull and duramater after obtaining the regularity how the viscoelasticity influences the deformation of cranial cavity.

2.2.1 The finite-element analysis of strains by ignoring the viscoelasticity of human skull and duramater

The craniospinal cavity may be considered as a balloon. For the purpose of our analysis, we adopted the model of hollow sphere (Fig2.6). We presented the development and validation of a 3D finite-element model intended to better understand the deformation mechanisms of human skull corresponding to the ICP change. The skull is a layered sphere constructed in a specially designed form with a Tabula externa, Tabula interna, and a porous Diploe sandwiched in between. Based on the established knowledge of cranial cavity importantly composed of skull and duramater, a thin-walled structure was simulated by the composite shell elements of the finite-element software [39].

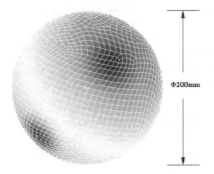

Φ200mm

Fig. 2.6. The sketch of 3D cranial cavity and grid division

Of course, the structure, dimension and characteristic parameter of human skull must be given before the calculation. The thickness of calvaria [40] varies with the position, age, gender and individual, so does dura mater [41]. Tabula externa and interna are all compact bones and the thickness of Tabula externa is more than that of Tabula interna. Diploe is the cancellous bone between Tabula externa and Tabula interna [42]. The parietal bone is the transversely isotropic material, namely it has the mechanical property of rotational symmetry in the axially vertical planes of skull [43]. The important mechanical characteristic of cancellous bone is viscoelasticity, which is generally considered as the semi-closed honeycomb structure composed of bone trabecula reticulation. The main composition of cerebral dura mater, a thick and tough bilayer membrane, is the collagenous fiber which has the characteristic of linear viscoelasticity [44]. And the thickness of dura mater obviously varies with the changing ICP [45]. The mechanical performance of skull is isotropic along the tangential direction on the surface of skull bone [46], in which the performance of compact bone in the Tabula externa is basically the same as that in the Tabula interna [47]. Thus both cancellous bone and dura mater can be regarded as isotropic materials. And the elastic modulus of fresh dura mater varies with the delay time [48].

Next we need determine the fluctuant scope of human ICP. ICP is not a static state, but one that influenced by several factors. It can rise sharply with coughing and sneezing, up to 50 or 60mmHg to settle down to normal values in a short time. It also varies according to the activity the person is involved with. For these reasons single measurement of ICP is not a true representation. ICP needs to be measured over a period. Measured by means of a lumbar puncture, the normal ICP in adults is 8 mmHg to 18 mmHg. But so far there are almost no records of the actual human being's ICP in clinic. The geometry and structure of monkey's skull, mandible and cervical muscle are closer to those of human beings than other animals'. So the ICP of monkeys [49] can be taken as the reference to that of human beings'. The brain appears to be mild injury when ICP variation is about 2.5 kPa, moderate injury when ICP variation is about 3.5 kPa and severe injury when ICP variation is about or more than 5 kPa. Therefore, we carried out the following theoretical analysis with the ICP scope from 1.5 kPa to 5 kPa.

In this paper, the finite-element software MSC_PATRAN/NASTRAN and ANSYS are applied to theoretically analyze the deformation of human skull with the changing ICP. The external diameter of cranial cavity is about 200 mm. The thickness of shell is the mean thickness of calvarias. The average thickness of adult's calvaria is 6.0 mm, that of Tabula externa is 2.0 mm, diploe is 2.8 mm, Tabula interna is 1.2 mm and, dura mater in the parietal position is 0.4 mm.

Considering the characteristic of compact bone, cancellous bone and dura mater, we adopt their elastic modulus and Poisson ratios as 1.5×10^4 MPa, 4.5×10^3 MPa [50], 1.3×10^2 MPa [51] and 0.21, 0.01, 0.23 respectively.

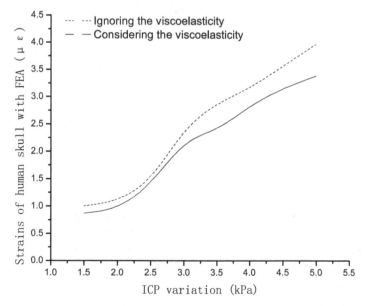

Fig. 2.7. The strain curves of finite-element simulation under the conditions of ignoring and considering the viscoelasticity of human skull and duramater with the changing ICP from 1.5 kPa to 5 kPa

After ignoring the viscoelasticity of human skull and dura mater, the strains of cranial cavity are shown in Table 1 with the finite-element software MSC_PATRAN/NASTRAN as ICP changing from 1.5 kPa to 5.0 kPa (Fig2.7). There is the measurable correspondence between skull strains and ICP variation. The strains of human skull can reflect the ICP change. When ICP variation is raised up to 2.5 kPa, the stress and strain graphs of skull bone are shown in Fig2.8~Fig2.13.

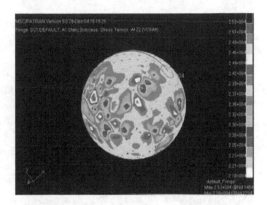

Fig. 2.8. Stress distribution
The scope of stress change on the outside surface is from 22.1 kPa to 25.3 kPa when ICP variation is raised up to 2.5 kPa by ignoring the viscoelasticity of human skull and dura mater.

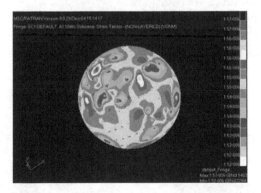

Fig. 2.9. Strain distribution
The scope of strain change on the outside surface is from 1.52 $\mu\varepsilon$ to 1.57 $\mu\varepsilon$ when ICP variation is raised up to 2.5 kPa by ignoring the viscoelasticity of human skull and dura mater.

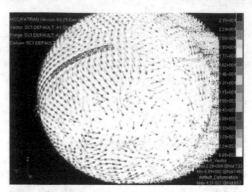

Fig. 2.10. The maximal stress vector distribution
The maximal main stress is about 22.4 kPa when ICP variation is raised up to 2.5 kPa by ignoring the viscoelasticity of human skull and dura mater.

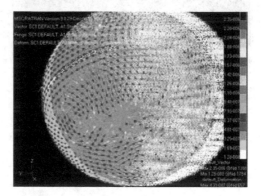

Fig. 2.11. The maximal strain vector distribution
The maximal main strain is about 2.2 $\mu\varepsilon$ when ICP variation is raised up to 2.5 kPa by ignoring the viscoelasticity of human skull and dura mater.

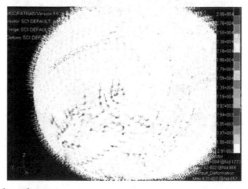

Fig. 2.12. Stress vector distribution
The main stress vector is about 21.8 kPa when ICP variation is raised up to 2.5 kPa by ignoring the viscoelasticity of human skull and dura mater.

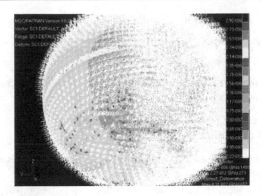

Fig. 2.13. Strain vector distribution

The main strain vector is about 2.14 $\mu\varepsilon$ when ICP variation is raised up to 2.5 kPa by ignoring the viscoelasticity of human skull and dura mater.

2.2.2 The finite-element analysis of strains by considering the viscoelasticity of human skull and dura mater

Human skull has the viscoelastic material [52]. Considering the viscoelasticity of human skull and dura mater, we use the viscoelastic option of the ANSYS finite-element program to analysis the strains on the exterior surface of human skull as ICP changing. According to the symmetry of 3D model of human skull, the preprocessor of the ANSYS finite-element program is used to construct a 1/8 finite-element model of human skull and dura mater consisting of 25224 nodes and 24150 three-dimensional 8-node isoparametric solid elements, shown in Fig2.14.

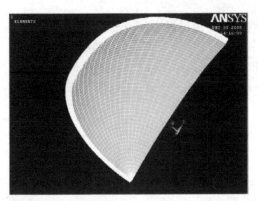

Fig. 2.14. Finite element model of 1/8 cranial cavity shell

The three-dimensional stress-strain relationships for a linear isotropic viscoelastic material are given by:

$$\sigma_{ij} = \int_0^t \left[2G(t-\tau)\frac{\partial e_{ij}(\tau)}{\partial \tau} + \delta_{ij}K(t-\tau)\frac{\partial \theta(\tau)}{\partial \tau} \right] d\tau \; ; \; (i,j=1,2,3) \tag{2.33}$$

Here, σ_{ij} — the Cauchy stress tensor; e_{ij} — the deviatoric strain tensor; δ_{ij} — the Kronecker delta; $G(t)$ — the shear relaxation function; $K(t)$ — the bulk relaxation function; $\theta(t)$ — the volumetric strain; t — the present time; τ — the past time.

Before the theoretical analysis of the minitraumatic strain-electrometric method, we need to set up the viscoelastic models to describe the relevant mechanical properties of human skull and dura mater.

(1) Viscoelastic Model of human skull

Under the constant action of stress, the strain of ideal elastic solid is invariable and that of ideal viscous fluid keeps on growing at the equal ratio with time. However, the strain of actual material increases with time, namely so-called creep. Generally, Maxwell and Kelvin models are the basic models to describe the performance of viscoelastic materials. Maxwell model represents in essence the liquid. Despite the representative of solid, Kelvin model can't describe stress relaxation but only stress creep (Fig2.15). So the combined models made up of the primary elements are usually adopted to describe the viscoelastic performance of actual materials. The creep of linear viscoelastic solid can be simulated by the Kelvin model of three parameters or the generalized Kelvin model.

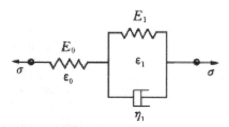

Fig. 2.15. Three parameters Kelvin model of human skull.

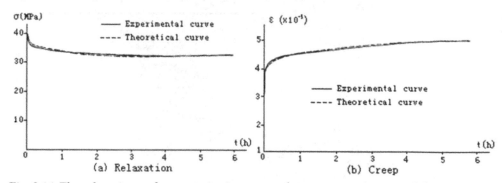

(a) Relaxation   (b) Creep

Fig. 2.16. The relaxation and creep train-time curves between experiment and three parameters Kelvin theoretical model of human skull.

Kelvin model of three parameters is shown in Fig2.15. Fig2.16 (a) is the relaxation curves of human skull and Kelvin model of three parameters in the compressive experiment. Fig2.16(b) is the creep curves of human skull and Kelvin model of three parameters. It shows that the theoretical Kelvin model of three parameters can well simulate the mechanical properties of human skull in the tensile experiments. Thus the Kelvin model of three parameters is adopted to describe the viscoelasticity of human skull in this paper.

For the Kelvin model of three parameters, the stress and strain of human skull are shown in equation (2.34),

$$\begin{cases} \varepsilon = \varepsilon_0 + \varepsilon_1 \\ \sigma = E_1\varepsilon_1 + \eta\dot{\varepsilon}_1 \\ \sigma = E_0\varepsilon_0 \end{cases} \tag{2.34}$$

After the calculation based on the equation (1), the elastic modulus of human skull is the Fig2.16,

$$E = \left(\frac{E_0 E_1}{E_0 + E_1} \right) + \left(\frac{E_0^2}{E_0 + E_1} \right) e^{\frac{t}{P_1}} \tag{2.35}$$

Here, σ — Direct stress acted on elastic spring or impact stress acted on viscopot; ε — Direct strain of elastic spring; E — Elastic modulus of tensile compression; η — Viscosity coefficient of viscopot; $\dot{\varepsilon}$ — strain ratio; $P_1 = \dfrac{\eta}{E_0 + E_1}$.

(2) Viscoelastic Model of human duramater

The generalized Kelvin model is shown in Fig2.17 (c). Fig2.17 (a) is the creep experimental curves of human duramater. Fig2.17 (b) is the curves of creep compliance for the generalized Kelvin model. It shows that the tendency of creep curve in the experiment is coincident with that of creep compliance for the generalized Kelvin model. Creep is the change law of material deformation with time under the invariable stress, so here σ is constant. For the generalized Kelvin model, the stress-strain relationship is $\varepsilon(t) = J(t)\sigma$. Thus the tendency of theoretical creep curve is totally the same as that of experimental one for human duramater. So in this paper, the generalized Kelvin model composed of three Kelvin-unit chains and a spring is adopted to simulate the viscoelasticity of human dura mater in this paper.

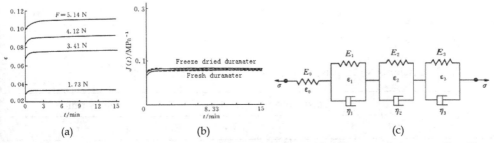

(a)   (b)   (c)

Fig. 2.17. Creep train-time curves under different loads for fresh human duramater (L_0=23 mm, θ=37 °C). Creep compliance curves of human duramatar Kelvin model. And the Kelvin model of the duramater.

For the viscoelastic model of human dura mater composed of the three Kelvin-unit chains and a spring, the stress and strain of human dura mater are shown in equation (2.36),

$$\begin{cases} \varepsilon = \varepsilon_0 + \varepsilon_1 + \varepsilon_2 + \varepsilon_3 \\ \varepsilon_0 = \dfrac{\sigma}{E_0} \\ \sigma = E_1\varepsilon_1 + \eta_1\dot{\varepsilon}_1 = E_2\varepsilon_2 + \eta_2\dot{\varepsilon}_2 = E_3\varepsilon_3 + \eta_3\dot{\varepsilon}_3 \end{cases} \qquad (2.36)$$

After the calculation based on the equation (2.36), the creep compliance of human dura mater is equation (2.37),

$$J(t) = E_0^{-1} + E(1 - e^{-t/\tau_1}) + E_2^{-1}(1 - e^{-t/\tau_2}) + E_3^{-1}(1 - e^{-t/\tau_3}) \qquad (2.37)$$

Then the elastic modulus of human dura mater is equation (2.38),

$$E = \left[E_0^{-1} + E_1^{-1}\left(1 - e^{-t/\tau_1}\right) + E_2^{-1}\left(1 - e^{-t/\tau_2}\right) + E_3^{-1}\left(1 - e^{-t/\tau_3}\right) \right]^{-1} \qquad (2.38)$$

Here, σ, ε, E, η, $\dot{\varepsilon}$ $--$Ditto mark; τ_1, τ_2, τ_3 $--$Lag time, that is $\tau_1 = \eta_1 / E_1$, $\tau_2 = \eta_2 / E_2$, $\tau_3 = \eta_3 / E_3$.

In the finite-element software ANSYS, there are three kinds of models to describe the viscoelasticity of actual materials, in which the Maxwell model is the general designation for the combined Kelvin and Maxwell models. Considering the mechanical properties of human skull and dura mater, we adopt the finite-element Maxwell model to simulate the viscoelasticity of human skull-dura mater system. The viscoelastic parameters of human skull and dura mater are respectively listed in Table 2.1 and Table 2.2.

	Elastic Modulus (GPa)		Viscosity (GPa/s)	Delay time τ (s)	
	E0	E1	η	τ_γ	* τ_d *
Compression	5.69±0.26	42.24±2.09	26.9±1.5	2022±198	2292±246
Tension	13.64±0.59	51.45±2.54	57.25±4.27	3180±300	4026±372

* $\tau_r = \eta / {E_1 + E_2}$, $\tau_d = \eta / {E_2}$

Table 2.1. Coefficients for the viscoelastic properties for human skull

	Elastic modulus (MPa)				Delay time τ (s)		
	E_0	E_1	E_2	E_3	τ_1	τ_2	τ_3
Duramater	16.67	125.0	150.0	93.75	40	10^4	10^6

Table 2.2. Creep coefficients for the viscoelastic properties for fresh human duramater

(3) The stress and strain distribution by the finite-element analysis

Fig2.18 (a) ~ (e) are the analytic graphs of stress and strain with finite-element software ANASYS when ICP variation is raised up to 2.5 kPa. After considering the viscoelasticity of human skull and duramater, the stresses and strains of cranial cavity are shown in Fig2.18 as the ICP changing from 1.5 kPa to 5kPa with the finite-element software ANSYS. It shows that the stress and strain distributions on the exterior surface of human skull are well-proportioned and that the stress and strain variation on the exterior surface of cranial cavity is relatively small corresponding to the ICP change. The strains of cranial cavity are coincident with ICP variation. The deformation scope of human skull is theoretically from 0.9 με to 3.4 με as the ICP changing from 1.5 kPa to 5.0 kPa. Corresponding to ICP of 2.5 kPa, 3.5 kPa and 5.0 kPa, the strain of skull deformation separately for mild, moderate and severe head injury is 1.5 με , 2.4 με , and 3.4 με or so.

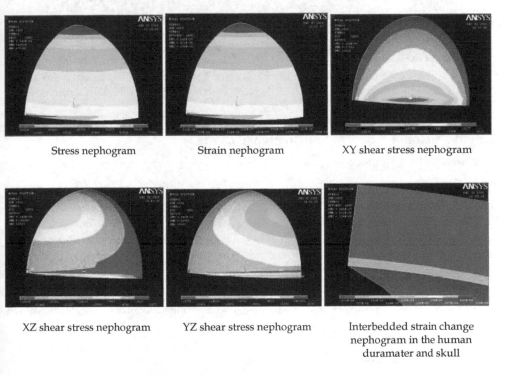

| Stress nephogram | Strain nephogram | XY shear stress nephogram |

| XZ shear stress nephogram | YZ shear stress nephogram | Interbedded strain change nephogram in the human duramater and skull |

Fig. 2.18. The stress and strain distribution considering viscoelasticity of human skull and duramater

From the relationships about total, elastic and viscous strains of human skull and dura mater in Fig2.19, the viscous strains account for about 40% and the elastic strains are about 60% of total strains with the increasing ICP.

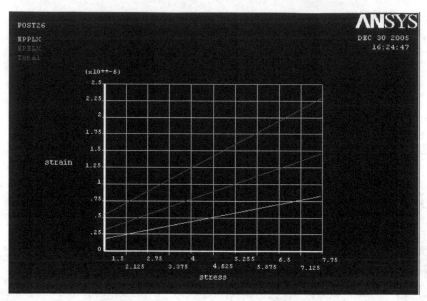

Fig. 2.19. Curves among total, elastic and viscous strain when the ICP increment is 2.5 kPa. Here EPELX is elastic strain curve, EPPLX is viscous strain curve. The viscous strain is about 40% of total strain.

3. Finite-element model of human cranial cavity

3.1 Materials and methods

3.1.1 CT scan

A healthy male volunteer, aged 40 years old, with body height 176 cm, weighing 75 kg, was included in this study. The volunteer explained no history of cranium brain. Common projections (posterior-anterior, lateral, dual oblique, hyperextension and hyperflexion) were made to exclude cranium brain degenerative disorders, cranial instability, and brain destruction.

Spiral CT scans (1 mm thickness) were output in the JPG image file format and saved in the computer. Prior to experiment, informed consent was obtained from this volunteer, shown in Fig3.1.

3.1.2 Experimental equipment

High performance computer (Lenovo, X200) and mobile storage equipment were used. Solid modeling software Mimics 13.0 (Materiaise's interactive medical image control system) was used in this study. As a top software in computer aided design, Mimics 13.0 provides many methods of precise modeling and has been widely used for precise processing. Its equipped Ansys, Partron finite element analysis module sequence were used for finite element analysis, and then the strain and deformation regularity of the real human cranial cavity were simulated with the changing ICP.

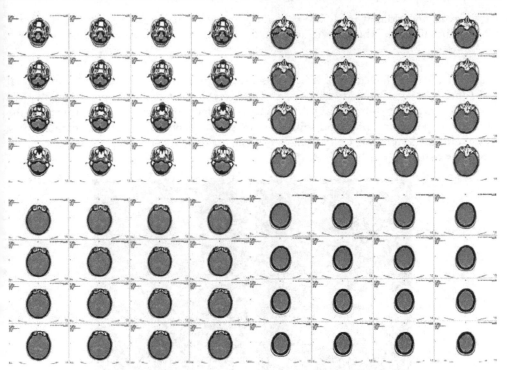

Fig. 3.1. CT scan image of cranial cavity

3.1.3 Flowchart of finite element method

Methods

Image boundary tracking

A self-programmed image boundary automatic recorder was used to acquire cranial information along the superior border to inferior border of cranium brain. The spatial boundary was recorded layer by layer. Following point selection from interior and exterior border of images, two-dimensional space coordinates were automatically recorded and saved in the .CDB form. After conversion, this file can be directly input into Mimics or CAD software.

Location

A coordinate system was determined. The CT scanning image starting from the lowest layer of cranium brain was set as the working background. CT scan was performed based on a fixed coordinate axis, with a know layer interval and magnification proportion. The spatial three-dimensional coordinate of each point in the image could be determined through drawing the horizontal coordinate of each point and referencing scanning interval. When CT machines recorded each layer of images, all images were in the same scanning range, which equaled to cranial location of two-dimensional CT images from each layer in the scanning direction. Calibration of two-dimensional images could be performed if the scar bar of CT scan images were given. In addition, each layer of image was scanned with some interval in the longitudinal direction, which was equivalent to calibration in the third dimensional direction.

Image reconstruction

In accordance with the sequence of CT scans and according to the scale bar and scan interval of CT faulted image, geometry data of each layer were input into the pre-processing module of finite element software to establish a geometry model in the rectangular coordinate in the sequence of point, line, area, and solid. The transverse plane of CT scan was parallel to xy plane, and the longitudinal plane was along the z axis. Three-dimensional reconstruction process is shown in Fig3.2.

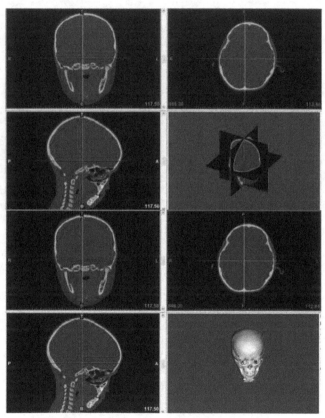

Fig. 3.2. 3d model of cranial cavity

3.2 Results

3.2.1 Reconstruction results

The fitted curves were assigned into different layers to construct the solid structure of bone (Tabula externa, Tabula interna, Diploe sandwiched in between), spongy durameter. During reconstruction of structure of Tabula externa, Tabula interna, Diploe sandwiched in between, dura mater, solid of each part was generated independently, and the union of all parts was collected. The 3d finite-element models in each direction of cranial cavity are shown in Fig3.3.

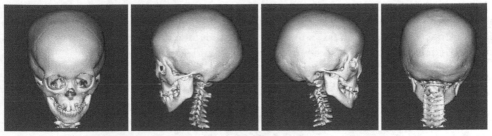

Fig. 3.3. Each view drawing on the 3d finite-element model in of cranial cavity

3.2.2 Model validation

Following type selection, finite element mesh generation was performed in the above-mentioned models which were given material characteristics. Then, through simulating practical situation, boundary condition was exerted and proper numerical process. And the three-dimensional analysis was performed. Based on previously published manuscripts, the elastic modulus and the Poisson's ratio of compact bone, cancellous bone and dura mater, we adopt their elastic modulus and Poisson ratios as 1.5×10^4 MPa, 4.5×10^3 MPa, 1.3×10^2 MPa and 0.21, 0.01, 0.23 respectively. 3d finite-element model of cranial cavity is meshed in Fig3.4. Simulation analysis of cranium brain three-dimensional finite element model is shown in Fig3.5.

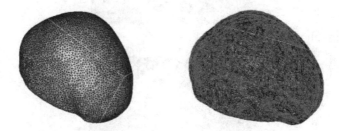

Fig. 3.4. 3d finite-element model of cranial cavity is meshed

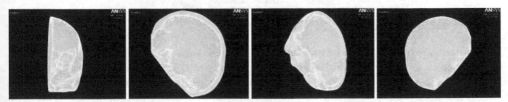

Fig. 3.5. Finite element model of 1/2 cranial cavity

Biomechanical model has been shown to play a key role in study of cranium brain, because it can be used to investigate the pathogenesis through model observation, thereby to propose the strategy of diagnosis and treatment.

Owing to irregular geometry and non-uniform composition of cervical spine cranium bone as well as impossible human mechanical tests, increasing attention has been recently paid to finite element method included in the biological study of cranium brain injury because this method exhibits unique advantages in analysis of complex structure.

Fig. 3.6. Loads on the finite element model of 1/2 cranial cavity

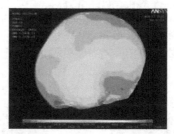

Fig. 3.7. Strain graph when the ICP is 3.0 kPa

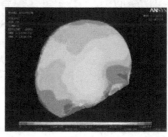

Fig. 3.8. Strain graph when the ICP is 5.0 kPa

Experimental results are the best method to verify model accuracy. When exerting persistent pressure to vertebral spine, non-linear computation is supplemented to the two-dimensional unit calculation of ligament structure, which more corresponds to human mechanical structure. Statics solver exhibits the self-testing function and can automatically analyze computation process, report errors, and control error range. The displacement graph, stress graph, and isogram drawn by post-processor visualizing the distribution ranges of stress or strain loaded on each part of cranium brain with the changing ICP. When loads are vertically added, the stress on the posterior wall of cranium brain, as well as on the end plate and the posterior part of intervertebral discs, relatively centralizes.

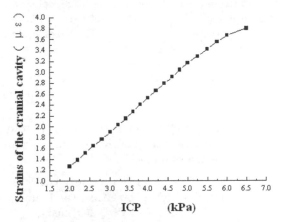

Fig. 3.9. Strains curve of cranial cavity with the ICP variation

3.2.3 Strains on the external surface of cranial cavity

Finite element analysis is an important mean to simulate human structural mechanical function in the field of biomechanics. A human finite element model with physical material characteristics under proper simulated in vivo condition can be used to effectively analyze the physical characteristics of human structure, for example, stress/strain of structure, modal analysis, exterior impact response, and fatigue test. With further understanding of cranium brain diseases, some complex models have not been developed, for example, finite element models of head and cranial cavity used to study the physio-pathological influences cervical spine loading in some complex exercises on cranium brain and soft tissue. Finite element analysis exhibits unexampled advantages in the biomechanical study of a severe medical brain problem. Theoretically, finite element method can simulate nearly all biomechanical experiments. Moreover, this method can better describe the interior changes of living body than practical study. Finite element method, as an emerging technique, has a broad developing space. However, it is a theoretical simulation analyses, only in conjunction with clinical detection and observation can it truly reflect the occurrence and progression of cranium brain disease and provide evidence for predicting curative effects, thereby exhibiting a synergic effect with clinical outcomes. The present model is only a represent. It can not reflect the changes between individual interior parts and between individuals in terms of bone contour and material characteristics. The present model is only a cranium brain motion segment. Its simulation analysis results might differ from the results from multiple motion segments. Actually, when much difference and many uncertainties exist between individuals, model simplification and idealization is to strengthen some research aspect, which removes experimental inherent difference. Of course, establishment of a finite element mechanical model is to provide mechanical methods for clinical and experimental studies. The present model needs further improvements due to some limitations, i.e., unable to reflect some complex condition, but it ensures the geometry data and material characteristic approximation for application of multiple toolsequipped by various softwares during the process of model establishment. In addition, finite element method, as one of tools used in the biomechanical field, can qualitatively analyze the stress change of cranium brain interior parts when bearing forces. Only by changing local structure or materials can the present model established by finite element method simulate the common clinical situation and the effects of intervention on ICP force. The present model should be further improved in the clinical and experimental processes.

4. Conclusion and discussion

4.1 Conclusion

We develop a new minitraumatic method for measuring ICP with strain-electrometric technology. The strains of skull bone can reflect the ICP change. The surgical procedures for this new method are easy, simple, safe and reliable.

This paper carries respectively on the stress and strain analysis on both conditions of ignoring and considering the viscoelasticity of human skull and duramater by finite-element software MSC_PPATRAN/NASTRAN and ANSYS as ICP changing from 1.5 kPa to 5 kPa. The three-dimensional finite element model of cranial cavity and the viscoelastic models of human skull and duramater are constructed in this paper. At the same time, the ANSYS finite-element software is in this paper used to reconstruct the three-dimensional cranial cavity of human being with the mild hypothermia treatment. The conclusion is as follows:

1. The human skull and duramater are deformed as ICP changing, which is corresponding with mechanical deformation mechanism.
2. When analyzing the strain of human skull and duramater as ICP changing by the finite-element software ANSYS, the strain by considering the viscoelasticity is about 14% less than that by ignoring the viscoelasticity of human skull and duramater. Because the viscoelasticity analysis by finite-element software ANSYS is relatively complex and the operation needs the huge memory and floppy disk space of computer, it is totally feasible to ignore the viscoelasticity while calculating the FEA strain of human skull and duramater as ICP changing.
3. The viscosity plays an important role in the total deformation strain of human skull and duramater as ICP changing. In the strains analysis of human skull and duramater with the changing ICP by the finite-element software ANASYS, the viscous strain accounts for about 40% of total strain, and the elastic strain is about 60% of total strain.
4. Because the strains of human skull are proportional to ICP variation and the caniocerebra characteristic symptoms completely correspond to different deformation strains of human skull, ICP can be completely obtained by measuring the deformation strains of human skull. That is to say, the minitraumatic method of ICP by strain electrometric technique is feasible. Furthermore, ICP variation is respectively about 2.5 kPa when the strain value of human skull is about 1.4 $\mu\varepsilon$, about 3.5 kPa when the strain value of human skull is about 2.1 $\mu\varepsilon$, and about 5 kPa when the strain value of human skull is about 3.9 $\mu\varepsilon$.
5. The strains decreased under the mild hypothermia environment about 0.56% than those under the normal temperature conditions during the same circumstance of ICP changes.

6.   The deformation scope of human skull is theoretically from 1.50 $\mu\varepsilon$ to 4.52 $\mu\varepsilon$ as the ICP changing from 2.0 kPa to 6.0 kPa under the normal situation, and from 1.50 $\mu\varepsilon$ to 4.49 $\mu\varepsilon$ under the mild hypothermia environment. Accordingly, the strains of skull deformation for mild, moderate and severe head injuries are separately 1.87 $\mu\varepsilon$, 2.62 $\mu\varepsilon$ and 3.74 $\mu\varepsilon$ or so corresponding to ICP of 2.5 kPa, 3.5 kPa and 5.0 kPa.

4.2 Discussion

In neurosurgery, one of the principle axes of treatment for neurosurgical disease is to control ICP. Because the skull bone is outside of and close to the brain, the surgical procedure in the strain-ICP monitoring system is relatively invasive and may affect experimental results from brain tissue. The strain-ICP monitoring has several advantages. First, the strain foil is far from the brain, and will not affect the surgery or experiments in the brain. Second, the wound surface on the parietal bone is very small and just about 11 mm^2. Third, the surgical procedure is not extremely invasive for patients compared to the conventional monitoring. Fourth, it is possible to keep the strain foil for a longer time, the fixation of strain foil to the periosteum is much easier than other methods. Fifth, the operation is performed in the cephalic skin, the risk, difficulty, infection and trauma to patients are relatively small. Sixth, no special posture of patients is demanded, skull bone can be hardly influenced by any diseases and will be deformed as long as ICP is fluctuant in brain. ICP can be synchronously and continuously monitored based on the dynamic measurement of skull strains. Thus, this system is relatively safe, and it is easier to keep the strain foils in the cranial cavity for a longer period of time.

In this paper, the finite-element simulation was carried out to analyze the deformation of cranial cavity. Many complex relationships and influencing factors lie in the actual deformation of cranial cavity with the changing ICP. Therefore, in order to obtain the accurate deformation tendency of cranial cavity, the precise simulation to the finite-element model and further experimental studies in vivo and clinic need to be carried on.

5. References

[1] Gregson BA, Banister K, Chambers IR. Statistics and analysis of the Camino ICP monitor. J Neurol Neurosurg Psychiatry 1995; 70: 138.

[2] Hilton G. Cerebral oxygenation in the traumatically brain-injured patient: are ICP and CPP enough? J Neurosci Nurs 2000; 32: 278-282.

[3] Richard KE, Block FR, Weiser RR. First clinical results with a telemetric shunt-integrated ICP-sensor. Neurol Res 1999; 21: 117-120.

[4] Rosner MJ, Becker DP. ICP monitoring: complications and associated factors. Clin Neurosurg 1976; 23: 494-519.

[5] Schmidt B, Klingelhofer J. Clinical applications of a non-invasive ICP monitoring method. Eur J Ultrasound 2002; 16: 37-45.

[6] North B, Reilly P. Comparison among three methods of intracranial pressure recording. Neurosurgery 1986; 18: 730-732.

[7] Powell MP, Crockare HA. Behavior of an extradural pressure monitor in clinical use. Comparison of extradural with intraventricular pressure in patients with acute and chronically raised intracranial pressure. J Neurosurg 1985; 53: 745-749.

[8] Hayes KC, Corey J. Measurement of cerebrospinal fluid pressure in the rat. J Appl Physiol 1970; 28: 872-873.

[9] Melton JE, Nattie EE. Intracranial volume adjustments and cerebrospinal fluid pressure in the osmotically swollen rat brain. Am J Physilo 1984; 246: 533-541.

[10] Andrews BT, Levy M, M cIntosh TK, Pitts LH. An epidural intracranial pressure monitor for experimental use in the rat. Neurol Res 1988; 10: 123-126.

[11] Yamane K, Shima T, Okada Y, Takeda T, Uozumi T. Acute brain swelling in cerebral embolization model of rats. Part I. Epidural pressure monitoring. Surg Neurol 1994; 41: 477-481.

[12] Shah JL. Positive lumbar extradural space pressure. Br J Anaesthesia 1994; 73: 309-314. Sutherland WG. The cranial bowl. Mankato, Minn: Free Press Company 1939.

[13] Zhong J, Dujovny M, Park H, Perez E, et al. Neurological Research 2003; 25: 339-350.

[14] Schmiedek P, Bauhuf C, Horn P, Vajkoczy P, Munch E. International Congress Series 2002; 1247: 605-610.

[15] Allocca JA. Method and apparatus for noninvasive monitoring of intracranial pressure. U.S. Patent 4204547, 1980.3.27.

[16] Rosenfeld JG. Method and apparatus for intracranial pressure estimation. U.S. Patent 4564022, 1986.1.14.

[17] Marchbanks RJ. Method and apparatus for measuring intracranial fluid pressure. U.S. Patent 4841986, 1989.6.27.

[18] Mick EC. Method and apparatus for the measurement of intracranial pressure. U.S. Patent 5074310, 1991.11.24.

[19] Mick EC. Method and apparatus for the measurement of intracranial pressure. U.S. Patent 5117835, 1992.1.2.

[20] Shepard S. Heat trauma. eMedicine 2004; 8: 20.

[21] Czosnyka M, Pickard JD. Monitoring and interpretation of intracranial pressure. J Neurol Neurosurg Psychiatry 2004; 75: 813-821.

[22] Monro A. Observations on the structure and function of the nervous system. Edinburgh, Creech & Johnson 1823; 5.

[23] Kellie G. An account of the appearances observed in the dissection of two of the three individuals presumed to have perished in the storm of the 3rd, and whose bodies were discovered in the vicinity of Leith on the morning of the 4th November 1821 with some reflections on the pathology of the brain. Edinburgh: Trans Med Chir Sci 1824; 1: 84-169.

[24] Retzlaff EW, Jones L, Mitchell Jr FL, Upledger J. Possible autonomic innervation of cranial sutures of primates and other animals. Brain Res 1973; 58: 470-477.

[25] Lakin WD, Stevens SA, Trimmer BI, Penar PL. A whole-body mathematical model for intracranial pressure dynamics. J Math Biol, 2003Apr;46 (4):347-38.

[26] R. M. Jones. Composite Material. 1st Eds. Shanghai Science and Technology Publisher, 1981. 6: 41-73.

[27] Zhimin Zhang. Structural Mechanics of Composite Material. 1st Eds. Beijing: Publish of BUAA, 1993. 9: 85-86.

[28] Willinger, R., Kang, H.S., Diaw, B.M. Development and validation of a human head mechanical model. Comptes Rendus de l'Academie des Sciences Series IIB Mechanics, 1999, 327: 125-131.

[29] Pithioux M, Lasaygues P, Chabrand P. An alternative ultrasonic method for measuring the elastic properties of cortical bone. Journal of Biomechanics 2002; 35: 961-968.

[30] Hakim S, Watkin KL, Elahi MM, Lessard L. A new predictive ultrasound modality of cranial bone thickness. IEEE Ultrason Sympos 1997; 2: 1153-1156.

[31] Hatanaka M. Epidural electrical stimulation of the motor cortex in patients with facial neuralgia. Clinical Neurology and Neurosurgery 1997; 99: 155.

[32] Odgaard A. Three-Dimensional methods for quantification of cancellous bone architecture. Bone 1997; 20: 315-328.

[33] Kabel J., Rietbergenvan B., Dalstra M., Odgaard A., Huiskes R. The role of an elective isotropic tissue modulus in the elastic properties of cancellous bone. Journal of Biomechanics1999; 32: 673-680.

[34] Noort van R., Black M.M., Martin T.R. A study of the uniaxial mechanical properties of human dura mater preserved in glycerol. Biomaterials 1981; 2: 41-45.

[35] Kuchiwaki H., Inao S., Ishii N., Ogura Y., Sakuma N. Changes in dural thickness reflect changes in intracranial pressure in dogs. Neuroscience Letters 1995; 198: 68-70.

[36] Cattaneo P.M., Kofod T., Dalstra M., Melsen B. Using the finite element method to model the biomechanics of the asymmetric mandible before, during and after skeletal correction by distraction osteogenesis. Computer Methods in Biomechanics & Biomedical Engineering 2005; 8(3): 157-165.

[37] Amit G., Nurit G., Qiliang Z., Ramesh R., Margulies S.S. Age-dependent changes in material properties of the brain and braincase of the rat. Journal of Neurotrauma 2003; 20(11): 1163-1177.

[38] Andrus C. Dynamic observation and nursing of ICP. Foreign Medical Sciences (Nursing Foreign Medical Science) 1992; 11(6): 247-249.

[39] Min W. Observation and nursing for fever caused by Acute Cerebrovascular Disease. Contemporary Medicine 2008; 143: 100-101.

[40] Shiozaki T., Sugimoto H., Taneda M., et al. Selection of Severely Head Injured Patients for Mild Hypothermia Therapy. Journal of Neurosurgery 1998; 89: 206–211.

[41] Lingjuan C. Current situation and development trend at home and abroad of Hypothermia Therapy Nursing. Chinese Medicine of Factory and Mine 2005; 18(3): 268-269.

[42] Yue X.F., Wang L., Zhou F. Experimental Study on the strains of skull in rats with the changing Intracranial Pressure.Tianjin Medicine Journal, 2007, 35(2):140-141.

[43] Yue X.F., Wang L., Zhou F. Strain Analysis on the Superficial Surface of Skull as Intracranial Pressure Changing.Journal of University of Science and Technology Beijing, 2006, 28(12):1143-1151.

[44] Qiang X., Jianuo Z. Impact biomechanics researches and finite element simulation for human head and neck.Journal of Clinical Rehabilitative Tissue Engineering Research, 2008, 12(48):9557-9560.

[45] Zong Z., Lee H., Lu C. A three-dimensional human head finite element model and power in human head subject to impact.Journal of Biomechanics, 2006, 39(2):284-292.

[46] Ren L., Hao L., Shuyuan L., et al. A study of the volume of cranial cavity calculating from the dimension of cranial outer surface in X-ray films—its stepw ise regressive equation and evaluation. Acta Anthropological Sinica, 1999, 18(1):17-21.

[47] Yunhong L., Yanbo G., Yunjian W., et al. Experiment study on shock resistance of skull bone. Medical Journal of the Chinese People's Armed Police Forces, 1998, 9(7): 408-409.

[48] Haiyan L., Shijie R., Xiang P., et al. The thickness measurement of alive human skull based on CT image. Journal of Biomedical Engineering, 2007, 24(5):964-968.

[49] Zhou L.F., Song D.L., Ding Z.R. Biomechanical study of human dura and substitutes. Chinese Medical Journal, 2002, 115(11):1657-1659.

[50] Willinger R., Kang H.S., Diaw B.M. Development and validation of a human head mechanical model. Competes Rendus de l'Academie des Sciences Series IIB Mechanics, 1999, 327:125-131.

[51] Odgaard A., Linde F. The Underestimation of young's modulus in compressive testing of cancellous bone specimens. Journal of Biomechanics, 1991, 24, 691-698.

[52] Wenjun S., Jialin H. Handbook of critical illness care. People's Military Medical Press, 1994:75-77.

Permissions

The contributors of this book come from diverse backgrounds, making this book a truly international effort. This book will bring forth new frontiers with its revolutionizing research information and detailed analysis of the nascent developments around the world.

We would like to thank David Moratal, for lending his expertise to make the book truly unique. He has played a crucial role in the development of this book. Without his invaluable contribution this book wouldn't have been possible. He has made vital efforts to compile up to date information on the varied aspects of this subject to make this book a valuable addition to the collection of many professionals and students.

This book was conceptualized with the vision of imparting up-to-date information and advanced data in this field. To ensure the same, a matchless editorial board was set up. Every individual on the board went through rigorous rounds of assessment to prove their worth. After which they invested a large part of their time researching and compiling the most relevant data for our readers. Conferences and sessions were held from time to time between the editorial board and the contributing authors to present the data in the most comprehensible form. The editorial team has worked tirelessly to provide valuable and valid information to help people across the globe.

Every chapter published in this book has been scrutinized by our experts. Their significance has been extensively debated. The topics covered herein carry significant findings which will fuel the growth of the discipline. They may even be implemented as practical applications or may be referred to as a beginning point for another development. Chapters in this book were first published by InTech; hereby published with permission under the Creative Commons Attribution License or equivalent.

The editorial board has been involved in producing this book since its inception. They have spent rigorous hours researching and exploring the diverse topics which have resulted in the successful publishing of this book. They have passed on their knowledge of decades through this book. To expedite this challenging task, the publisher supported the team at every step. A small team of assistant editors was also appointed to further simplify the editing procedure and attain best results for the readers.

Our editorial team has been hand-picked from every corner of the world. Their multi-ethnicity adds dynamic inputs to the discussions which result in innovative outcomes. These outcomes are then further discussed with the researchers and contributors who give their valuable feedback and opinion regarding the same. The feedback is then collaborated with the researches and they are edited in a comprehensive manner to aid the understanding of the subject.

Apart from the editorial board, the designing team has also invested a significant amount of their time in understanding the subject and creating the most relevant covers. They scrutinized every image to scout for the most suitable representation of the subject and create an appropriate cover for the book.

The publishing team has been involved in this book since its early stages. They were actively engaged in every process, be it collecting the data, connecting with the contributors or procuring relevant information. The team has been an ardent support to the editorial, designing and production team. Their endless efforts to recruit the best for this project, has resulted in the accomplishment of this book. They are a veteran in the field of academics and their pool of knowledge is as vast as their experience in printing. Their expertise and guidance has proved useful at every step. Their uncompromising quality standards have made this book an exceptional effort. Their encouragement from time to time has been an inspiration for everyone.

The publisher and the editorial board hope that this book will prove to be a valuable piece of knowledge for researchers, students, practitioners and scholars across the globe.

List of Contributors

Wirley Gonçalves Assunção, Valentim Adelino Ricardo Barão, Érica Alves Gomes, Juliana Aparecida Delben and Ricardo Faria Ribeiro
Univ Estadual Paulista (UNESP), Aracatuba Dental School, Univ of Sao Paulo (USP), Dental School of Ribeirao Preto, Brazil

Carlos José Soares, Andréa Dolores Correia Miranda Valdivia, Aline Arêdes Bicalho, Crisnicaw Veríssimo, Bruno de Castro Ferreira Barreto and Marina Guimarães Roscoe
Federal University of Uberlândia, Brazil

Antheunis Versluis
University of Tennessee, USA

Erika O. Almeida
Department of Biomaterials and Biomimetics, New York University College of Dentistry, New York, NY, USA
Department of Dental Materials and Prosthodontics, Sao Paulo State University College of Dentistry, Araçatuba, SP, Brazil

Nick Tovar and Paulo G. Coelho
Department of Biomaterials and Biomimetics, New York University College of Dentistry, New York, USA

Eduardo P. Rocha
Department of Dental Materials and Prosthodontics, São Paulo State University College of Dentistry, Araçatuba, SP, Brazil

Amilcar C. Freitas Júnior
Postgraduate Program in Dentistry, Potiguar University, College of Dentistry – UnP, Natal, RN, Brazil

Roberto S. Pessoa
Department of Mechanical Engineering - Federal University at Uberlândia, FEMEC - Uberlândia, Uberlândia, MG, Brazil

Nikhil Gupta
Department of Mechanical and Aerospace Engineering, Polytechnic Institute of New York University, New York, USA

Paulo Roberto R. Ventura, Isis Andréa V. P. Poiate and Adalberto Bastos de Vasconcellos
Federal Fluminense University, Brazil

Edgard Poiate Junior
Pontifical Catholic University, Brazil

Rafael Yagüe Ballester
University of São Paulo, Brazil

Ching-Chang Ko
Department of Orthodontics, University of North Carolina School of Dentistry, USA
Department of Material Sciences and Engineering, North Carolina State University Engineering School, Raleigh, USA

Eduardo Passos Rocha
Department of Orthodontics, University of North Carolina School of Dentistry, USA
Faculty of Dentistry of Araçatuba, UNESP, Department of Dental Materials and Prosthodontics, Araçatuba, Saõ Pauló, Brazil

Matt Larson
Department of Orthodontics, University of North Carolina School of Dentistry, USA

Márta Kurutz
Budapest University of Technology and Economics, Hungary

László Oroszváry
Knorr-Bremse Hungaria Ltd, Budapest, Hungary

Silvia Schievano, Claudio Capelli, Daria Cosentino, Giorgia M. Bosi and Andrew M. Taylor
UCL Institute of Cardiovascular Science, London, UK

Luis Gracia, Elena Ibarz, José Cegoñino, Sergio Gabarre, Sergio Puértolas and Enrique López
Engineering and Architecture Faculty, University of Zaragoza, Spain

Antonio Lobo-Escolar, Jesús Mateo and Antonio Herrera
Medicine School, University of Zaragoza, Spain
Miguel Servet University Hospital, Zaragoza, Spain

Xianfang Yue, Li Wang, Ruonan Wang, Yunbo Wang and Feng Zhou
Mechanical Engineering School, University of Science and Technology Beijing, Beijing, China